Bayesian Methods for Nonlinear Classification and Regression

Bayesian Methods for Nonlinear Classification and Regression

David G. T. Denison and **Christopher C. Holmes**

Imperial College of Science, Technology and Medicine, UK

Bani K. Mallick

Texas A&M University, USA

Adrian F. M. Smith

Queen Mary, University of London, UK

JOHN WILEY & SONS, LTD

Telephone (+44) 1243 779777

Email (for orders and customer service enquiries): cs-books@wiley.co.uk
Visit our Home Page on www.wileyeurope.com or www.wiley.com

Reprinted February 2004

Other Wiley Editorial Offices

John Wiley & Sons Inc., 111 River Street, Hoboken, NJ 07030, USA

Jossey-Bass, 989 Market Street, San Francisco, CA 94103-1741, USA

Wiley-VCH Verlag GmbH, Boschstr. 12, D-69469 Weinheim, Germany

John Wiley & Sons Australia Ltd, 33 Park Road, Milton, Queensland 4064, Australia

John Wiley & Sons (Asia) Pte Ltd, 2 Clementi Loop #02-01, Jin Xing Distripark, Singapore 129809

John Wiley & Sons Canada Ltd, 22 Worcester Road, Etobicoke, Ontario, Canada M9W 1L1

British Library Cataloguing in Publication Data

A catalogue record for this book is available from the British Library

ISBN 0471 49036 9

Produced from files supplied by the authors, typeset by T&T Productions Ltd, London
Printed and bound in Great Britain by Antony Rowe Ltd, Chippenham, Wiltshire
This book is printed on acid-free paper responsibly manufactured from sustainable forestry in which at least two trees are planted for each one used for paper production.

Contents

Preface

This book covers the relatively new field of nonlinear Bayesian modelling. The need for models which are more flexible than those with linear assumptions has been known for some time. However, their routine use has become possible only recently due to significant increases in computational power. Also, Bayesian methodology has been increasing in popularity for the past half-century for two basic reasons. Firstly, Bayesian methods take an axiomatic view of uncertainty allowing the user to make coherent inference. Secondly, Bayesian modelling is particularly well suited to incorporating prior information, which is often available.

We bring together the ideas of nonlinear modelling and Bayesian theory to provide methods for producing probabilistically sound inferences. The basic ideas behind the models are shown to be relatively straightforward. The main difficulties lie in the implementation of the methods, as well as understanding which model would be expected to perform well in what situation. This discussion of the strengths and weaknesses of each model is an important contribution of the book.

The work in this book has its roots in the theses of D.G.T.D. (supervised by B.K.M. and A.F.M.S.) and C.C.H. (supervised by B.K.M.). It brings together all this previous work, as well as that from other researchers, in an up-to-date and consistent manner. It is intended as a text for researchers in nonlinear modelling, whether Bayesian or not, and would be particularly well suited for a graduate level course in the subject. For this reason we have included problems at the end of most of the chapters.

Acknowledgements

As is always the case, many people have contributed significantly to this book. We owe collective thanks to the Statistics Section of the Department of Mathematics at Imperial College, London, where we first met and which provided a wonderfully stimulating research environment.

Other people we would like to especially thank are those who read through, and made many perceptive comments on, earlier drafts of the book. These include Petros Dellaportas, Arnaud Doucet, Simon Godsill, Peter Müller, Dave Stephens. Other help was also provided by Niall Adams, Rob McCulloch, Tomé Ferreira, Ed George, Leo Knorr-Held and Stephen Roberts. Thanks also to the staff at Wiley for being so efficient in preparing this monograph, especially to Rob Calver, who was very patient with us.

Finally, most heartfelt thanks go to Maria, who had to listen to many hours of detailed discussion on how the book was progressing, always with the same conclusion: not very quickly.

Nomenclature

$\mathbb{R}$	The set of real numbers
Y	The column vector of response values
p	The number of predictors
$X_1, \ldots, X_p$	The predictor variables
$(y, \boldsymbol{x})$	A general datum
n	The number of points in the dataset
k	The number of basis functions in a model (not including the constant one)
K	The maximum number of basis functions a model may contain
$\mathcal{D}$	The dataset $\mathcal{D} = \{y_i, \boldsymbol{x}_i\}_1^n$
$\mathcal{X}$	The predictor space, $\boldsymbol{x} \in \mathcal{X}$
$\boldsymbol{B}$	The $n \times k$ basis function matrix
$\boldsymbol{\beta}$	The $k \times 1$ vector of coefficients
σ^2	The regression variance
$\boldsymbol{\epsilon}$	The error vector
$\boldsymbol{\theta}$	The parameter vector that defines the model
$\boldsymbol{\phi}$	The parameters of the model that are assigned conjugate priors
η	The linear predictor
$B_1, \ldots, B_k$	The basis functions that combine linearly to form the model
C	The number of classes in a classification problem
f	The true functional relationship between $\boldsymbol{x}$ and y
g	An approximation to f that constitutes our model

1

Introduction

1.1 Regression and Classification

Quantifying the relationship between a response variable of interest and measurements taken on a set of possibly related observations is one of the most fundamental problems in statistics: a problem which conventionally is split into two distinct topics, regression and classification. Both focus on approximating the relationship between a set of input variables, or predictors, and an output variable, or response. In short, regression involves the case when the response variable is continuous, whereas classification is used when the response is categorical.

Figure 1.1 displays a dataset for which regression is appropriate. The data are taken from the Great Barrier Reef dataset described in detail in Poiner *et al.* (1997). The interest lies in determining the relationship between the longitude and the weight of fauna, given in terms of a score, captured at that longitude. By convention, we plot the response variable (score) on the y-axis with the predictor (longitude) on the x-axis. The aim of the regression analysis is to provide a good approximation to the true functional relationship between these two variables. The simplest such approximation widely used in data analyses assumes that the relationship is a straight line such that

$$\text{Score} = \beta_0 + \beta_1 \times \text{Longitude},$$

where β_0 and β_1 give the intercept and slope of the line. The line with β_0 and β_1 set so that they best fit the data, in least squares terms, is shown in Figure 1.1. However, this straight line seems to oversimplify the true relationship between the variables, and completely fails to model the particularly rapid drop in score for values of longitude greater than 143.3.

Figure 1.2 displays a dataset for which classification is appropriate. The dataset was originally described by Spyers-Ashby (1996) and was collected to determine how arm tremor measurements on an individual were related to the presence or absence of Parkinson's disease. The two axes represent measurements of two different types of arm tremor. We label these predictors X_1 and X_2 and write the set of all predictors as $\boldsymbol{X} = (X_1, X_2)$. Each sampled data point is represent by a cross if that particular patient had Parkinson's disease and a circle otherwise. The problem of interest is to split up the two-dimensional space in which the observed predictor values lie into regions, where the patients whose predictors lie in each one have a particular probability of

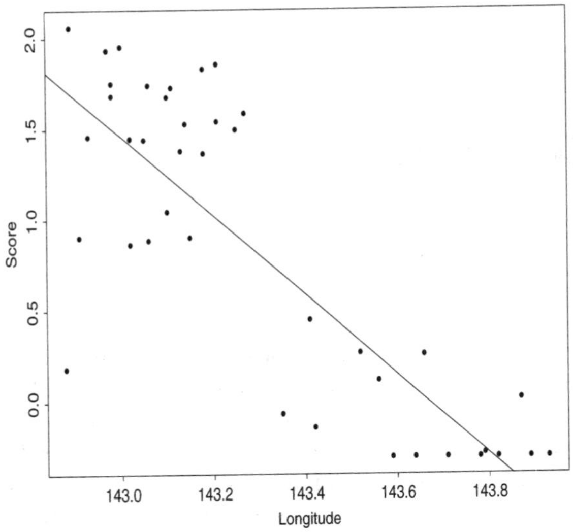

Figure 1.1 The Great Barrier Reef dataset together with the straight line which best fits the data in terms of sum of squared residuals.

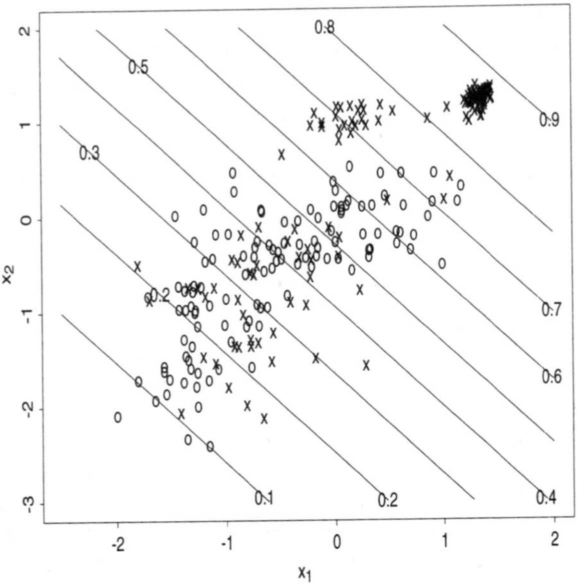

Figure 1.2 The arm tremor dataset with the crosses marking the location of individuals with Parkinson's disease and the circles representing those without it. The lines are the contours of probability found by fitting a straight line to $\text{logit}(p(y = 1 \mid \boldsymbol{x}))$.

having Parkinson's disease. One way to do this is to estimate a surface that captures how the probability of the presence of Parkinson's disease is related to $\boldsymbol{X}$.

A standard way to solve this classification problem is to use another model based on a straight line. That is, if $\boldsymbol{x} = (x_1, x_2)$ represents a general vector of measurements of the predictors, we take

$$\text{logit}\{p(y = 1 \mid \boldsymbol{x})\} = \log\left\{\frac{p(y = 1 \mid \boldsymbol{x})}{p(y = 0 \mid \boldsymbol{x})}\right\} = \beta_0 + \beta_1 x_1 + \beta_2 x_2,$$

where the response y, equals one if the patient has Parkinson's disease, zero otherwise, and β_0, β_1 and β_2 are coefficients that define the probability surface. Using this underlying model, known as logistic regression, we find that the probability surface fitted is the simple two-dimensional plane represented by the contours in Figure 1.2. However, it appears that this plane does not fit the data adequately, especially failing to capture the near certainty of Parkinson's disease for those patients for which X_2 was greater than 0.5.

In both regression and classification problems the underlying focus is on determining some form of functional relationship between the predictors and responses. Despite their widespread use and popularity, we see from the two examples given above that linear models (i.e. those based on straight lines, or planes) will typically be too restrictive to accurately capture the actual underlying relationship. We need classes of more general, nonlinear models that are flexible enough to capture the complexity of the data.

The search for appropriate models for classification and regression was once the exclusive domain of the statistician but now it has been taken up by a wide range of other researchers, especially those in the more modern, computer-based research fields. Although linear methods still have uses in modern statistics, it has become apparent that more sophisticated approaches are required to model accurately a wide class of datasets. This need was recognised some time ago by various pioneering, far-sighted researchers (see, for example, Halpern 1973; Rosenblatt 1956; Whittaker 1923) who suggested models that would have advantages over the conventional linear model but, at the time, had no practical possibility of being implemented routinely using the computing methods and technology of their time. However, with the substantial increase in currently available computational power these, so-called nonlinear, methods have gained popularity.

The other consequence of readily available computing power is the ease with which data can now be stored and collected. A knowledge of data analysis, and how to extract information from data, is therefore increasingly seen as important. This has implications in a wide variety of areas, from the social sciences (responses to questionnaires in a marketing campaign), medical sciences (classification of EEG traces, detection of relevant gene sequences), financial markets (understanding the volatility of a stock, assigning optimal portfolios) and commerce (predicting the value a client might be to a company over their lifetime, determining a person who is a poor credit risk). It is now commonplace for researchers from electrical engineering, computer science and bioinformatics, amongst others, to be involved with problems of classification and regression, which have also become known by a bewildering collection of new names such as pattern recognition, machine learning, intelligent data analysis, knowledge discovery, data mining and artificial intelligence. All of which essentially describe an approach to either regression of classification.

1.2 Bayesian Nonlinear Methods

Before any data analysis takes place, the true functional form between the input and output variables is almost always unknown (this is why we are trying to determine it!). We may have some clues about the relationship, or we may have some ideas about what we think our approximation should look like, but apart from this we may know very little.

Among the obvious questions which may influence how we proceed are the following.

1. What type of approximating functions should we use, or do we have available?
2. How do we know when we have found the 'best' approximation to the truth?
3. How can we incorporate any quantitative or qualitative knowledge we have about the relationship?

We shall address these questions in turn.

1.2.1 Approximating functions

There are a huge number of ways to approximate the truth and no one single specific approach can be uniformly better than any other in terms of predictive ability (see, for example, the discussion in Friedman (1993)). Commonly used approximating functions include linear and generalised linear models, smoothing splines, neural networks, Fourier bases, wavelets, decision trees and kernel smoothers. All of these provide explicit models for the relationship between the responses and predictors. Making inference about the relationship between y and $\boldsymbol{x}$ by defining such models is the focus of this book. We shall introduce a range of such models, describing their strengths and weaknesses, and demonstrate how use them to provide probabilistically sound inferences.

1.2.2 The 'best' model

A model is necessarily an approximation to the truth. In any real data analysis situation no single model will completely capture the true relationship between the inputs and outputs. We may think of the 'best' model among a set of alternatives as the one which most closely captures the true relationship for the particular purpose we have in mind. For example, if our aim is to predict well as judged by minimising the squared error between our predictions and the actual outcome, we shall see that a 'supermodel', corresponding to an average over all the specific models we might have considered, is the 'best'. This process of averaging over models means that the resulting approximation is drawn from a wider, more flexible, class of functions than is provided by any single specific model. We shall learn how to construct algorithms based on Markov chain Monte Carlo methods that identify good models and, where required, combine them in a sensible way.

1.2.3 Bayesian methods

We should clearly be able to improve the quality of the models we develop by incorporating whatever *a priori* qualitative or quantitative knowledge we have available. For example, knowledge of the degree of smoothness of a regression relationship might lead us to favour classes of models which provide smoothness in the second derivative (e.g. smoothing splines). Such approaches naturally lead us to Bayesian methods. These allow us to assign prior distributions to the parameters in the model which capture known qualitative and quantitative features, and then to update these priors in the light of the data, yielding a posterior distribution via Bayes' Theorem

$$\text{Posterior} \propto \text{Likelihood} \times \text{Prior}.$$

The ability to include prior information in the model is not only an attractive pragmatic feature of the Bayesian approach, it is theoretically vital for guaranteeing coherent inferences. Our aim in this book is to provide the reader with an insight into the field of Bayesian nonlinear modelling, detailing the models and demonstrating the wealth of information we can glean from them through determination of appropriate posterior distributions. Further, Bayesian methods perform well in relation to other approaches and numerous papers (which we shall reference as appropriate in the later chapters) demonstrate their empirical effectiveness.

To whet the appetite, Figures 1.3 and 1.4 display the outcomes of Bayesian nonlinear modelling approaches to the problems considered earlier in Figures 1.1 and 1.2. Here we have used standard Bayesian models to estimate the true relationship between the response and predictors. We shall consider the form and implementation of such models later on in the book. For now, we content ourselves with noting how these nonlinear models capture the observed features in the data dramatically better than the linear models described earlier.

1.3 Outline of the Book

The next chapter underpins the understanding of the book as a whole. It begins by giving an outline of why we choose to adopt a Bayesian approach to predictive inference and then goes on to demonstrate how we can implement such an approach. This is described with reference to the regression problem which dominates the first few chapters. In particular, Chapter 2 provides familiarity with analytic posterior inference for the Bayesian linear model (Lindley and Smith 1972) and with simulation algorithms.

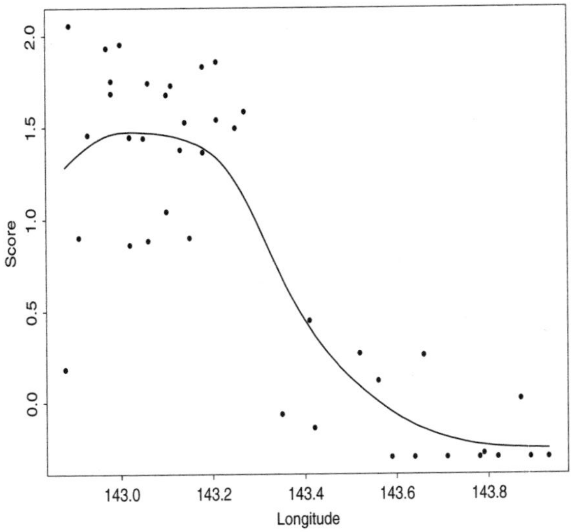

Figure 1.3 The Great Barrier Reef dataset together with the posterior mean Bayesian spline fit.

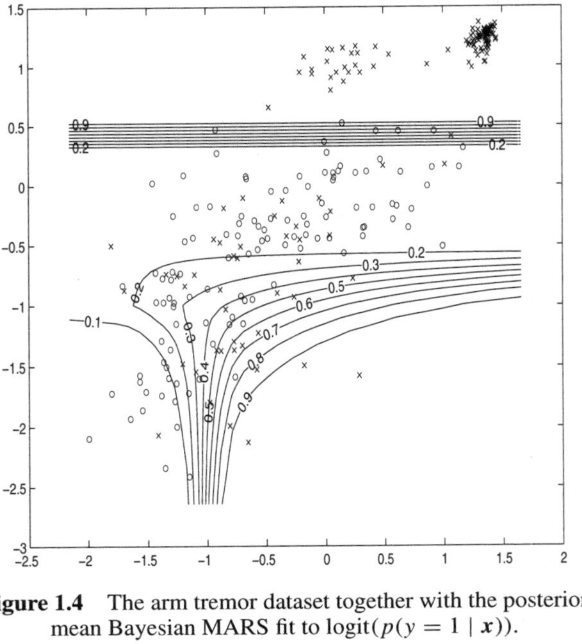

Figure 1.4 The arm tremor dataset together with the posterior mean Bayesian MARS fit to $\text{logit}(p(y = 1 \mid \boldsymbol{x}))$.

Chapter 3 goes through, in detail, the process of fitting simple Bayesian models, both analytically and via posterior simulation. We use the curve fitting problem to demonstrate simple applications of the methods with a view to aiding the understanding of readers new to the field.

Chapter 4 then introduces many models for fitting regression surfaces to data. It makes a distinction between models that are more suited to problems when there are many predictors and those that work well in fewer dimensions.

Bayesian generalised linear models form the focus of Chapter 5. Here, we show how to make posterior inference in these more difficult problems when analytic simplification is not available and simulation methods for high-dimensional posteriors are required. Generalised linear models provide a powerful toolkit for modelling in a wide variety of situations when the errors in the observations are not additive and normally distributed.

Chapter 6 covers Bayesian extensions to tree modelling. These are mainly used for the classification problem, although we include discussion of regression trees.

Chapter 7 extends the general method to partition models. These are similar to trees but allow for more flexible partitioning of the predictor space into regions where the data are assumed homogeneous.

Chapter 8 goes on to describe Bayesian nearest-neighbour modelling. Again, this method uses local structure in the data to lead us to a good way of making predictions.

Finally, Chapter 9 covers multiple response modelling when we wish to make inference about a vector of response variables given the predictors. This generalises the single response case, which is studied extensively throughout the rest of the book.

2

Bayesian Modelling

2.1 Introduction

This chapter describes the basic ideas that will run throughout the book. The aim is to introduce Bayesian methods through the Bayesian linear model. This is the basic parametric model we refer to throughout the first few chapters, and, despite its name, will be seen to be highly relevant to later material on nonlinear modelling. We do not go into great detail about the theoretical background to subjective probability as this is outside the scope of this applied text. Instead, we take as our starting point the idea that Bayesian inference is the coherent approach to statistics when some form of decision-making is (explicitly or implicitly) the ultimate objective – as here. There is a large literature that shows that if we want to act rationally so as not to contradict a set of reasonable axioms, we are naturally led to using the Bayesian paradigm. For more details see de Finetti (1937/1964, 1963, 1964/1972), Cox (1961), DeGroot (1970), Savage (1971) and Bernardo and Smith (1994).

An important theorem underpinning the work presented here is the representation theorem of de Finetti (1930). This provides the link between subjective probabilities and parametric models. It simplifies the way we need to place priors over observed responses by providing a constructive decomposition of the problem, revealing the role of Bayes' Theorem in the context of parametric modelling and demonstrating its relationship to predictive inference.

The Bayesian linear model (Lindley and Smith 1972) is central to the first few chapters of the book and we initially focus on making inference with this model. This involves determining the comparative merits of competing models as well as considering how to search for models that describe the data well. We also review the approach of averaging over models. We go on to describe methods that allow us to approximate Bayesian inferences and predictions when exact inference is not computationally feasible.

2.2 Data Modelling

2.2.1 The representation theorem for classification

We recall the arm tremor dataset displayed in Figure 1.2. Here we wish to determine the relationship between the two measurements on arm tremor, represented by the random vector $\boldsymbol{x} = (x_1, x_2)$, and the response variable, y, which shall define as taking the value 1 if a patient has Parkinson's disease (the crosses in Figure 1.2) and 0 otherwise. To help us to determine this relationship we make use of the dataset which contains measurements on the predictors and response for the n (= 357) data points; this is an example of binary classification.

First of all let us step back and consider the problem of making predictions for a new response, y_{n+1}, without any measurements on the predictors for any of the individuals so that the dataset is just given by $\mathcal{D} = (y_1, \ldots, y_n)$. That is, we want to determine $p(y_{n+1} \mid \mathcal{D})$, the density of the new response y_{n+1} conditioned on the observed responses. For the arm tremor example this involves predicting the probability that a new patient has Parkinson's disease before taking their arm tremor measurements and given only the responses (Parkinson's disease or not) of the other patients in the study. Conditional probability tells us that

$$p(y_{n+1} \mid y_1, \ldots, y_n) = \frac{p(y_{n+1} \cap \mathcal{D})}{p(\mathcal{D})} = \frac{p(y_1, \ldots, y_n, y_{n+1})}{p(y_1, \ldots, y_n)}, \qquad (2.1)$$

whenever the mass function in the denominator is non-zero.

To find the distribution we require (the left-hand side of (2.1)) we need to find expressions for both the joint mass functions in the right-hand side of (2.1). It is not easy to do this directly but we find that we can simplify the problem using the general representation theorem of de Finetti (1930).

2.2.2 *The general representation theorem*

Before we give the general representation theorem we must define the important concept of *exchangeability*, which is central to it (see Bernardo and Smith 1994, Section 4.2.2). A sequence of random variables $y_1, \ldots, y_n$ is said to be exchangeable if the joint density,

$$p(y_1, \ldots, y_n) = p(y_{\pi(1)}, \ldots, y_{\pi(n)}),$$

for all permutations π defined on $\{1, \ldots, n\}$. That is, the joint density of the y_is is unchanged whichever way we choose to label each individual y_i. Going back to the Parkinson's disease example this means that whichever individual we label as the first, the second, etc., does not alter the joint density of all the responses $p(y_1, \ldots, y_n)$.

The concept of exchangeability can also be straightforwardly extended to infinite sequences of random variables and this is used in the following theorem.

General Representation Theorem. *If $y_1, y_2, \ldots$ is an infinitely exchangeable sequence of real-valued random quantities, there exists a probability distribution F over $\mathcal{F}$, the space of all density functions, such that the joint density function of $y_1, \ldots, y_n$ has the form,*

$$p(y_1, \ldots, y_n) = \int_{\mathcal{F}} \prod_{i=1}^{n} p(y_i \mid F) \, \mathrm{d}Q(F),$$

where $p(y_i \mid F)$ is the density of y_i given that it distributed according to F, and

$$Q(F) = \lim_{n \to \infty} P(F_n),$$

where $P(F_n)$ is a distribution function evaluated at the empirical distribution function defined by

$$F_n(y) = \frac{1}{n} \sum_{i=1}^{n} I(y_i \leqslant y).$$

The general representation theorem underpins all parametric Bayesian modelling. Although, at first glance, it does not appear readily accessible it does tell us that we can decompose the joint density of the y_is by conditioning on a distribution F and then integrating over the range of F, $\mathcal{F}$ the space of all distribution functions. Note that the distribution Q encodes our beliefs about the empirical distribution, F_n.

Integrating over the space of all distribution functions to find the joint density of the responses is not as difficult as it might, at first, appear. Such work lies in the field of Bayesian *nonparametric* modelling (Lavine 1992, 1994; Walker *et al.* 1999). The main aim in this situation is to make inference about the unknown distribution, F.

In contrast to nonparametric modelling, we wish to make inference about the actual observed responses by approximating F using some vector of parameters, $\boldsymbol{\theta}$, that lie in some parameter space $\boldsymbol{\Theta}$. Thus, we choose to decompose the joint density using the *models* defined by $\boldsymbol{\theta} \in \boldsymbol{\Theta}$. Hence, from the general representation theorem, we find (Bernardo and Smith 1994, Corollary 1, p. 180) that

$$p(y_1, \ldots, y_n) = \int_{\boldsymbol{\Theta}} \left\{ \prod_{i=1}^{n} p(y_i \mid \boldsymbol{\theta}) \right\} p(\boldsymbol{\theta}) \, \mathrm{d}\boldsymbol{\theta}. \tag{2.2}$$

It is up to the individual modeller to choose which parameter vector, $\boldsymbol{\theta}$, is appropriate, as well as the form of $p(y_i \mid \boldsymbol{\theta})$. This is the main difference between parametric and nonparametric modelling.

Throughout the book we shall highlight suitable parameter vectors, $\boldsymbol{\theta}$, to use in various situations. This places us firmly in the parametric modelling framework. Nevertheless, such models have occasionally been called nonparametric, perhaps because of their relationship to the nonparametric models in the non-Bayesian literature.

2.2.3 *Bayes' Theorem*

Having placed ourselves in a parametric modelling framework via (2.2) we find that the predictive density of a new response is

$$p(y_{n+1} \mid \mathcal{D}) = \int_{\boldsymbol{\Theta}} p(y_{n+1} \mid \boldsymbol{\theta}) p(\boldsymbol{\theta} \mid \mathcal{D}) \, \mathrm{d}\boldsymbol{\theta}, \tag{2.3}$$

where $\mathcal{D} = (y_1, \ldots, y_n)$ and

$$p(\boldsymbol{\theta} \mid \mathcal{D}) = \frac{p(\mathcal{D} \mid \boldsymbol{\theta})p(\boldsymbol{\theta})}{p(\mathcal{D})} \tag{2.4}$$

$$= \frac{\{\prod_1^n p(y_i \mid \boldsymbol{\theta})\}p(\boldsymbol{\theta})}{\int_{\boldsymbol{\Theta}}\{\prod_1^n p(y_i \mid \boldsymbol{\theta})\}p(\boldsymbol{\theta})\,\mathrm{d}\boldsymbol{\theta}}.$$

So, in general, we find that inference about a new response y_{n+1}, given the data $\mathcal{D} = (y_1, \ldots, y_n)$ is made via the *posterior predictive density* given in (2.3). Further, this density can be written as an integral over the parameter space, $\boldsymbol{\Theta}$. To evaluate this integral we need to determine, $p(\boldsymbol{\theta} \mid \mathcal{D})$, which is known as the *posterior distribution* of $\boldsymbol{\theta}$. This is found by Bayes' Theorem, which, in its simplest form, is given by (2.4). We can even simplify (2.4) further by writing it as a function in terms of $\boldsymbol{\theta}$, i.e.

$$p(\boldsymbol{\theta} \mid \mathcal{D}) \propto p(\mathcal{D} \mid \boldsymbol{\theta})p(\boldsymbol{\theta}),$$

or

$$\text{Posterior} \propto \text{Likelihood} \times \text{Prior},$$

where $\propto$ denotes that two quantities are proportional to each other. Hence, the posterior distribution is found by combining the prior distribution for the parameters, $p(\boldsymbol{\theta})$, with the probability of observing the data given the parameters, $p(\mathcal{D} \mid \boldsymbol{\theta})$ (the likelihood).

We have seen how integration is ubiquitous in Bayesian modelling. We have found that we can model joint densities which are difficult to specify via parameters over which we must integrate. Hence, Bayesian modelling involves integrals over the parameters, whereas non-Bayesian methods often rely on optimisation of the parameters. The main difference between these methods is that optimisation fails to take into account the inherent uncertainty we have in the parameters. There is no 'true' value for $\boldsymbol{\theta}$ which can be found by optimisation. Instead, there is a range of plausible values for $\boldsymbol{\theta}$, each with some associated density.

2.2.4 *Modelling with predictors*

So far we have restricted our attention to modelling the distribution of a new response, y_{n+1}, given a dataset made up of n other responses $y_1, \ldots, y_n$ ($y_1, \ldots, y_{n+1}$ exchangeable). However, we can see that for the arm tremor dataset (Figure 1.2), the probability of being in either class varies over the space spanned by the measurements of arm tremor (i.e. the space spanned by predictors X_1 and X_2), which we shall call $\mathcal{X}$. A similar situation occurs for the regression example seen earlier; the Great Barrier Reef dataset (Figure 1.1). Here, the mean level of the response variable, Score, changes with the value of the predictor, Longitude. We need to model this dependence between the responses $y_1, \ldots, y_n, y_{n+1}$ and their corresponding measurements on the predictor variables $\boldsymbol{x}_1, \ldots, \boldsymbol{x}_n, \boldsymbol{x}_{n+1}$.

When we need to determine the conditional distribution of y_{n+1} given $\boldsymbol{x}_{n+1}$ we cannot appeal to the general representation theorem as the sequence of random variables, $y_1 \mid \boldsymbol{x}_1, y_2 \mid \boldsymbol{x}_2, \ldots$, is not an exchangeable sequence unless all of the $\boldsymbol{x}_i$ are identical. As $p(y_i \mid \boldsymbol{x}_i, \boldsymbol{\theta})$ is not equal for every value $\boldsymbol{x}_i$, to apply the representation theorem we make $\boldsymbol{\theta}$ a function of the predictor values $\boldsymbol{x} = (\boldsymbol{x}_1, \ldots, \boldsymbol{x}_n)$. Hence, we assign a prior over the extended set of parameters $\boldsymbol{\theta}(\boldsymbol{x}) = (\boldsymbol{\theta}(\boldsymbol{x}_1), \ldots, \boldsymbol{\theta}(\boldsymbol{x}_n))$ so that

$$p(y_1, \ldots, y_n \mid \boldsymbol{x}_1, \ldots, \boldsymbol{x}_n) = \int_{\Theta} \left\{ \prod_{i=1}^{n} p(y_i \mid \boldsymbol{\theta}(\boldsymbol{x}_i)) \right\} p(\boldsymbol{\theta}(\boldsymbol{x})) \, \mathrm{d}\boldsymbol{\theta}(\boldsymbol{x}).$$

To see how this works consider again the Great Barrier Reef dataset. In Chapter 1 we considered fitting a function which related the Score linearly with Longitude. This suggests taking

$$E(y_i \mid x_i, \mathcal{D}) = \beta_0 + \beta_1 x_i := \mu(\boldsymbol{x}_i). \tag{2.5}$$

Having used this model for the mean of the sampling distribution $p(y_i \mid \boldsymbol{\theta}(\boldsymbol{x}_i))$, we may also want to specify its variance. This is unlikely to be known before the modelling begins but is likely to be similar for all values of $\boldsymbol{x}_i$. Hence, we take the variance to be given by the unknown constant σ^2. The normal model is often used for such regression problems so we could make the extra assumption that

$$p(y_i \mid \boldsymbol{\theta}(\boldsymbol{x}_i)) = N(\mu(\boldsymbol{x}_i), \sigma^2),$$

where the normal density for a random variable $X \sim N(\mu, \sigma^2)$ is given by

$$p(x \mid \mu, \sigma^2) = \frac{1}{\sqrt{2\pi\sigma^2}} \exp\left\{ \frac{(x-\mu)^2}{2\sigma^2} \right\},$$

for $x \in \mathbb{R}$.

The assumptions on the mean and variance of the sampling distribution, as well as the assumption of normality, are all choices left to the data modeller. They lead to us being able to write down

$$p(y_1, \ldots, y_n \mid \boldsymbol{x}_1, \ldots, \boldsymbol{x}_n) = \int_{\Theta} \left\{ \prod_{i=1}^{n} N(y_i \mid \mu(\boldsymbol{x}_i), \sigma^2) \right\} p(\boldsymbol{\mu}(\boldsymbol{x}), \sigma^2) \, \mathrm{d}\boldsymbol{\theta}(\boldsymbol{x}), \tag{2.6}$$

where $\boldsymbol{\mu}(\boldsymbol{x}) = (\mu(\boldsymbol{x}_1), \ldots, \mu(\boldsymbol{x}_n))$, but they are all subject to being changed where appropriate. It is the focus of this book to highlight good choices for the mean function $\mu(\boldsymbol{x}_i)$, in particular, but also for other parameters such as the variance and the choice of sampling distribution to use.

When fitting a linear model to the mean of the regression function, as in (2.5), we see that each $\mu(\boldsymbol{x}_i)$ depends only on the choice of the intercept parameter, β_0, and the slope, β_1. So, instead of (2.6), we can write the model down in terms of these parameters

$$p(y_1, \ldots, y_n \mid \boldsymbol{x}_1, \ldots, \boldsymbol{x}_n)$$
$$= \int_{\mathbb{R}^+} \int_{\mathbb{R}} \int_{\mathbb{R}} \left\{ \prod_{i=1}^{n} N(y_i \mid \beta_0 + \beta_1 x_i, \sigma^2) \right\} p(\beta_0, \beta_1, \sigma^2) \, \mathrm{d}\beta_0 \, \mathrm{d}\beta_1 \, \mathrm{d}\sigma^2.$$

This means that we can place our priors directly on the parameters that define the model β_0, β_1 and σ^2.

We saw in Figure 1.1 that using a simple linear model for the mean did not provide us with enough parameters to vary to capture the true underlying curve adequately. This is in contrast to the estimated mean level in Figure 1.3, which shows much more fidelity to the data. The improved results were due to the fact that Figure 1.3 was produced using a generalisation of the linear model which is much more flexible. This is done by using more parameters to define the mean level, in what we shall call basis function models. We now go on to provide a detailed outline of how to perform regression with the Bayesian linear model, and show how this relates to the nonlinear basis function models we shall encounter in later chapters.

2.3 Basics of Regression Modelling

2.3.1 The regression problem

The regression problem involves determining the relationship between some response variable Y and a set of p predictor variables $\boldsymbol{X} = (X_1, \ldots, X_p)$. The most common form of structural assumption is that the responses are assumed to be related to predictors through some deterministic function f and some additive random error component ϵ so that

$$Y = f(\boldsymbol{X}) + \epsilon, \tag{2.7}$$

where ϵ is a zero-mean error distribution.

In most situations the predictor variables, $\boldsymbol{X}$, are assumed to be observed without error so they are not considered random. Thus, by taking expectations of (2.7) we find that the true relationship between Y and $\boldsymbol{X}$ is, in fact, just the conditional expectation of Y given $\boldsymbol{X} = \boldsymbol{x}$, i.e. $f(\boldsymbol{x}) = E(Y \mid \boldsymbol{X} = \boldsymbol{x})$. Our aim is to determine f so that we can uncover the true relationship between the response y at predictor location $\boldsymbol{x}$, given by $y = f(\boldsymbol{x})$. Typically, we are only interested in estimating f over some range of plausible predictor values which we shall denote by $\mathcal{X}$.

The true regression function f is unknown and we have no way of determining its analytic form exactly, even if one actually exists. We must content ourselves with finding approximations to it which are close to the truth. To do this we must make use of the observed dataset, $\mathcal{D}$, which consists of n observed responses at some known predictor locations so $\mathcal{D} = \{y_i, \boldsymbol{x}_i\}_{i=1}^n$.

We now consider approaches to modelling this regression problem and the way in which Bayes' Theorem enables us to update our beliefs about aspects of the model in the light of the dataset.

2.3.2 Basis function models for the regression function

Without prior knowledge as to the exact form of f we need to approximate it. A simple, and historically much used, solution is to make direct linear assumptions about the estimating function and take

$$g(\boldsymbol{x}) = \beta_0 + \beta_1 x_1 + \cdots + \beta_p x_p, \tag{2.8}$$

as a good proxy for f. We have seen (Figure 1.1) that this form for g (the estimating function) is restrictive and, in Great Barrier Reef example, does not have the flexibility to model the dataset adequately. Instead, we shall make use of the more general *basis function* models that assume that g is made up of a linear combination of basis functions and corresponding coefficients. Hence g can be written as

$$g(\boldsymbol{x}) = \sum_{i=1}^{k} \beta_i B_i(\boldsymbol{x}), \qquad \boldsymbol{x} \in \mathcal{X}, \tag{2.9}$$

where $\boldsymbol{\beta} = (\beta_1, \ldots, \beta_k)'$ is the set of coefficients corresponding to basis functions $B = (B_1, \ldots, B_k)$. We see that the linear model in (2.8) is just a special case of (2.9), where $k = p + 1$, $B_1(\boldsymbol{x}) = 1$ and $B_i(\boldsymbol{x}) = x_{i-1}$ for $i = 2, \ldots, p + 1$.

Typically, the basis functions used in (2.9) are nonlinear transformations of the data vector $\boldsymbol{x}$, thus extending enormously the class of functions defined by (2.8). Later on in the book we shall describe different basis sets that can be used for regression, and demonstrate how basis functions models can also be used for classification. For now, we proceed generally and just need to think of each basis function B_i as defining a map from the predictor space $\mathcal{X}$ to the real line $\mathbb{R}$.

2.4 The Bayesian Linear Model

The mechanisms of the Bayesian approach to model fitting to make inferences consists of three basic steps:

1. assign priors to all the unknown parameters;
2. write down the likelihood of the data given the parameters;
3. determine the posterior distribution of the parameters given the data using Bayes' Theorem.

From (2.7) and (2.9), we see that when we approximate f by g we get

$$y_i = \sum_{i=1}^{k} \beta_i B_i(\boldsymbol{x}_i) + \epsilon_i, \qquad i = 1, \ldots, n, \tag{2.10}$$

or, in matrix notation,

$$\boldsymbol{Y} = \boldsymbol{B}\boldsymbol{\beta} + \boldsymbol{\epsilon}, \tag{2.11}$$

where $\boldsymbol{Y} = (y_1, \ldots, y_n)'$, $\boldsymbol{\epsilon} = (\epsilon_1, \ldots, \epsilon_n)'$ and

$$\boldsymbol{B} = \begin{pmatrix} B_1(\boldsymbol{x}_1) & \cdots & B_k(\boldsymbol{x}_1) \\ \vdots & \ddots & \vdots \\ B_1(\boldsymbol{x}_n) & \cdots & B_k(\boldsymbol{x}_n) \end{pmatrix},$$

where $\boldsymbol{B}$ is known as the design matrix of the regression. However, we shall tend to refer to this as the *basis function matrix*.

We now follow through steps 1–3 for the standard Bayesian linear regression model (Lindley and Smith 1972), assuming that the errors are independent, normal random variables with common unknown regression variance σ^2.

2.4.1 The priors

Assuming that we have fully specified the set of basis functions, $B = (B_1, \ldots, B_k)$, the only unknown parameters in the model are the set of coefficients, $\boldsymbol{\beta} = (\beta_1, \ldots, \beta_k)'$, and the regression variance σ^2. To proceed in a Bayesian framework we must put priors over these unknowns.

The choice of prior distributions represents information available about unknown parameters. Provided it does not overly distort the representation of such functions, it is convenient to choose mathematically convenient forms of prior distributions which result in computationally tractable posterior distributions. In general, this is achieved through the use of *conjugate* prior distributions (see, for example, Bernardo and Smith 1994; Broemeling 1985; O'Hagan 1994).

For the Bayes linear model we find that the conjugate choice of (joint) prior for $\boldsymbol{\beta}$ and σ^2 is the normal inverse-gamma (NIG), which we shall denote by $p(\boldsymbol{\beta}, \sigma^2) = \mathrm{NIG}(\boldsymbol{m}, \boldsymbol{V}, a, b)$. This distribution is defined by four parameters and via its conditional representation. Hence the probability density function (PDF) of the prior is given by

$$\begin{aligned} p(\boldsymbol{\beta}, \sigma^2) &= p(\boldsymbol{\beta} \mid \sigma^2) p(\sigma^2) \\ &= N(\boldsymbol{m}, \sigma^2 \boldsymbol{V})\, \mathrm{IG}(a, b) \\ &= \frac{b^a}{(2\pi)^{k/2} |\boldsymbol{V}|^{1/2} \Gamma(a)} (\sigma^2)^{-(a+(k/2)+1)} \\ &\quad \times \exp[-\{(\boldsymbol{\beta} - \boldsymbol{m})' \boldsymbol{V}^{-1} (\boldsymbol{\beta} - \boldsymbol{m}) + 2b\}/(2\sigma^2)], \end{aligned} \tag{2.12}$$

for $\sigma^2 > 0$ and where $\Gamma(\cdot)$ represents the standard gamma function, i.e. $\Gamma(a) = \int_0^\infty t^{a-1} \mathrm{e}^{-t}\, \mathrm{d}t$. As $p(\boldsymbol{\beta}, \sigma^2) = p(\boldsymbol{\beta} \mid \sigma^2) p(\sigma^2)$ we see that the prior is made up of a normal prior on the coefficients given the regression variance, and an inverse-gamma prior on the regression variance. Thus $\boldsymbol{m}$ is the prior mean of the coefficients and their prior variance is $\sigma^2 \boldsymbol{V}$. Note that we parametrise the inverse-gamma distribution so that the PDF of $\sigma^2 \sim \mathrm{IG}(a, b)$ is

$$p(\sigma^2) = \frac{b^a}{\Gamma(a)} (\sigma^2)^{-(a+1)} \exp(-b/\sigma^2), \qquad \sigma^2 > 0, \tag{2.13}$$

where $a, b > 0$.

2.4.2 *The likelihood*

The likelihood for a model is defined, up to proportionality, as the joint probability of observing the data regarded as a function of the model parameters. For the linear model with a fixed design matrix $\boldsymbol{B}$, we shall write this as just $p(\mathcal{D} \mid \boldsymbol{\beta}, \sigma^2)$, or alternatively $p(Y \mid \boldsymbol{X}, \boldsymbol{\beta}, \sigma^2)$. As mentioned earlier, we assume that the errors are normally distributed so that $\boldsymbol{\epsilon} \sim N(\boldsymbol{0}, \sigma^2 \boldsymbol{I})$, where $\boldsymbol{I}$ is the identity matrix of suitable dimension (n in this case). From (2.11) we see that $\boldsymbol{\epsilon} = Y - \boldsymbol{B\beta}$, hence

$$p(\mathcal{D} \mid \boldsymbol{\beta}, \sigma^2) = N(\boldsymbol{B\beta}, \sigma^2 \boldsymbol{I}),$$

so that the complete likelihood is found with reference to the density of a multivariate normal distribution, and is given by

$$p(\mathcal{D} \mid \boldsymbol{\beta}, \sigma^2) = (2\pi\sigma^2)^{-n/2} \exp\left\{-\frac{(Y - \boldsymbol{B\beta})'(Y - \boldsymbol{B\beta})}{2\sigma^2}\right\}. \tag{2.14}$$

We note the similarities between this expression and the prior in (2.12) which motivates the choice of the conjugate prior.

2.4.3 *The posterior*

We know from Bayes' Theorem that the posterior distribution of the model parameters satisfies

$$p(\boldsymbol{\beta}, \sigma^2 \mid \mathcal{D}) = \frac{p(\mathcal{D} \mid \boldsymbol{\beta}, \sigma^2) p(\boldsymbol{\beta}, \sigma^2)}{p(\mathcal{D})}. \tag{2.15}$$

By multiplying together the expressions in (2.12) and (2.14) we can determine the form of the posterior up to a constant term, $p(\mathcal{D})$, that does not depend on either $\boldsymbol{\beta}$ or σ^2. Thus, we find that by using the identity

$$\begin{aligned}(\boldsymbol{Y} - \boldsymbol{B\beta})'(Y - \boldsymbol{B\beta}) + (\boldsymbol{\beta} - \boldsymbol{m})'\boldsymbol{V}^{-1}(\boldsymbol{\beta} - \boldsymbol{m}) + 2b \\ \equiv (\boldsymbol{\beta} - \boldsymbol{m}^*)'(\boldsymbol{V}^*)^{-1}(\boldsymbol{\beta} - \boldsymbol{m}^*) + 2b^*,\end{aligned}$$

we can write

$$p(\boldsymbol{\beta}, \sigma^2 \mid \mathcal{D}) \propto (\sigma^2)^{-(a^*+(k/2)+1)} \exp\left\{-\frac{(\boldsymbol{\beta} - \boldsymbol{m}^*)'(\boldsymbol{V}^*)^{-1}(\boldsymbol{\beta} - \boldsymbol{m}^*) + 2b^*}{2\sigma^2}\right\}, \tag{2.16}$$

where

$$\boldsymbol{m}^* = (\boldsymbol{V}^{-1} + \boldsymbol{B}'\boldsymbol{B})^{-1}(\boldsymbol{V}^{-1}\boldsymbol{m} + \boldsymbol{B}'Y), \tag{2.17}$$

$$\boldsymbol{V}^* = (\boldsymbol{V}^{-1} + \boldsymbol{B}'\boldsymbol{B})^{-1}, \tag{2.18}$$

$$a^* = a + n/2, \tag{2.19}$$

$$b^* = b + \{\boldsymbol{m}'\boldsymbol{V}^{-1}\boldsymbol{m} + Y'Y - (\boldsymbol{m}^*)'(\boldsymbol{V}^*)^{-1}\boldsymbol{m}^*\}/2. \tag{2.20}$$

We see that the functional form of the posterior (2.16) is the same as that for the prior (2.12) except that the four parameters in the prior have been updated. Hence $p(\boldsymbol{\beta}, \sigma^2 \mid \mathcal{D}) = \text{NIG}(\boldsymbol{m}^*, \boldsymbol{V}^*, a^*, b^*)$ with the normalising constant in (2.16), found by comparison with (2.12), given by

$$\frac{(b^*)^{a^*}}{(2\pi)^{k/2}|\boldsymbol{V}^*|^{1/2}\Gamma(a)}.$$

This standard updating result for the normal inverse-gamma model is given in many textbooks on Bayesian statistics (see Bernardo and Smith 1994; Broemeling 1985; O'Hagan 1994) and allows us to determine the posterior distribution of the parameters $\boldsymbol{\beta}$ and σ^2 simply.

The updated parameters in (2.17)–(2.20) can be interpreted without difficulty as $\boldsymbol{m}^*$ is the posterior mean estimate to the coefficients $\boldsymbol{\beta}$ with their variance dependent on $\boldsymbol{V}^*$. Further, as $Y'Y - (\boldsymbol{m}^*)'(\boldsymbol{V}^*)^{-1}\boldsymbol{m}^*$ equals the sum of squared residuals of the fit using the posterior mean, b^*/a^* is approximately equal to the residual sum of squares in the case where $a, b \to 0$ and $\boldsymbol{V} \to \boldsymbol{0}$.

2.5 Model Comparison

It is often the case that a number of competing theories (or models) exist to describe the process that generated Y. Suppose now that we have M such competing models $\mathcal{M}_1, \ldots, \mathcal{M}_M$ and we wish to compare them. These models may have different basis sets, as well as different priors, but we wish to see which one models the data 'best', in some sense. To make this explicit, let model $\mathcal{M}_i$ be defined by the linear regression of $\boldsymbol{X}$ on Y according to

$$Y = \boldsymbol{B}_i\boldsymbol{\beta}_i + \boldsymbol{\epsilon},$$

where $\boldsymbol{\epsilon} \sim N(\boldsymbol{0}, \sigma^2\boldsymbol{I})$. Further, the basis function matrix $\boldsymbol{B}_i$ is fixed and known and the prior over the unknowns is given by $p(\boldsymbol{\beta}_i, \sigma^2) = \text{NIG}(\boldsymbol{m}_i, \boldsymbol{V}_i, a, b)$. Thus we allow for a different prior mean and variance of the coefficients in each model but have the same prior on the regression variance throughout. We shall now see how to make model comparison in this particular situation.

Suppose for the time being that we know that the set of models under consideration contains the true one. This is sometimes known as the $\mathcal{M}_{\text{closed}}$ modelling perspective (Bernardo and Smith 1994). Then we know that the actual data we observe have been generated according to one of the $\mathcal{M}_i$. An obvious question is then 'Which one is the true model?'. This involves comparing the relative merits of each model after we have observed the data, and to make a choice between the available models it is natural to utilise a decision theoretic approach (Bernardo and Smith 1994).

This requires us to define a loss function that encodes our ideas about the relative desirability of the different actions (choices of model) we can take. In this case, as we are only interested in finding the true model, and we are assuming it is in the set of models we are considering, the appropriate loss function has zero-one form (i.e. zero loss if we determine the true model correctly, and loss one otherwise).

Let d_i represents the decision 'choose model $\mathcal{M}_i$' and $\ell(m_i, \mathcal{M}_j)$ represent the loss associated with making decision m_i when the true model is $\mathcal{M}_j$. Now, we can write down the loss function mathematically as

$$\ell(d_i, \mathcal{M}_j) = \begin{cases} 0, & \text{if } i = j, \\ 1, & \text{otherwise.} \end{cases} \tag{2.21}$$

Hence the expected loss of the decision d_i (choosing $\mathcal{M}_i$), given the data, is given by

$$\begin{aligned} E\{\ell(d_i \mid \mathcal{D})\} &= \sum_{j=1}^{M} \ell(d_i, \mathcal{M}_j) p(\mathcal{M}_j|\mathcal{D}) \\ &= 1 - p(\mathcal{M}_i \mid \mathcal{D}). \end{aligned}$$

So, to minimise the expected posterior loss, we should choose the model with the largest posterior probability, an intuitively sensible choice.

2.5.1 Bayes' factors

If we wish to make a comparison of just two models, $\mathcal{M}_i$ and $\mathcal{M}_j$, the above analysis leads us to look at the posterior odds ratio

$$\frac{p(\mathcal{M}_i \mid \mathcal{D})}{p(\mathcal{M}_j \mid \mathcal{D})} = \frac{p(\mathcal{D} \mid \mathcal{M}_i)}{p(\mathcal{D} \mid \mathcal{M}_j)} \times \frac{p(\mathcal{M}_i)}{p(\mathcal{M}_j)}, \tag{2.22}$$

where $p(\mathcal{M}_i)$ is the prior probability of $\mathcal{M}_i$ being the true model. To determine this ratio we need to multiply the prior odds ratio by what is known as the marginal likelihood ratio (also known as the integrated likelihood ratio or prior predictive).

The marginal likelihood of model $\mathcal{M}_i$ gives a measure of the probability of observing the data given that $\mathcal{M}_i$ is true. To account for the uncertainty in the unknowns associated with each model we determine the marginal likelihood by integrating out the model parameters. So, for the linear model example already described, we know that

$$p(\mathcal{D} \mid \mathcal{M}_i) = \int\int p(\mathcal{D} \mid \boldsymbol{\beta}_i, \sigma^2) p(\boldsymbol{\beta}_i, \sigma^2)\, \mathrm{d}\boldsymbol{\beta}_i\, \mathrm{d}\sigma^2,$$

for $i = 1, \ldots, M$. These integrals can be performed analytically when we use the conjugate prior specification. This proceeds by first integrating out $\boldsymbol{\beta}_i$ by writing the integrand as a normal density, and then by eliminating σ^2 by comparison with an inverse-gamma distribution. However, a simpler way to determine $p(\mathcal{D} \mid \mathcal{M}_i)$ is given by first rearranging (2.15) so that

$$p(\mathcal{D} \mid \mathcal{M}_i) = \frac{p(\mathcal{D} \mid \boldsymbol{\beta}_i, \sigma^2) p(\boldsymbol{\beta}_i, \sigma^2)}{p(\boldsymbol{\beta}_i, \sigma^2 \mid \mathcal{D})}.$$

Then, using equations (2.12), (2.14) and (2.16) we find that for the Bayes linear model with conjugate priors,

$$p(\mathcal{D} \mid \mathcal{M}_i) = \frac{|\boldsymbol{V}_i^*|^{1/2}(b)^a \Gamma(a_i^*)}{|\boldsymbol{V}_i|^{1/2}\pi^{n/2}\Gamma(a)}(b_i^*)^{-a_i^*}, \tag{2.23}$$

for $i = 1, 2$, and where the updated parameters are given in (2.17)–(2.20).

From (2.22) we can see the important role that the marginal likelihood plays in transforming relative prior beliefs into relative posterior beliefs. For this reason an important quantity for choosing between models is the Bayes factor (Aitkin 1991; Bernardo and Smith 1994; Good 1988; Kass and Raftery 1995). This is defined for the comparison of two competing models and, if we wish to consider the relative merits of $\mathcal{M}_i$ over $\mathcal{M}_j$, is given by

$$\text{BF}(\mathcal{M}_i, \mathcal{M}_j) = \frac{p(\mathcal{M}_i \mid \mathcal{D})}{p(\mathcal{M}_j \mid \mathcal{D})} \bigg/ \frac{p(\mathcal{M}_i)}{p(\mathcal{M}_j)},$$

the posterior to prior odds ratio.

The Bayes factor measures whether the data have increased or decreased the relative odds of $\mathcal{M}_i$ over $\mathcal{M}_j$. For values greater than one these odds have been increased, whereas for values less than one the relative odds of $\mathcal{M}_j$ have increased. Further, in the case where the prior probabilities of each model have been taken to be equal, we find that the Bayes factor is exactly the same as the posterior odds ratio. In this case, choosing the model with the highest posterior probability is equivalent to picking the model whose Bayes' factor with respect to any other model is greater than one. Kass and Raftery (1995) suggest that, if the Bayes factor for $\mathcal{M}_i$ over $\mathcal{M}_j$ is between 1 and 3, then there is little perceived difference between the models, between 3 and 20 there is positive evidence in favour of $\mathcal{M}_i$, 20 to 150 strong evidence and, if the Bayes factor is over 150, there is very strong evidence in favour of $\mathcal{M}_i$.

Returning to the linear regression example we see that, if there is no prior preference between $\mathcal{M}_i$ and $\mathcal{M}_j$, then the Bayes factor in favour of $\mathcal{M}_i$ over $\mathcal{M}_j$ is given by

$$\text{BF}(\mathcal{M}_i, \mathcal{M}_j) = \frac{|\boldsymbol{V}_j|^{1/2}|\boldsymbol{V}_i^*|^{1/2}(b_j^*)^{a^*}}{|\boldsymbol{V}_i|^{1/2}|\boldsymbol{V}_j^*|^{1/2}(b_i^*)^{a^*}}, \tag{2.24}$$

where $a^* = a_i^* = a_j^* = a + n/2$.

2.5.2 *Occam's razor*

Bayesian analysis provides a coherent framework for model comparison. However, for this to be a sensible method we need to know that when we compare only two models, say

$$\mathcal{M}_1: \quad g(\boldsymbol{x}) = \sum_{i=1}^{k} \beta_i B_i(\boldsymbol{x}) \qquad \text{and} \qquad \mathcal{M}_2: \quad g(\boldsymbol{x}) = \sum_{i=1}^{k+1} \beta_i B_i(\boldsymbol{x}), \tag{2.25}$$

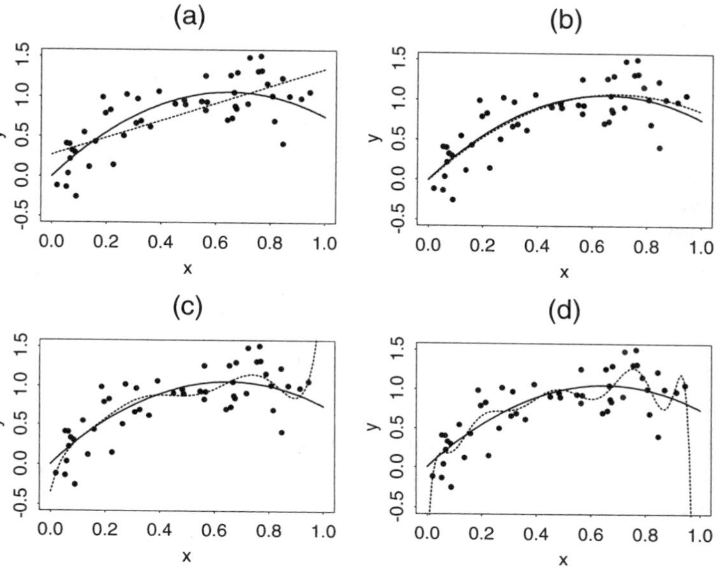

Figure 2.1 The effect of overfitting. In each plot we display a 50-point dataset simulated from a quadratic function (the solid lines) with added independent normal errors. The dotted lines give fits to the data with polynomials of order k: (a) $k = 1$; (b) $k = 2$; (c) $k = 7$; (d) $k = 12$.

we are not always led to choosing $\mathcal{M}_2$ just because it has one more basis function than $\mathcal{M}_1$. This extra basis function will allow us to explain the data better but will not necessarily lead to better predictions by the model as the extra basis function may give rise to the model *overfitting* the data.

Combating overfitting is vital to the performance of flexible models. For example, consider the data in Figure 2.1. Here we have fitted, by least squares, the polynomial functions,

$$g(x) = \beta_0 + \sum_{i=1}^{k} \beta_i x^i,$$

for various values of k, to the simulated dataset shown. We see that just fitting larger and larger models is not a good strategy to use, even though they are only generalisations of the smaller ones. As k increases the fitted functions vary more rapidly, but picking k too small (i.e. $k = 1$ in this case) does not allow enough flexibility for the fitted function to adequately approximate the truth. In particular, we see that at the edges of the data and for large values of k, the fitted curves can be significantly different from the truth. This is obviously undesirable and is a major reason why too flexible, or overfitted, models perform particularly poorly in terms of prediction.

On the face of it we might be concerned that the flexible modelling strategy we advocate might be prone to overfitting the data by adding too many basis functions. Indeed, many papers found in the literature advocate explicit priors on the model space that penalise the dimension of the model. However, throughout this book we argue that such a measure is unnecessary. The Bayesian framework contains a natural penalty against over complex models, sometimes called *Occam's razor*, which essentially states that a simpler theory is to be favoured over a more complex one, all other things being equal.

We illustrate the Bayesian embodiment of this philosophy, adapted from the excellent review by MacKay (1995) (see also the review of Jeffreys and Berger 1992), in the context of models $\mathcal{M}_1$ and $\mathcal{M}_2$ given earlier in (2.25). We know that the relative plausibility of the models having observed data $\mathcal{D}$ can be determined by the Bayes factor in favour of $\mathcal{M}_1$ over $\mathcal{M}_2$. When the priors on each model are assumed equal this is simply given by

$$\mathrm{BF}(\mathcal{M}_1, \mathcal{M}_2) = \frac{p(\mathcal{D} \mid \mathcal{M}_1)}{p(\mathcal{D} \mid \mathcal{M}_2)} = \frac{p(\mathcal{M}_1 \mid \mathcal{D})}{p(\mathcal{M}_2 \mid \mathcal{D})}.$$

The first thing to note is that simpler models tend to make more precise prior predictions. We can see this by considering the prior predictive distributions at $\boldsymbol{x} = \boldsymbol{0}$ for both the models. Without loss of generality we can fix the basis functions so that $B_1(\boldsymbol{0}) = \cdots = B_{k+1}(\boldsymbol{0}) = 0$. Then, if we take the prior on each of the coefficients to be independent with $\beta_i \sim N(0, v)$, for some known constant v. Thus the prior predictions at $\boldsymbol{x} = \boldsymbol{0}$ are given by

$$p(y \mid \boldsymbol{x} = \boldsymbol{0}, \mathcal{M}_1) \sim N(0, kv) \quad \text{and} \quad p(y \mid \boldsymbol{x} = \boldsymbol{0}, \mathcal{M}_2) \sim N(0, (k+1)v).$$

So, in this case, the more complex the model the less precise the prior prediction is. We find that this is illustrative of a more general rule demonstrated in Figure 2.2. A direct result of this principle is that if the observed data are reasonably supported by a simple model then the simple model will have higher posterior probability, because its prior is more concentrated around the observed data. Conversely, the more complex model spreads the marginal likelihood of the data over a wider region of data-space.

2.5.3 *Lindley's paradox*

We have seen how the Bayes factor naturally penalises unnecessarily complex models. Although this is an appealing reason to use Bayes' factors for model comparison, it is important to understand that the use of very vague priors can lead to misleading results when comparing models of different dimension.

Again let us consider comparing the relative merits of models $\mathcal{M}_1$ and $\mathcal{M}_2$ given in (2.25). Commonly the prior means for the coefficients in these models are taken to be zero, reflecting prior uncertainty in the sign of each coefficient. Also, if we assume that the coefficients are independent and identically distributed *a priori* this leads us to choosing a NIG($\boldsymbol{0}$, $v\boldsymbol{I}$, a, b) prior, where v is a prior parameter that we need to set. With this particular prior we find the Bayes factor in favour of $\mathcal{M}_1$ over $\mathcal{M}_2$ is

$$\sqrt{v}\frac{|v^{-1}\boldsymbol{I} + \boldsymbol{B}_1'\boldsymbol{B}_1|^{-1/2}}{|v^{-1}\boldsymbol{I} + \boldsymbol{B}_2'\boldsymbol{B}_2|^{-1/2}}\left(\frac{b_2^*}{b_1^*}\right)^{a^*}, \tag{2.26}$$

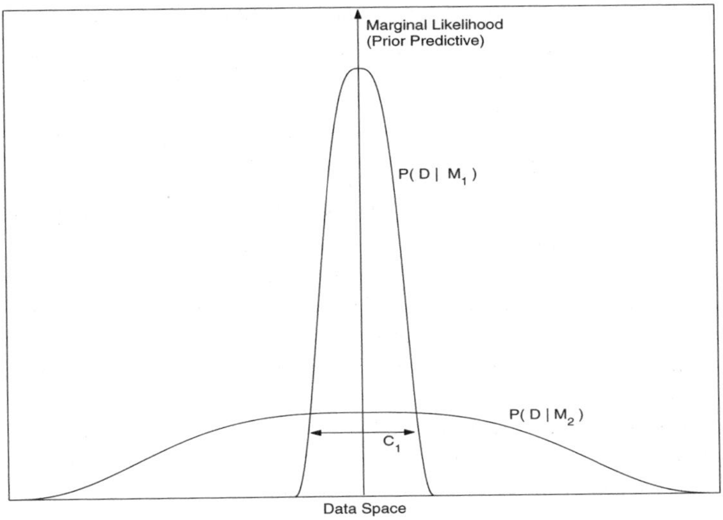

Figure 2.2 An illustration of how Bayesian inference adheres to the principle of Occam's razor. The horizontal axis represents the data-space with the vertical axis giving the marginal likelihood of the data. Suppose there are two competing models $\mathcal{M}_1$ and $\mathcal{M}_2$, where model $\mathcal{M}_1$ covers only a subspace of that covered by $\mathcal{M}_2$. As both $p(\mathcal{D} \mid \mathcal{M}_1)$ and $p(\mathcal{D} \mid \mathcal{M}_2)$ must integrate to one over the data-space, if data are observed that lie in the region C_1, then the simpler model $\mathcal{M}_1$ is preferred as it has a higher probability in this region.

where the updated values a^* and b_1^* and b_2^* are found with reference to (2.19) and (2.20). Now, if we fix a and b we see that the Bayes factor in (2.26) is solely a function of v, the prior variance of the coefficients and the data.

In most situations knowledge about the spread of coefficient values is not directly available. We could think of encoding this prior ignorance by taking a 'flat' prior for $\boldsymbol{\beta}_i$, i.e. a normal with variance tending to infinity. If we take the limit of (2.26) as $v \to \infty$, we find that the Bayes factor also tends to infinity. Hence, $\mathcal{M}_1$ will always be the favoured model for large settings of v. It turns out that this is the case only because $\mathcal{M}_1$ has fewer coefficients than $\mathcal{M}_2$. This is an example of Lindley's paradox (Bartlett 1957; Lindley 1957), which asserts that when comparing models with different numbers of parameters and using diffuse priors, the simpler model is always favoured over the more complex one. In this example $\mathcal{M}_1$ contains only k basis functions, whereas $\mathcal{M}_2$ has $k + 1$. This is why, as $v \to \infty$, the simpler (i.e. smaller) model $\mathcal{M}_1$ is favoured over $\mathcal{M}_2$, irrespective of the data.

There has been some criticism of the Bayesian paradigm in reaction to Lindley's paradox (see, for example, Lavine and Schervish 1999; Shafer 1982), mainly because flat priors on coefficients are mistakenly thought to be uninformative in this context. In fact they are highly informative, as pointed out by DeGroot (1982). He noted that by assigning a diffuse prior over a coefficient value, say β, we are putting a lot of prior weight on values for which $|\beta|$ is large. Further, assigning a diffuse prior does not

Table 2.1 The number of possible models, M, for various values of T.

T	2	5	10	25	50	200	1000
M	4	32	1024	3.3×10^7	$\approx 10^{15}$	$\approx 10^{60}$	$\approx 10^{301}$

correspond to ignorance about functions of β. For instance DeGroot (1982) noticed that if $\eta = \exp(\beta)/\{1 + \exp(\beta)\}$, then a diffuse prior on β suggests that nearly all the prior weight on η is around 0 or 1.

We feel that Lindley's paradox teaches us that there is no natural way to encode complete ignorance about the spread of the coefficients. Instead, we must find good ways to choose v as it controls the complexity of the favoured model. In some sense it is a smoothing parameter as large values of it lead to choosing models with fewer parameters.

Further discussion on Lindley's paradox can be found in Edwards *et al.* (1963), Leamer (1978), Berger (1980) and Robert (1995b). In the next section we highlight how it is an important principle in Bayesian statistics.

2.6 Model Selection

In the previous sections we have described the Bayesian linear model with conjugate priors and how to choose between a set of competing models $\mathcal{M}_1, \ldots, \mathcal{M}_M$ with different priors and basis function matrices. Further, if the prior probability of each model is identical we have seen how using the Bayes factor to choose between models is equivalent to picking the model with the highest posterior probability. For a moderate number of competing models such methods are straightforward and relatively fast. In terms of computation, they only require M marginal likelihood evaluations followed by a comparison between the marginal likelihoods to determine the largest. The most time consuming aspect of determining the marginal likelihoods is in calculating the posterior variance $\boldsymbol{V}^*$ from (2.18) as this involves inverting a square matrix with dimension equal to the number of basis functions. The computations involved can be made faster by using the Cholesky decomposition (see, for example, Thisted 1988) as the matrices involved are symmetric.

Consider the situation where we have T possible basis functions which we could include in the model, $B_1, \ldots, B_T$. Our aim is to determine which combination of these T bases gives us the largest marginal likelihood. As each basis function can either be included or excluded from the model, we see that there are $2^T (= M)$ possible combinations of basis functions. Hence to determine the preferred model we must evaluate the likelihood for all these 2^T possibilities and find which is the largest. For small T this can be achieved in a reasonable amount of computational time but, as the number of models grows exponentially, for medium to large T the set becomes too large to perform an exhaustive search of the model space. In Table 2.1 we show how this model space grows, but note for large T we also have to invert matrices which can be up to size $T \times T$ so things are even worse than suggested by the table. We feel that with current computational power for $T < 25$ evaluation of every possible model is feasible but for larger T we need other methods to try and search for models with high marginal likelihoods.

2.6.1 *Searching for models*

As we have seen in the last subsection, selection of models by an exhaustive search over the model space is impractical in all but the most trivial problems. We must restrict our attention to searching only over a subspace of models. Obviously, we would wish this selective search to include many of the models with the highest marginal likelihood. We now list some methods we could use for determining exactly, or approximately, the best model amongst those we consider. We give references for completeness, but we shall not use these in what follows. Note that throughout this section we shall denote by H some lack-of-fit criterion and $\delta H = H_{\text{prop}} - H_{\text{curr}}$, the difference in this criterion between the proposed and current model.

Greedy searches. The most obvious technique for searching for good models is to start with the model with no basis functions and then look among the potential basis functions we could add to this current model and add the best one, according to some prespecified criterion, H. That is, we add the model for which δH is a minimum amongst all the candidate models with one more basis function than the current one. We can continue to do this until some stopping rule suggests we do not add any more bases. This method is known as forward selection. Of course, we could do the same in reverse, starting with the full model and removing basis functions one at a time. This is known as backwards deletion and does not necessarily give the same final model as forwards selection.

Obvious difficulties with these, so-called greedy, algorithms are determining the criterion, H, as well as determining a good stopping rule. A well-motivated method is given by taking H as the generalised cross-validation criterion of Craven and Wahba (1979). However, Friedman (1979) noted that choosing a suitable stopping rule would be difficult so instead suggested adding basis functions one at a time until the model was much larger than desired and then performing backwards deletion until H is at a minimum.

The greedy search strategy (see, for example, Aho *et al.* 1983) is a quick way to find a reasonable model but the one-step-ahead search for basis functions can cause problems. The fitting criterion is only minimised with respect to one basis function at a time. However, we would much rather the complete set of basis functions were chosen together so they could jointly minimise the lack-of-fit criterion. Nevertheless, the greedy search strategy is the most widely used model search technique for the non-Bayesian equivalents of the models we present in this book (see, for example, Breiman *et al.* 1984; Friedman 1991; Hastie and Tibshirani 1990).

Leaps and bounds. The leaps and bounds algorithm proposed by Furnival and Wilson (1974) is an efficient search technique that can be used to find the set of bases with minimum residual sum of squares (RSS), amongst all the 2^k possible combinations. It proceeds by searching down 'trees' that successively remove basis functions from the full model with k bases. When a model is reached for which the RSS is greater than that for another model evaluated with less bases, no more splits are made down the branch of the tree with greater RSS. This is because we know that all the models in that branch are worse, in terms of RSS, than another model that we have already discovered. This method is a considerable improvement over exhaustive search and can be used to determine the 'best' set of basis by repeating the procedure and using a criterion, H, that is monotone in RSS to choose between the models like the C_p statistic of Mallows (1973). In the Bayesian literature, this method has been used by, amongst others, Volinsky (1997) and Volinsky *et al.* (1997). However, in our case we wish to rank the models by the marginal likelihood, which does not necessarily increase when we add a basis function to the current model. This makes this method unsuitable for our purpose of finding the best model in terms of marginal likelihood, rather than some other criterion which is a simple function of the residual sum of squares.

EM algorithms. Dempster *et al.* (1977a) introduced the expectation–maximisation, or EM, algorithm. This is an iterative procedure aimed at producing maximum likelihood estimators. The algorithm is often written down as a way of providing maximum likelihood estimators to a set of parameters in the presence of missing data. The algorithm then proceeds, after the initial setting of the parameters, by cycling over the two steps:

1. E-step: compute the conditional expectation of the missing parameters given the observed data and the current parameter settings;
2. M-step: update the current parameter vector by maximising the conditional expectation determined in the E-step.

The algorithm is run until convergence to a single model, within some acceptable toleration, has been detected.

Although the method was developed to handle incomplete datasets we find that the EM algorithm can also be used to find good models in many other contexts. In fact, any model where auxiliary, or latent, variables can be introduced to simplify the form of likelihood fall into this class. The most frequently used situation in which the EM algorithm is used is the mixture model which is used to estimate unknown densities (see, for example, Titterington *et al.* 1985).

The main problem of the EM algorithm is that the starting values can sometimes have a significant influence on the final model. This is because the algorithm is deterministic by nature so can get stuck in local modes or saddle points of the likelihood surface. To overcome this stochastic versions of the algorithm have been suggested (see, for example, Celeux and Diebolt 1985; Diebolt and Ip 1996). These proceed similarly to the usual EM algorithm except that the E-step is replaced by a stochastic step which draws a realisation from the conditional distribution of the missing data (or parameters). This allows the algorithm the possibility to escape local modes and has been shown to have better convergence properties (Diebolt and Celeux 1993) than the simple EM algorithm.

Simulated annealing. A weakness of greedy search procedures is that they only optimise the set of bases in a one-step-ahead fashion, rather than jointly. By always taking the basis function which decreases the lack-of-fit the most we can get trapped in locally good models, rather than the globally good ones we are searching for. So, instead of performing such deterministic 'hill-climbs', a better strategy is to introduce a random component into the algorithm. This is designed to allow the algorithm to escape local modes by actually moving to models with larger lack-of-fit than the current one. This is to ensure that the search can jump out of local modes and hopefully, later on, find better models. Simulated annealing (Geyer and Thompson 1995; Kirkpatrick *et al.* 1983; Otten and van Ginneken 1989) is one such strategy which proceeds by proposing jumps from the current model according to some user-designed proposal mechanism. The proposed model is then always accepted if it lowers the lack-of-fit criterion, and accepted with probability $\exp(-\delta H/T)$ otherwise, where T is known as the temperature. For $T = 0$ we only accept models which decrease H and the method is just another example of a greedy algorithm. However, when T is large we nearly always move to the proposed model and essentially pay no attention to the lack-of-fit criterion. Both these methods are, by themselves, unacceptable so simulated annealing involves running the algorithm at a variety of temperatures and taking T to be a function of the number of iterations, or proposals, made. Thus, at the beginning of the algorithm many moves are made but as the number of iterations increases the algorithm tends to the greedy approach. The way that T is decreased as a function of time is known as the 'cooling schedule'. Asymptotic results suggest that slow cooling using the logarithmic cooling schedule ensures that the global optimum is reached with certainty (Hajek 1988).

Genetic algorithms. Another popular search strategy is the genetic algorithm (Goldberg 1989; Holland 1975; Mitchell 1996). This bases the search for a good model on biological evolution by maintaining a population of models which are used to form new generations of models through combining and mutating the current population. The models most likely to be combined are the ones which have the lowest values of the lack-of-fit criterion, H, biasing the next generation to contain similarities to the better models of the previous one. This mimics the evolutionary premise of 'survival of the fittest', as the worst models in the population are unlikely to 'parent' models in the next generation. For example, when searching for basis functions the offspring from two-parent models might include each individual basis function with probability 0 if neither parents include it, $\frac{1}{2}$ if one parent includes it, and with certainty otherwise. The mutation step is included to locally move among individual models, mimicking genetic mutation. Thus, a particular member of the population might be mutated so that in the next generation it has one less, or one more, basis function. Generations are successively borne until some suitable stopping rule is satisfied.

We have outlined a variety of search techniques. The method used is often a matter of personal preference but the appropriateness of each one does depend on the search space to be explored. The fact that so many search methods have been proposed demonstrates the difficulties involved in the search for the 'optimal' set of model parameters. However, we shall see in the next section that such a search is often not necessary.

2.7 Model Averaging

2.7.1 Predictive inference

We have seen how decision theory leads us to picking the model with the highest posterior probability, based on the assumption that one of the models under consideration exactly represents the true relationship between the response Y and the predictors $\boldsymbol{X}$ and using a zero-one loss function. However, practically speaking we can rarely expect to exactly replicate the truth with any model and, even if we could, we would be extremely fortunate to include it in our set of models under consideration, $\mathcal{M}_1, \ldots, \mathcal{M}_M$. This makes the $\mathcal{M}_{\text{closed}}$ perspective and the use of the zero-one loss function in (2.21) unsuitable in most sensible data analyses. An alternative approach is the $\mathcal{M}_{\text{open}}$ modelling perspective (Bernardo and Smith 1994). This accepts that none of the models is true with the prior probability on each reflecting our relative degrees of belief on each model. Thus, the $\mathcal{M}_{\text{open}}$ perspective provides answers closest to those that would be derived from the true model.

Again we use decision theory to aid us in making a sensible decision, giving an operational meaning to 'closest answers' by seeking to minimise the difference between a model-based estimate, $\tilde{y}$, and the true response, y, at some given values of the predictors $\boldsymbol{x} (\in \mathcal{X})$. This gives rise to the predictive squared-error loss function, $(\tilde{y} - y)^2$. Thus, we can define our estimate to y using this loss function as

$$\tilde{y} = \arg\min_d \int (d - y)^2 p(y \mid \boldsymbol{x}, \mathcal{D})\, \mathrm{d}y.$$

Straightforward differentiation with respect to d yields that

$$\tilde{y} = \int y\, p(y \mid \boldsymbol{x}, \mathcal{D})\, \mathrm{d}y = E(y \mid \boldsymbol{x}, \mathcal{D}), \tag{2.27}$$

the expectation of the posterior predictive distribution for y. Now, this distribution can be written as

$$p(y \mid \boldsymbol{x}, \mathcal{D}) = \sum_{i=1}^{M} p(y \mid \boldsymbol{x}, \mathcal{D}, \mathcal{M}_i) p(\mathcal{M}_i \mid \mathcal{D}), \tag{2.28}$$

which is simply a mixture of the individual predictive distributions for y given each model, weighted by the posterior probability of each model. So, to minimise the predictive loss we are naturally led to posterior averaging across the model space. In this way the Bayesian framework naturally circumvents the problems associated with selecting a single model. Throughout this book we adopt this predictive approach to inference, using model averaging in the majority of situations.

For the linear model we can write down the distribution in (2.28) explicitly. First, Bayes' Theorem gives us

$$p(\mathcal{M}_i \mid \mathcal{D}) = \frac{p(\mathcal{D} \mid \mathcal{M}_i)p(\mathcal{M}_i)}{\sum_j p(\mathcal{D} \mid \mathcal{M}_j)p(\mathcal{M}_j)}, \tag{2.29}$$

with $p(\mathcal{D} \mid \mathcal{M}_i)$ given by (2.23). Further, we find that (see Problem 2.10)

$$p(y \mid \boldsymbol{x}, \mathcal{D}, \mathcal{M}_i) = \mathrm{St}(\boldsymbol{b}'\boldsymbol{m}^*, b^*(\boldsymbol{I} + \boldsymbol{b}'\boldsymbol{V}^*\boldsymbol{b}), a^*), \tag{2.30}$$

where $\boldsymbol{b}' = (B_1(\boldsymbol{x}) \cdots B_k(\boldsymbol{x}))$ is the output of the basis functions in model $\mathcal{M}_i$ at location $\boldsymbol{x}$. The distribution in (2.30) is a Student distribution, where for $y \sim \mathrm{St}(\mu, v, c)$, the PDF is given by

$$p(y) = \frac{\Gamma(\frac{1}{2}(c+1))}{\Gamma(\frac{1}{2}c)\sqrt{v\pi}}\left\{1 + \frac{(y-\mu)^2}{v}\right\}^{-(c+1)/2}, \tag{2.31}$$

which has mean μ and variance $v/(c-2)$ for $c > 2$, where c is known as the degrees of freedom of the distribution.

From (2.30) we see that the distribution of y is Student with a^* $(= a + n/2)$ degrees of freedom. The Student distribution is well known to have more probability in its tails than a normal distribution with the same location and scale parameters. It is the uncertainty in the regression variance, σ^2, that leads to y being distributed in this way. Hence, by using the linear model with unknown regression variance we are naturally led to relating the response and predictors via a *Student process*, rather than the much more widely known Gaussian process (see, for example, Neal 1999; Rasmussen 1996). A possible advantage of this is the greater robustness we would expect from a Student process. However, for reasonable sample sizes ($n > 100$) the difference between the two processes is minimal as when $a^* \to \infty$ the Student process tends to a Gaussian process. A comparison of Student and Gaussian processes is outside the scope of this book, but it is worth pointing out that by assigning Gaussian process priors, or adopting the model priors we shall suggest, we find predictive distributions of similar form (Gaussian or Student). However, one significant difference remains between the two methods. Gaussian processes require the inversion of an $n \times n$ matrix to provide predictions whereas in the above we only require the inversion of a $k \times k$ matrix (where the number of basis functions k is nearly always much smaller than the sample size n).

In summary, from (2.28) and (2.30) we see that the predictive distribution is a weighted mixture of Student distributions. Further, as expectation is a linear operator we find from (2.27) and (2.30) that

$$\begin{aligned}\tilde{y} &= \sum_{i=1}^{M} E(y \mid \boldsymbol{x}, \mathcal{D}, \mathcal{M}_i) p(\mathcal{M}_i \mid \mathcal{D}) \\ &= \sum_{i=1}^{M} (\boldsymbol{b}'\boldsymbol{m}^*) p(\mathcal{M}_i \mid \mathcal{D}).\end{aligned} \tag{2.32}$$

Although these results have only been shown in relation to the prediction at one new location similar ones can be found if we are interested in the joint response at a series of points in $\mathcal{X}$ (O'Hagan 1994, see p. 262). We also leave as an exercise for the reader (Problem 2.10) the derivations of the estimates for some other common loss functions. However, we note that when making predictive inference no matter what sensible loss function is used, we are never led to choosing a single model. Instead estimates are based on the mixture density given in (2.28).

2.7.2 Problems with model selection

Although decision theory naturally leads us, under reasonable assumptions, to using model averaging to make predictions, we find that this method also makes sense for more abstract reasons.

First, individual model selection fails to take into account the inherent uncertainty associated with using only the set of bases associated with the best model. Even the most optimistic modeller, when analysing real data, would never suspect that the model selected exactly mimics the true unknown relationship, f. There is uncertainty present and, especially when the basis functions themselves are poorly aligned with the truth, this can be significant. Even though in this book we shall propose using flexible basis functions which can capture many types of true relationship, potential for *model misspecification* still exists. By admitting that no model we can propose will ever be exactly true we are adopting the $\mathcal{M}_{\text{open}}$ view of modelling discussed in Bernardo and Smith (1994) and Key *et al.* (1999). This leaves the Bayesian modeller in a dilemma as we place priors on models we have accepted as not being true. Practically speaking we feel that the priors only give a relative prior weight amongst the models under consideration and do not represent our true beliefs. These ideas are taken further in the discussion of Raftery *et al.* (1996) and by Holmes *et al.* (1999b).

The second problem with model selection is that the data are used to choose both the best model, and to account for the uncertainty in the model parameters. This leads to selection bias as some of the parameters are estimated conditional on the model chosen being true, e.g. in a regression set-up we use the data both to estimate the coefficients and regression variance as well as choosing the basis functions to include in the model. See Copas (1983) for a discussion of some of the problems associated with selection bias.

We saw in Section 2.5.1 how, for the linear model, we could account for the uncertainty in $\boldsymbol{\beta}$ and σ^2 by marginalising out these parameters. Similarly, model averaging accounts for the model misspecification by marginalising over the set of possible models. From the theorem of total probability, we know

$$p(\mathcal{D}) = \sum_{i=1}^{M} p(\mathcal{D} \mid \mathcal{M}_i) p(\mathcal{M}_i), \tag{2.33}$$

so that marginalisation only requires the calculation of this sum (or integral if the model space is continuous). For the linear model this can be calculated from (2.23) once a prior distribution has been assigned to the individual models.

2.7.3 Other work on model averaging

The idea of model averaging in a Bayesian context appears to originate with Roberts (1965), who combines two models to include information from two 'experts'. Leamer (1978) develops the ideas further and more recent work includes Hodges (1987), George and McCulloch (1993), Chatfield (1995), Draper (1995), Volinsky (1997), Raftery *et al.* (1997), Clyde (1999) and Hoeting *et al.* (1999). Outside the Bayesian framework, the idea arises in bootstrapping (Efron 1982; Efron and Tibshirani 1993), bagging (Breiman 1996) and boosting (Freund 1995; Freund and Schapire 1996, 1997).

We shall, for the most part, take the view that combining forecasts is a good idea and typically use such estimates to make predictions. Nevertheless, in certain situations it may not be appropriate to use this estimate. Averaged models do not necessarily lie in the set of considered models and they are usually more difficult to interpret. Single models can sometimes be more appropriate summaries of the data analysis although they usually predict less well.

2.8 Posterior Sampling

We saw in Section 2.6 how, when we wish to select one model from a large number of possible models, we need to adopt either a clever optimal search strategy or an approximate search of the model space. Similar problems face us when we try to model average over a large candidate set of models. In these cases, summation (or integration) in the denominator of (2.29) cannot be implemented exactly and we are forced to employ methods that approximate the posterior distribution.

Now, instead of thinking about choosing between models $\mathcal{M}_1, \ldots, \mathcal{M}_M$ let us parametrise this collection of models via a one-to-one mapping to some vector of parameters $\boldsymbol{\theta} = (\theta_1, \ldots, \theta_k)$, where $\boldsymbol{\theta}$ lies in the model space $\boldsymbol{\Theta}$. Here k is the total number of parameters in the model class and can be varying or constant, hence the parameter vector $\boldsymbol{\theta}$ is not necessarily of fixed dimension. With this representation we can write

$$p(\mathcal{D}) = \sum_{i=1}^{M} p(\mathcal{D} \mid \mathcal{M}_i) p(\mathcal{M}_i)$$
$$= \int_{\boldsymbol{\Theta}} p(\mathcal{D} \mid \boldsymbol{\theta}) p(\boldsymbol{\theta}) \, \mathrm{d}\boldsymbol{\theta}.$$

Again we see how integration is a central feature in Bayesian methods. It is required that we not only evaluate posterior distributions, but also the predictive distributions we are especially interested in, for example,

$$p(y \mid \boldsymbol{x}, \mathcal{D}) = \int_{\boldsymbol{\Theta}} p(y \mid \boldsymbol{x}, \boldsymbol{\theta}, \mathcal{D}) p(\boldsymbol{\theta} \mid \mathcal{D}) \, \mathrm{d}\boldsymbol{\theta}. \tag{2.34}$$

These integrals are rarely in an analytically tractable form. Sometimes, when the integrals can be expressed as summations, they can be written down explicitly but even then cannot be evaluated in a reasonable amount of computation time as they have prohibitively many terms.

In the integrand of (2.34) is the posterior density of $\boldsymbol{\theta}$ which we cannot evaluate. The other density is known as, when we choose which model to use, we explicitly give a form for $p(y \mid \boldsymbol{x}, \boldsymbol{\theta})$ and the prediction y is typically conditionally independent of the data, given $\boldsymbol{\theta}$. When k is small the integral may be only over a small number of parameters so well-known approximate integration techniques such as quadrature exist to first determine $p(\boldsymbol{\theta} \mid \mathcal{D})$ and then (2.34) (see, for example, Lindley 1980). However, when k is large such methods break down and alternatives must be found.

Even without exact knowledge of $p(\boldsymbol{\theta} \mid \mathcal{D})$, if we can sample from this distribution then we know that

$$p(y \mid \boldsymbol{x}, \mathcal{D}) \approx \sum_{t=1}^{N} p(y \mid \boldsymbol{x}, \boldsymbol{\theta}^{(t)}, \mathcal{D}) p(\boldsymbol{\theta}^{(t)} \mid \mathcal{D})$$
$$= \frac{1}{N} \sum_{t=1}^{N} p(y \mid \boldsymbol{x}, \boldsymbol{\theta}^{(t)}, \mathcal{D}), \tag{2.35}$$

where $\boldsymbol{\theta}^{(1)}, \ldots, \boldsymbol{\theta}^{(N)}$ are draws from the posterior distribution of $\boldsymbol{\theta}$. This is the basis of the Monte Carlo method for approximating integrals. The accuracy of the approximation increases with N but it is also related to the dependence of the draws. If the draws in the generated sample of models are dependent, then the greater the dependence the longer the approximation takes to converge to the truth.

There are general classes of methods that allow us to sample from $p(\boldsymbol{\theta} \mid \mathcal{D})$ which only require knowledge of the distribution up to the constant of proportionality. These are collectively known as Markov chain Monte Carlo (MCMC) simulation algorithms. The basis of the methods is that, starting from an initial set of parameters $\boldsymbol{\theta}^{(0)}$, a Markov chain is run with transition density from $\boldsymbol{\theta}^{(t)}$ to $\boldsymbol{\theta}^{(t+1)}$ given by $q(\boldsymbol{\theta}^{(t+1)} \mid \boldsymbol{\theta}^{(t)})$. Given the transition density q, a Markov chain may, or may not, have a stationary distribution

$\pi(\boldsymbol{\theta})$, where $\pi(\boldsymbol{\theta})$, if it exists, is a distribution to which the Markov chain converges from any initial state, $\boldsymbol{\theta}^{(0)}$. Remarkably, we can assign transition densities q in such a way that the stationary distribution of the Markov chain, $\pi(\boldsymbol{\theta})$, is identical to the posterior distribution we are interested in sampling from, $p(\boldsymbol{\theta} \mid \mathcal{D})$. Thus, to generate dependent random samples from $p(\boldsymbol{\theta} \mid \mathcal{D})$ we need to run the Markov chain until it has converged to its stationary distribution and then we can use output from the chain to approximate required integrals using expressions like (2.35).

To give the algorithm a chance to converge to $\pi(\boldsymbol{\theta})$ the output (i.e. the $\boldsymbol{\theta}^{(t)}$) from the first few iterations of the algorithm, known as the burn-in, are discarded. After this time convergence is assumed and yet some of the models after this time are discarded too. This is to make the generated sample of models less dependent, so the approximation to the integral requires a smaller value of N to obtain reasonable accuracy.

In the next few subsections we shall provide examples of MCMC algorithms. These provide methods for specifying transition densities that ensure that, given a sufficient number of iterations, the chain reaches the correct stationary distribution.

A pragmatic difficulty with MCMC methods is that it is difficult to know when convergence has been reached, so we do not know at what point to start collecting the generated sample. We defer further discussion of this point until after we have introduced the sampling algorithms that we shall employ extensively in the forthcoming chapters.

2.8.1 The Gibbs sampler

The Gibbs sampler described below requires the knowledge of the conditional posterior distributions $p(\theta_i \mid \mathcal{D}, \boldsymbol{\theta}_{-i})$, where $\boldsymbol{\theta}_{-i}$ is the elements of $\boldsymbol{\theta}$ with the ith one removed.

Gibbs-sampler:

Assign starting values to $\boldsymbol{\theta}^{(0)}$;
Set $t = 0$;
REPEAT
  Draw $\theta_1^{(t+1)}$ from $p(\theta_1 \mid \mathcal{D}, \theta_2^{(t)}, \ldots, \theta_k^{(t)})$;
  Draw $\theta_2^{(t+1)}$ from $p(\theta_2 \mid \mathcal{D}, \theta_1^{(t+1)}, \theta_3^{(t)}, \ldots, \theta_k^{(t)})$;
  etc;
  Draw $\theta_k^{(t+1)}$ from $p(\theta_k \mid \mathcal{D}, \theta_1^{(t+1)}, \theta_2^{(t+1)}, \ldots, \theta_{k-1}^{(t+1)})$;
  $t = t + 1$;
  Store every mth value of $\boldsymbol{\theta}$ after
    an initial burn-in period;
END REPEAT;

Other ways of 'blocking' the variables are also acceptable, rather than just simulating from each univariate conditional distribution in turn. Also, we are not restricted to using this deterministic scan of the elements of $\boldsymbol{\theta}$ and can use a random method to decide which one to update next. However, the Gibbs sampler is most often used in this form. For further information see Geman and Geman (1984), Gelfand and Smith (1990) and Smith and Roberts (1993).

The linear model with unknown basis functions introduced before demonstrates a simple application of the Gibbs sampler. Suppose again, as in Section 2.6, that there are T possible basis functions $B_1, \ldots, B_T$ and the model space consists of all $M = 2^T$ possible combinations of these. Now, let θ_i be an indicator random variable which is one when B_i is included in the model and zero otherwise. Every model can now be described by the parameter vector $\boldsymbol{\theta} = (\theta_1, \ldots, \theta_T)$ and all the possible models lie in the T-dimensional model space, $\boldsymbol{\Theta} = \{0, 1\}^T$. This leads to a redefinition of the estimating function given in (2.9), so now we take

$$g(\boldsymbol{x}) = \sum_{i=1}^{T} \theta_i \beta_i B_i(\boldsymbol{x}). \tag{2.36}$$

To implement the Gibbs sampler for this model all we need to do is write down the conditional posteriors of the θ_i. As θ_i is binary this is straightforward as

$$p(\theta_i = 0 \mid \mathcal{D}, \boldsymbol{\theta}_{-i}) = \frac{p(\mathcal{D} \mid \theta_i = 0, \boldsymbol{\theta}_{-i}) p(\theta_i = 0 \mid \boldsymbol{\theta}_{-i})}{\sum_{j=0}^{1} p(\mathcal{D} \mid \theta_i = j, \boldsymbol{\theta}_{-i}) p(\theta_i = j \mid \boldsymbol{\theta}_{-i})}, \tag{2.37}$$

for $i = 1, \ldots, T$ and where $\boldsymbol{\theta}_{-i} = \boldsymbol{\theta} \backslash \theta_i$. Note that all the terms above can be calculated once a prior on $\boldsymbol{\theta}$ has been assigned and by using (2.23). Obviously,

$$p(\theta_i = 1 \mid \mathcal{D}, \boldsymbol{\theta}_{-i}) = 1 - p(\theta_i = 0 \mid \mathcal{D}, \boldsymbol{\theta}_{-i})$$

so sampling from this distribution is very simple.

2.8.2 *The Metropolis–Hastings algorithm*

This Metropolis–Hastings sampler (Hastings 1970; Metropolis *et al.* 1953; Tierney 1994) does not require knowledge of the conditional posteriors to update the parameters so is more general than the Gibbs sampler. However, to implement the algorithm we must choose proposal distributions with which to update each element of $\boldsymbol{\theta}$. In the following description of the algorithm we have not written it in generality but in a simpler form corresponding to how we intend to use it later in the book. As in the algorithm for the Gibbs sampler we choose to update the elements singly and in a fixed order. Also, we use q_i to denote the user-set (univariate) proposal density which is used to update the ith element of the parameter vector.

MH-sampler:

```
Assign starting values to θ^(0);
Set t = 0;
REPEAT
  FOR i = 1, ..., k
    Set θ_{-i} = (θ_1^(t+1), ..., θ_{i-1}^(t+1), θ_{i+1}^(t), ..., θ_k^(t));
    Draw θ'_i from the proposal density q_i(θ'_i | θ_i^(t), θ_{-i});
    Set u to a draw from a U(0, 1) distribution;
    IF u < Acceptance-prob(θ_i^(t), θ'_i, θ_{-i}, i)
      Set θ_i^(t+1) = θ'_i;                    /*Accept proposal*/
    ELSE
      Set θ_i^(t+1) = θ_i^(t);                 /*Reject proposal*/
    END IF;
  END FOR;
  t = t + 1;
  Store every mth value of θ^(t) after
    an initial burn-in period;
END REPEAT;
```

Acceptance-prob$(\theta_i^{(t)}, \theta_i', \boldsymbol{\theta}_{-i}, i)$**:**

$$\alpha = \min\left\{1, \frac{p(\theta_i', \boldsymbol{\theta}_{-i} \mid \mathcal{D})}{p(\theta_i^{(t)}, \boldsymbol{\theta}_{-i} \mid \mathcal{D})} \frac{q_i(\theta_i^{(t)} \mid \theta_i', \boldsymbol{\theta}_{-i})}{q_i(\theta_i' \mid \theta_i^{(t)}, \boldsymbol{\theta}_{-i})}\right\};$$

RETURN(α);

Having to set the proposal densities makes this method less attractive for sampling from posteriors where we actually know the full conditionals. In this case Gibbs sampling is often preferred due to its greater simplicity. Also, as might be expected, if we take the q_i to be the conditional posteriors, $p(\theta_i \mid \boldsymbol{\theta}_{-i}, \mathcal{D})$ (i.e. those used in the Gibbs sampler), then the two methods are identical (Problem 2.10). Other possible proposal distributions are a symmetric one (e.g. a normal distribution centred around the current value of the parameter), which leads to an example of the original Metropolis algorithm given in Metropolis *et al.* (1953). For this sampler the acceptance probability is just

$$\begin{aligned}\alpha &= \min\left\{1, \frac{p(\theta_i', \boldsymbol{\theta}_{-i} \mid \mathcal{D})}{p(\theta_i^{(t)}, \boldsymbol{\theta}_{-i} \mid \mathcal{D})}\right\} \\ &= \min\left\{1, \frac{p(\mathcal{D} \mid \theta_i', \boldsymbol{\theta}_{-i}) p(\theta_i', \boldsymbol{\theta}_{-i})}{p(\mathcal{D} \mid \theta_i^{(t)}, \boldsymbol{\theta}_{-i}) p(\theta_i^{(t)}, \boldsymbol{\theta}_{-i})}\right\}. \qquad (2.38)\end{aligned}$$

Note that it is only through (2.38) that we can calculate the acceptance probability. The use of Bayes' Theorem to obtain this expression eliminates the need to determine $p(\mathcal{D})$, the intractable normalising constant. Remember we only resorted to simulation methods because of our inability to determine $p(\mathcal{D})$, so its algebraic cancellation in the above expression is crucial.

Other important choices of proposal densities are the conditional prior densities, so $q_i(\theta_i' \mid \theta_i, \boldsymbol{\theta}_{-i}) = p(\theta_i' \mid \boldsymbol{\theta}_{-i})$. This is a form of independence sampler as the proposed move to θ_i' is made independently of the current state of that parameter θ_i. Adopting these proposals, the acceptance probability can again be written straightforwardly as

$$\begin{aligned}\alpha &= \min\left\{1, \frac{p(\mathcal{D} \mid \theta_i', \boldsymbol{\theta}_{-i})p(\theta_i', \boldsymbol{\theta}_{-i})}{p(\mathcal{D} \mid \theta_i^{(t)}, \boldsymbol{\theta}_{-i})p(\theta_i^{(t)}, \boldsymbol{\theta}_{-i})} \frac{p(\theta_i^{(t)} \mid \boldsymbol{\theta}_{-i})}{p(\theta_i' \mid \boldsymbol{\theta}_{-i})}\right\} \\ &= \min\left\{1, \frac{p(\mathcal{D} \mid \theta_i', \boldsymbol{\theta}_{-i})}{p(\mathcal{D} \mid \theta_i^{(t)}, \boldsymbol{\theta}_{-i})}\right\},\end{aligned} \tag{2.39}$$

where we have used the relationship $p(\theta_i, \boldsymbol{\theta}_{-i}) = p(\theta_i \mid \boldsymbol{\theta}_{-i})p(\boldsymbol{\theta}_{-i})$ to simplify the expression. We see that this choice of proposal density leads to the acceptance probability depending on the ratio of model likelihoods. Hence, when the model proposed has a greater likelihood than the current one it is always accepted, otherwise it is accepted with a probability determined by the likelihood ratio. Thus the sampler performs a stochastic 'hill-climb' over the posterior distribution.

Again, we can use the Metropolis–Hastings method to sample from the posterior distribution of the model space for linear models. Defining the model space as in Section 2.8.1 and taking the proposal densities as the conditional priors we see from (2.39) and (2.24) that the acceptance probability is

$$\alpha = \min\{1, \mathrm{BF}([\theta_i', \boldsymbol{\theta}_{-i}], [\theta_i, \boldsymbol{\theta}_{-i}])\}, \tag{2.40}$$

which is just the minimum of 1 and the Bayes factor in favour of the proposed model over the current one. We have seen before how the Bayes factor is used for model choice, which is why we feel this simple form for the acceptance probability is particularly attractive. In fact, throughout the book we shall use the conditional priors as 'default' proposal distributions for this reason.

2.8.3 The reversible jump algorithm

Both the Gibbs and Metropolis–Hastings samplers we have outlined require a deterministic scan over the elements in the random vector of model parameters, $\boldsymbol{\theta}$. However, if $\boldsymbol{\theta}$ is of a very high dimension this method becomes impractical. For instance, later in the book we shall present methods for which the number of possible basis functions, T, is greater than one million. If every iteration required a cycle over one million parameters, the computational effort in approximating the posterior would be immense and impractical. To allow us to consider such large model spaces we must get round this problem. We can do this by defining the model space in a more parsimonious fashion.

Let us return to the linear model of (2.36). Suppose that there are 10 possible basis functions and we wish to represent the model with the first, fifth and seventh basis functions included. In this framework this is done through the parameter vector,

$$\boldsymbol{\theta} = (1, 0, 0, 0, 1, 0, 1, 0, 0, 0).$$

This model space requires redundant zeros for basis functions that are not included so, when T is very large, far too many random variables are needed to represent a model. We would rather be able to define $\boldsymbol{\theta}$ in such a way that it is only as long as the number of basis functions in the model. This is easy in this case as we could just take $\boldsymbol{\theta} = (1, 5, 7)$. So in general, we can write this new model formulation as

$$g(\boldsymbol{x}) = \sum_{i \in \boldsymbol{\theta}} \beta_i B_i(\boldsymbol{x}). \tag{2.41}$$

The model space, $\boldsymbol{\Theta}$, in the old representation of (2.36) is $\{0, 1\}^T$ and contains 2^T elements. In this new representation we think of the model space differently as being made up of a union of smaller subspaces so that $\boldsymbol{\Theta} = \bigcup_{k=0}^{T} \boldsymbol{\Theta}_i$, where each $\boldsymbol{\Theta}_k$ consists of all the combinations of k elements from $\{1, \ldots, T\}$. Hence each $\boldsymbol{\Theta}_i$ contains

$$\binom{T}{i}$$

members so the complete model space in this new formulation still has 2^T possible elements as

$$\binom{T}{0} + \binom{T}{1} + \cdots + \binom{T}{T} = 2^T.$$

The reason that this new formulation is so useful is that when T is very large we only need a few of the bases in the model to adequately approximate the true regression function. Hence, we define some maximum number of basis functions, say K, that the model is allowed to contain. Thus $\boldsymbol{\theta}$ is always of dimension less than K so it does not become too computationally inefficient to update in the sampling algorithm.

The two samplers already described are only adapted to work with fixed-dimensional posteriors. However, the dimension of $\boldsymbol{\theta}$ is no longer fixed, as before, so we need a different method to sample from its posterior distribution, $p(\boldsymbol{\theta} \mid \mathcal{D})$.

Green (1995) proposed the reversible jump MCMC method, which is a generalisation to the Metropolis–Hastings algorithm which can sample from posteriors of varying dimension. Again, we shall describe the methodology in the way that we use it to sample from linear models, rather than in complete generality. For more details see Green (1995) and Richardson and Green (1997). Basically, the algorithm proceeds similarly to **MH-sampler** except that $\boldsymbol{\theta}$ is now defined differently so we do not need to cycle through the elements individually. Also, the acceptance probability is modified to take into account the change in dimension.

Instead of cycling through the θ_i in the main loop of the algorithm, we make a random choice between adding a basis function, or taking one away. This corresponds to adding or taking away an element of $\boldsymbol{\theta}$. We define b_k to be the probability that we choose to add a basis function given that $\boldsymbol{\theta}$ is currently of dimension k, and d_k is defined similarly for the death probability. In this case we know that $b_k + d_k = 1$ for $k = 0, 1, \ldots, K$.

We see from the acceptance probability in **MH-sampler** that the fraction involved is just a posterior ratio for the current and proposed models multiplied by a proposal ratio. From Bayes' Theorem we see that this acceptance probability could also be thought of as

$$\min\{1, \text{likelihood} \times \text{prior} \times \text{proposal ratios}\}. \tag{2.42}$$

However, when we change between dimension, Green (1995) shows that the acceptance probability also includes a Jacobian term. This is needed because when we propose to add an element to $\boldsymbol{\theta} = (\theta_1, \ldots, \theta_k)$, we must generate a random variable, say u, which we use to determine a proposed new element, θ'_{k+1}. The Jacobian term then accounts for the change in variables from

$$(\boldsymbol{\theta}, u) \longleftrightarrow (\boldsymbol{\theta}, \theta'_{k+1}) =: \boldsymbol{\theta}'.$$

For a Jacobian to exist the transformation involved must be bijective, and this is equivalent to what Green (1995) calls detailed balance. Obviously, the above change of variables assumes that only the existing parameters of the model are unchanged when we make a birth. More complex mappings are possible but we shall not need to use these in this book.

Jacobians are used to account for transformations of continuous random variables. However, in the models we propose we ensure that for parameters of varying dimension the model space is discrete. Hence, the random variable u which we need to generate to find the new parameter in the model is also discrete. Then, providing the relationship between u and θ'_i is bijective, there is no need for a Jacobian. For example, suppose that u is drawn from a distribution defined on the integers $\{0, 1, \ldots, K\}$ and $\theta'_i = 2u$. The transformation is bijective and no Jacobian is required to account for this change of variables. However, in the continuous case, with the density of u non-zero on the interval $[0, K]$ and $\theta'_i = 2u$ still, the Jacobian turns out to be $\frac{1}{2}$. So, only when using continuous model spaces is the Jacobian required and for the models we shall present they are not needed.

Having seen that we do not need to include the extra Jacobian term introduced by Green (1995) our version of the reversible jump sampler becomes even more like the Metropolis–Hastings one. For a birth step the acceptance probability for the linear model, using the conditional prior as the proposal distribution, is just

$$\min\left\{1, \mathrm{BF}(\boldsymbol{\theta}', \boldsymbol{\theta})\frac{d_{k+1}}{b_k}\right\}, \tag{2.43}$$

which is the same as (2.39) apart from the fraction involving the birth and death probabilities. We choose to define $b_k = d_k = \frac{1}{2}$ for $k = 1, \ldots, K - 1$ with $b_0 = d_K = 1$ and $b_K = d_0 = 0$ so that in general (2.43) also accepts the model with the minimum of 1 and the Bayes factor. For the death step this probability is the same as (2.43) except that the fraction that multiplies the Bayes factor is b_{k-1}/d_k. Dellaportas *et al.* (2002) also suggest using 'simple' proposal distributions to sample from the posterior with reversible jump algorithms and find that in this form the reversible jump algorithm has very strong connections with the Gibbs sampling strategy for variable selection proposed by Carlin and Chib (1995).

Although we theoretically produce samples from the correct posterior using just birth and death moves this is often not the best strategy. Combining moves that change dimension with updates that keep the dimension fixed have been shown to work well (see, for example, Denison *et al.* 1998a; Richardson and Green 1997). Thus we may incorporate into the algorithm a new move step, or move steps, which just resample the $\boldsymbol{\theta}$ without altering the overall dimension of the model. These usually take the form of proposing to perturb a randomly chosen element of $\boldsymbol{\theta}$ and accepting this proposal using the Metropolis acceptance probability. We favour choosing the move step so that it is formed by first proposing a model with one less basis function (a death step) and then adding a basis function to this new model (a birth step). In this way the final model is of the same dimension as the original one and only one of the components will have been altered.

When this extra move step is included we must reduce the probabilities of birth and deaths by an arbitrary amount to ensure that this step is actually undertaken by the algorithm. Again we suggest a default choice that chooses each possible move type being chosen with equal probability. We go into more detail about these choices in the next chapter, as well as providing an algorithm for reversible jump sampling in the way we have described here.

2.8.4 *Hybrid sampling*

In practice, model spaces are not defined as simply as those laid out before. Usually, there is some subset of parameters that change dimension, but there are often also other parameters that are included in every model. For parameters that are ever present we may sometimes know their full conditional posteriors, but on other occasions we may not. Further, as we saw for the Bayesian linear model case, some unknowns may even be able to be integrated out (e.g. $\boldsymbol{\beta}, \sigma^2$). We need to devise a sampling strategy that can accommodate all these situations.

Let us now define some more notation:

$\boldsymbol{\theta}$ a varying-dimensional parameter vector relating to parts of the model space which are not fixed;

$\boldsymbol{\phi}$ a vector containing those parameters which can be analytically integrated out to form the marginal likelihood (could be of variable dimension or not);

$\boldsymbol{\psi}_{\mathrm{G}}$ the parameters which are always present in the model and for which we can easily sample from their known posterior conditional on the other parameters;

$\boldsymbol{\psi}_{\mathrm{H}}$ the same as $\boldsymbol{\psi}_{\mathrm{G}}$ except that for these parameters sampling from the required conditional posterior is not easy.

Essentially, a model is made up by different combinations of these types of parameters. For the simple linear model we have been looking at $\boldsymbol{\theta}$ is defined as in Section 2.8.3, $\boldsymbol{\psi}_{\mathrm{G}} = \boldsymbol{\psi}_{\mathrm{H}} = \emptyset$ and $\boldsymbol{\phi} = (\boldsymbol{\beta}, \sigma^2)$. We note that the parameters $\boldsymbol{\phi}$ do not need to be sampled directly to determine the model posterior as they can be marginalised over. However, even these parameters must have a prior distribution assigned to them.

As we can determine $p(\mathcal{D} \mid \boldsymbol{\theta}, \boldsymbol{\psi}_{\mathrm{G}}, \boldsymbol{\psi}_{\mathrm{H}})$ explicitly, by integrating out $\boldsymbol{\phi}$, we can make inference using the posterior distribution $p(\boldsymbol{\theta}, \boldsymbol{\psi}_{\mathrm{G}}, \boldsymbol{\psi}_{\mathrm{H}} \mid \mathcal{D})$. Further, by definition, we know that $p(\boldsymbol{\theta} \mid \boldsymbol{\psi}_{\mathrm{G}}, \boldsymbol{\psi}_{\mathrm{H}}, \mathcal{D})$ can be sampled using a reversible jump MCMC algorithm. Also, $p(\boldsymbol{\psi}_{\mathrm{G}} \mid \boldsymbol{\theta}, \boldsymbol{\psi}_{\mathrm{H}}, \mathcal{D})$ can be sampled using a Gibbs step as the required conditional posteriors are known. The posterior of the other parameters $\boldsymbol{\psi}_{\mathrm{H}}$ then requires Metropolis–Hastings updates to sample from $p(\boldsymbol{\psi}_{\mathrm{H}} \mid \boldsymbol{\theta}, \boldsymbol{\psi}_{\mathrm{G}}, \mathcal{D})$.

We can combine all the three samplers outlined before into what Tierney (1994) calls a *hybrid sampler*. This uses a combination of sampling strategies to provide draws from the required posterior. The form of this for the models we shall introduce later is given by the very simple algorithm.

Hybrid-sampler:

```
Assign starting values to (θ^(0), ψ_G^(0), ψ_H^(0));
Set t = 0;
REPEAT
  Use RJ-sampler to draw θ^(t+1) given (ψ_G^(t), ψ_H^(t), D);
  Use Gibbs sampler to draw ψ_G^(t+1) given (θ^(t+1), ψ_H^(t), D);
  Use MH-sampler to draw ψ_H^(t+1) given (θ^(t+1), ψ_G^(t+1), D);
  t = t + 1;
  Store every mth value of (θ^(t), ψ_G^(t), ψ_H^(t)) after
    an initial burn-in period;
END REPEAT;
```

When $\boldsymbol{\psi}_{\mathrm{G}}$ or $\boldsymbol{\psi}_{\mathrm{H}}$ are not themselves univariate, we usually update each element of them singly so the steps for updating these parameters in the above algorithm may involve a loop over each of their elements.

2.8.5 Convergence

Having just outlined the simulation methods which we shall use extensively in this book, it is worth remarking that these provide a (very useful) tool for making posterior inference. However, the focus of this book is on models for regression and classification that can be analysed with Bayesian methods, not the technicalities behind making posterior inferences. Hence, we concern ourselves more with eliciting appropriate priors over the model space for the various problems that we might be faced with and leave the detailed discussion of sampling methods used to make posterior inference to, for example, Gilks *et al.* (1996), Gamerman (1997), Robert and Casella (1999) and Chen *et al.* (2000).

The issue of convergence of the simulation algorithm, i.e. when to start collecting the sample of models and how many to collect, is the subject of current research. For simple parameter spaces, where the dimension of the model is fixed, various tests for non-convergence exist (see, for example, Brooks and Roberts 1998, 1999; Cowles and Carlin 1996; Gelman and Rubin 1992). Although none of these can actually detect convergence they give a good idea as to non-convergence of the sampling algorithm. However, $\boldsymbol{\theta}$ is of varying dimension in most of the situations that we shall be looking at. In these cases standard convergence diagnostics do not apply. Brooks and Giudici (1999) have offered one potential solution but its practicality for wide-ranging problems is still not fully developed at the time of writing.

We choose to offer some common-sense ideas about the problem of convergence. In the main these should involve looking at output for variables that do not change dimension, or by determining whether or not the predictive distributions of interest have settled down. An easy check for non-convergence which should always be undertaken is to plot the marginal likelihood of the models generated by the Markov chain and check that these are not 'drifting' in any direction. We shall come back to these ideas when we come onto looking at particular models (in particular, see Sections 3.3.4 and 7.5.6).

2.9 Further Reading

This book is a text for researchers interested primarily in classification and regression so we have only provided minimal coverage of foundational issues surrounding Bayesian inference. For those interested in these the original work of de Finetti (1937/1964, 1963, 1964/1972) as well as the alternative viewpoints on the Bayesian paradigm given by Cox (1961), DeGroot (1970), Savage (1971) and Bernardo and Smith (1994) are highly recommended.

This chapter has introduced the Bayesian linear model (Lindley and Smith 1972) and shown how it relates to the basis function models we shall meet in detail later on. We have also discussed and detailed strategies for choosing between competing models. Further discussion in the linear model literature includes Atkinson (1978), Smith and Spiegelhalter (1980), Spiegelhalter and Smith (1982), Pericchi (1984) and Mitchell and Beauchamp (1988).

More recently, Bayesian models have been combined with efficient search strategies to determine good sets of predictors to use in a linear regression context where there are only p candidate predictors (or basis functions) to chose from. However, the searches are usually undertaken to provide models with which to perform posterior averaging rather than to just pick a single set of predictors as the 'best'. Madigan and Raftery (1994) describes an approach known as Occam's window, which averages over models which have similar posterior probability to the best model found. To find the models to include in the window an efficient search for models with high posterior probability is undertaken (Madigan and Raftery 1994; Raftery *et al.* 1996). Another approximate method is the stochastic search variable selection method (George and Foster 2000; George and McCulloch 1993), which has already been mentioned. This essentially is just the Gibbs sampling approach outlined Section 2.8.1. They suggest not ever completely removing the influence of any predictor, but instead they restrict the coefficients for the bases 'out' of the model to be close to zero. Other authors have advocated a similar approach but have instead suggested that single models should either include or exclude predictors (Dellaportas *et al.* 2002; Geweke 1996; Hoeting *et al.* 1999; Madigan and York 1995). Other work that formulates the model as a varying-dimensional parameter vector, like the reversible jump algorithm, include Carlin and Chib (1995), Grenander and Miller (1994) and Phillips and Smith (1996).

2.10 Problems

2.1 Show that the posterior predictive distribution for the response y given predictor values $\boldsymbol{x}$, the data $\mathcal{D}$, and the model $\mathcal{M}_i$, is given by equation (2.30).

2.2 Suppose that we define the posterior loss between our estimate, $\tilde{y}$, and the true response, y, as either

(i) $\ell(\tilde{y}, y) = |\tilde{y} - y|$, or

(ii) $\ell(\tilde{y}, y) = 1$ if $\tilde{y} \neq y$ and 0 otherwise.

Show for each case that we should take $\tilde{y}$ as

(i) the median of $p(y \mid \boldsymbol{x}, \mathcal{D})$, or

(ii) the mode of $p(y \mid \boldsymbol{x}, \mathcal{D})$ (assuming one exists).

[Hint: *For (ii) assume the loss function is zero for* $\|\tilde{y} - y\| < \epsilon$ *and 1 otherwise. Then find the limit as* $\epsilon \to 0$.]

2.3 Show that if we take the proposal densities, q_i, to be the conditional posterior distributions so that

$$q_i(\theta_i' \mid \theta_i, \boldsymbol{\theta}_{-i}) = p(\theta_i' \mid \mathcal{D}, \boldsymbol{\theta}_{-i}),$$

then the Gibbs sampler and the Metropolis–Hastings sampler of Section 2.8 are identical.

2.4 Suppose that we wish to make inference in the usual linear model set-up but with known regression variance σ^2, i.e. the only unknown is $\boldsymbol{\beta}$ and we assign it an $N(\boldsymbol{m}, \boldsymbol{V})$ prior. Show that the posterior distribution of $\boldsymbol{\beta}$ is $N(\boldsymbol{m}^*, \boldsymbol{V}^*)$, where

$$\boldsymbol{V}^* = (\boldsymbol{V}^{-1} + \sigma^{-2}\boldsymbol{B}'\boldsymbol{B})^{-1},$$
$$\boldsymbol{m}^* = \boldsymbol{V}^*(\boldsymbol{V}^{-1}\boldsymbol{m} + \sigma^{-2}\boldsymbol{B}'Y).$$

Now determine the likelihood of the data, marginalised over $\boldsymbol{\beta}$.

2.5 Consider the Bayesian linear model (Lindley and Smith 1972) where $p(Y \mid \boldsymbol{\beta}) = N(\boldsymbol{B}\boldsymbol{\beta}, \boldsymbol{\Sigma})$, with the basis function and covariance matrices, $\boldsymbol{B}$ and $\boldsymbol{\Sigma}$, both known. Adopting a conjugate prior for $\boldsymbol{\beta}$ such that $p(\boldsymbol{\beta}) = N(\boldsymbol{A}, \boldsymbol{C})$, where $\boldsymbol{A}$ is the prior design matrix, $\boldsymbol{m}$ the prior coefficient vector and $\boldsymbol{C}$ the prior covariance matrix ($\boldsymbol{A}$, $\boldsymbol{m}$ and $\boldsymbol{C}$ are all known).

(i) Show that the posterior distribution of $\boldsymbol{\beta}$ is $N(\boldsymbol{Zd}, \boldsymbol{Z})$, where $\boldsymbol{Z}^{-1} = \boldsymbol{BC}'\boldsymbol{B} + \boldsymbol{C}^{-1}$ and $\boldsymbol{d} = \boldsymbol{B}'\boldsymbol{\Sigma}^{-1}Y + \boldsymbol{C}^{-1}\boldsymbol{Am}$.

(ii) Show that the marginally $Y \sim N(\boldsymbol{BAm}, \boldsymbol{\Sigma} + \boldsymbol{BC}'\boldsymbol{B})$.

(iii) Give the point estimate of $\boldsymbol{\beta}$ obtained when assuming a squared-error loss function.

(iv) If we now wish to include uncertainty in the specification of the covariance matrices by taking $\boldsymbol{\Sigma} = \sigma^2\boldsymbol{I}$ and $\boldsymbol{C} = \sigma^2\boldsymbol{V}$ ($\boldsymbol{V}$ known), determine the posterior distribution of $\boldsymbol{\beta}$ and σ^2 and compare it to the result given in (2.16).

2.6 If the basis function matrix $\boldsymbol{B}$ is orthonormal so that $\boldsymbol{B}'\boldsymbol{B} = \boldsymbol{I}$, find the posterior distribution of the regression parameters in the Bayesian linear model set-up of Section 2.4. Describe a method that can be used to transform an arbitrary design matrix to an orthonormal design matrix highlighting what advantages this will give when making posterior inference.

2.7 Consider a linear model of the form,

$$y_i = \begin{cases} \beta_0 + \epsilon_i, & x_i \leqslant c, \\ \beta_1 + \beta_2 x_i + \epsilon_i, & x_i > c, \end{cases}$$

for $i = 1, \ldots, n$, $\epsilon_i \sim N(0, \sigma^2)$ with all the ϵ_i independent and c a known constant.

(i) By first determining the relevant basis function matrix $\boldsymbol{B}$, express this set-up as a Bayesian linear model.

(ii) Derive the joint posterior of $\boldsymbol{\beta} = (\beta_0, \beta_1, \beta_2)$ when

$$p(\boldsymbol{\beta}, \sigma^2) = \text{NIG}(\boldsymbol{0}, \sigma^2\boldsymbol{I}, a, b)$$

and comment on the dependence structure between the individual elements of $\boldsymbol{\beta}$.

(iii) Suppose now that we allow for uncertainty in c and assign a discrete uniform prior on the elements $\{c_1, \ldots, c_m\}$ to it. Write down an expression for the posterior distribution of c, $p(c \mid \mathcal{D})$, marginalised over $(\boldsymbol{\beta}, \sigma^2)$.

(iv) Describe how you might sample from the posterior distribution of c found in (iii) using the Gibbs sampler and the Metropolis–Hastings algorithm. Relating your discussion to the number of possible values, m, of c comment on which sampling algorithm would be preferred.

2.8 Due to the violation of some assumptions in the usual linear model formulation given in Section 2.4 we can generalise it to the transformation model given by

$$h_\lambda(y_i) = \sum_{i=1}^{k} \boldsymbol{\beta}_i B_i(\boldsymbol{x}_i) + \epsilon_i,$$

for $i = 1, \ldots, n$, with the ϵ_i independent and identically distributed $N(0, \sigma^2)$ random variables, and

$$h_\lambda(y_i) = \begin{cases} \lambda^{-1} y_i^{\lambda-1}, & \lambda \neq 0, \\ \log(y_i), & \lambda = 0. \end{cases}$$

Denoting $Y_\lambda = (h_\lambda(y_1), \ldots, h_\lambda(y_n))$, derive the Bayes factor to compare two linear models with the same basis set but with two different transformations corresponding to λ_1 and λ_2.

2.9 Consider a linear model, $\mathcal{M}_1$, and a quadratic regression model, $\mathcal{M}_2$, given by

$$\begin{aligned} \mathcal{M}_1 : \quad & y_i = \beta_0 + \beta_1 h_1(x_i) + \epsilon_i, \\ \mathcal{M}_2 : \quad & y_i = \beta_0 + \beta_1 h_1(x_i) + \beta_2 h_2(x_i) + \epsilon_i, \end{aligned}$$

where h_1 and h_2 are arbitrary polynomial functions of order one and two, respectively, for which $\sum_i h_j(x_i) = 0$ for $j = 1, 2$. If we adopt the usual prior and likelihood assumptions as those in Section 2.4,

(i) derive the Bayes factor in favour of model $\mathcal{M}_1$ over model $\mathcal{M}_2$;

(ii) simplify this expression for the special case where h_1 and h_2 are orthonormal functions such that $\sum_{i=1}^n h_j(x_i) h_k(x_i) = 1$ if $j = k$ and zero otherwise;

(iii) suppose that we are now interested in comparing models with degree 1 to degree 20. What advantages, in terms of computational requirements, are there in ensuring that $h_1, h_2, \ldots, h_{20}$ are orthonormal polynomials as described above.

3

Curve Fitting

3.1 Introduction

This chapter reviews some nonlinear Bayesian methods for curve fitting, the simplest type of regression problem when there is only one predictor variable. We begin by describing curve fitting when there is considerable knowledge of the underlying regression function. In these cases we can tailor Bayesian methods to match features that we expect to see in the data. However, we usually do not have this luxury and therefore, when there is little knowledge of the underlying function, a more general (and usually much larger) set of models needs to be considered. We also explore the role of both the model and parameter priors. In addition we suggest prior formulations that can be used for default curve fitting, robust curve fitting and situations in which the errors are correlated.

We begin, in Section 3.2, with a changepoint problem, for which we believe, *a priori*, that the underlying curve takes the form of a *step function*. The assumption is firm; but we are uncertain as to whether the underlying curve is a constant, has a single break in its level, or has more than one break in its level. We shall illustrate our general modelling framework and model comparison approach with a reanalysis of the well-known Nile discharge data.

In Section 3.3, we consider the situation where we believe, *a priori*, in a certain level of smoothness of the underlying curve, but do not wish to go further to posit a specific functional form. Initial implementation of the Bayesian approach is given using the class of *spline* basis functions.

In Section 3.4, the alternative class of *wavelet* basis functions is presented. Then in Section 3.5 we discuss issues relating to prior specification and the robustness of the Bayesian analyses. In particular, we examine in detail how these relate to the spline basis case.

Section 3.6 outlines how to extend the modelling approach when we believe the errors to be overdispersed compared to a normal distribution. This involves using an error distribution than it more robust to outlying responses than the normal one.

Finally, in Sections 3.7 and 3.8 we discuss the approaches introduced and give ideas for related further reading.

Note that this chapter provides a broad base on which the rest of the book is heavily dependent as extensions to curve fitting presented in later chapters draw heavily on the material presented here.

3.2 Curve Fitting Using Step Functions

3.2.1 Example: Nile discharge data

In Figure 3.1 we display the Nile discharge dataset described in Cobb (1978). This consists of 100 measurements of the Nile discharge at Aswan, Egypt, every year from 1871 to 1970. The response measurements, y_i, are the annual volume of discharge in 10^8 m^3 for the years $x_i = 1870 + i$, which are the predictor locations. We write this complete dataset as $\mathcal{D} = \{y_i, x_i\}_1^{100}$.

The Nile data have attracted interest because of the apparent change in the mean level of the discharge sometime near the end of the 19th century. One possible explanation is the presence of the Aswan dam. This began functioning in 1902 and it was not known if the corrections made after its completion led to accurate measurements of the volume of discharge in subsequent years. Alternatively, tropical rainfall evidence suggests that the change in mean level may be real. This dataset will be used to illustrate some of the ideas of Bayesian model fitting and selection.

Given our understanding of the problem, the first model we might think of is one which allows for a single shift in the volume of discharge at an unknown time point. This is known as a mean-shift model and we shall refer to it as model $\mathcal{M}_1$. We denote the model parameters of $\mathcal{M}_1$ by $\boldsymbol{\theta}$, as in the general model descriptions in Chapter 2. For this model, $\boldsymbol{\theta} = (t)$, the location of a single changepoint. So the function we use to estimate the relationship between the discharge volume and year is

$$g(x) = \beta_1 I(x \leqslant t) + \beta_2 I(x > t),$$

where, in the framework of the previous chapter, we have a model with basis functions $B_1(x) = I(x \leqslant t)$ and $B_2(x) = I(x > t)$. Here $I(\cdot)$ is the indicator function which returns the value one if its argument is true, and zero otherwise. Note that we do not include the coefficients β_1 and β_2 in the parameter vector $\boldsymbol{\theta}$ as, using the linear model of Chapter 2, we can integrate them out.

For this simple model $\boldsymbol{\theta}$ and t are interchangeable but, for consistency of notation with subsequent usage in the book, we stick with $\boldsymbol{\theta}$ to denote the model parameters (a single parameter in this case).

It is computationally convenient to adopt standard conjugate priors for the coefficients and error variance given in (2.12) so that, after taking out the mean of the data (i.e. removing $\overline{y}$ from each response), we take

$$p(\boldsymbol{\beta} \mid \sigma^2) = N(\mathbf{0}, \sigma^2 \boldsymbol{V}),$$
$$p(\sigma^2) = \mathrm{IG}(a, b),$$

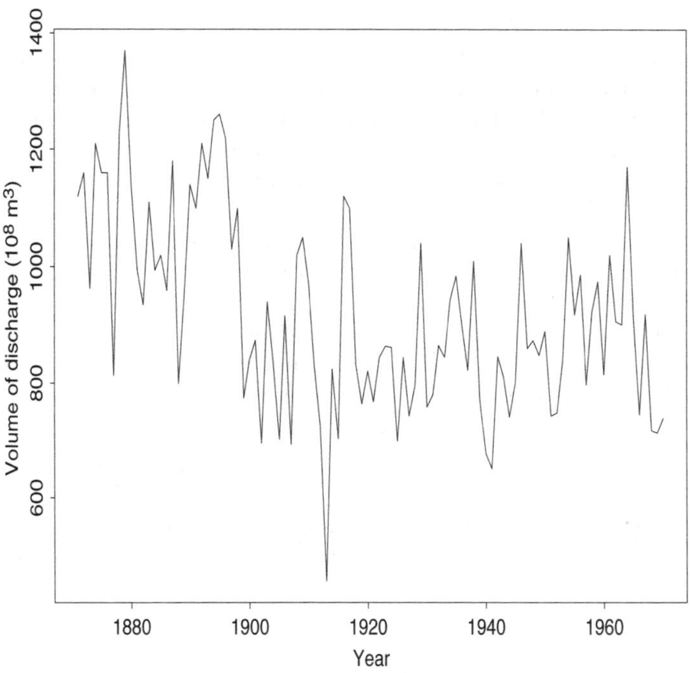

Figure 3.1 The Nile discharge data giving the annual volume (in 10^8 m^3) of discharge of the Nile at Aswan from 1871 to 1970.

with $\boldsymbol{V} = v\boldsymbol{I}$ (see Section 3.5.3 for discussion of this choice of $\boldsymbol{V}$). To reflect an assumption of ignorance about the position of the changepoint before we have seen the data we assign a discrete uniform prior for the position of the changepoint t, so that $p(t = x_i) = 1/(n-1)$ for $i = 1, \ldots, n-1$, where $n = 100$ in this example. Note that we do not allow a changepoint at the last predictor value because this would correspond to a model with no changepoints which we shall consider later.

We find that with this prior formulation the marginal likelihood is given by (2.23) with

$$\boldsymbol{m} = \boldsymbol{0};$$

$$\boldsymbol{V} = \begin{pmatrix} v & 0 \\ 0 & v \end{pmatrix};$$

$$\boldsymbol{B} = \begin{pmatrix} I(x_1 \leqslant t) & I(x_1 > t) \\ \vdots & \vdots \\ I(x_n \leqslant t) & I(x_n > t) \end{pmatrix}.$$

Further, the updates to the prior parameters are given by (2.17)–(2.20).

Initially, we wish to investigate the posterior distribution of the model parameters (or changepoint location), $p(\boldsymbol{\theta} \mid \mathcal{D}, \mathcal{M}_1) = p(t \mid \mathcal{D}, \mathcal{M}_1)$. The first step in determining the posterior is to find the marginal likelihood for each model. To do this we need to set a, b and v. Parameters of the prior, such as these, are commonly referred to as *hyperparameters*. For discussion on the setting of the hyperparameters, see Section 3.5, but for now we choose the broadly uninformative settings of $a = b = 0.01$ and $v = 1000$.

The natural logarithm of the marginal likelihood, $p(\mathcal{D} \mid \boldsymbol{\theta}, \mathcal{M}_1)$, is shown by the solid line in Figure 3.2. This has been calculated by evaluating (2.23) for every possible value of $\boldsymbol{\theta}$. We have drawn the log marginal likelihood as a line only for visual clarity although it is actually made up of 99 discrete points. By inspection of this curve we can read off the changepoint location most supported by the data, which we denote by $\tilde{\boldsymbol{\theta}} = 1898$. In terms of Bayes' factors (Section 2.5.1), this satisfies

$$\mathrm{BF}(\tilde{\boldsymbol{\theta}}, \boldsymbol{\theta}) = \frac{p(\mathcal{D} \mid \tilde{\boldsymbol{\theta}}, \mathcal{M}_1)}{p(\mathcal{D} \mid \boldsymbol{\theta}, \mathcal{M}_1)} \geqslant 1 \quad \text{for all } \boldsymbol{\theta} \in \boldsymbol{\Theta},$$

where $\boldsymbol{\Theta} = \{1871, \ldots, 1969\}$ denotes the set of possible values for $\boldsymbol{\theta}$. Note that under the uniform prior specification for $\boldsymbol{\theta}$, $\tilde{\boldsymbol{\theta}}$ is also the maximum *a posteriori* (MAP) estimate of $\boldsymbol{\theta}$, as

$$p(\boldsymbol{\theta} \mid \mathcal{D}, \mathcal{M}_1) \propto p(\mathcal{D} \mid \boldsymbol{\theta}, \mathcal{M}_1)$$

whenever $p(\boldsymbol{\theta}) \propto 1$.

So far we have only considered a single model, namely the mean-shift one, $\mathcal{M}_1$. However, one of the most attractive features of Bayesian analysis is the way in which competing models of the data can be assessed. All random variables, even if they are models themselves, can be treated within a single probabilistic framework.

Now, we wish to compare the single changepoint model with a number of alternatives. Firstly, it would be insightful to see the support the data give to model $\mathcal{M}_0$ which has no change in the mean level. Also, we introduce model $\mathcal{M}_2$ which suggests a change in both mean level and regression variance at the changepoint $\boldsymbol{\theta}$. This is known as a mean-variance shift model. The marginal likelihoods for both $\mathcal{M}_0$ and $\mathcal{M}_2$ are slight variations on the one for $\mathcal{M}_1$ which are straightforward to calculate using the same hyperparameter values as before (Problem 3.9).

In Figure 3.2 we also display the log marginal likelihoods of $\mathcal{M}_0$ and $\mathcal{M}_2$, given by the dashed and dotted lines, respectively. We can see that both models $\mathcal{M}_1$ and $\mathcal{M}_2$ have $\tilde{\boldsymbol{\theta}} = 1898$ as the MAP estimate for the changepoint. However, this figure can be misleading in assessing the true posterior support for the year 1898 over the others. As we have used the log-scale for displaying the results, small differences in the likelihood can appear insignificant. To illustrate this point in Figure 3.3 we plot the posterior probability for $\boldsymbol{\theta}$, assuming $\mathcal{M}_1$, on the natural scale. For only four values of $\boldsymbol{\theta}$ do the data suggest any discernible weight, and this figure ably demonstrates the great preference for $\tilde{\boldsymbol{\theta}} = 1898$ over the others (here $p(\tilde{\boldsymbol{\theta}} \mid \mathcal{D}, \mathcal{M}_1) = 0.77$).

From Figure 3.3 we can also see that there is little uncertainty about the position of the changepoint. For instance, there is negligible evidence in favour of a changepoint

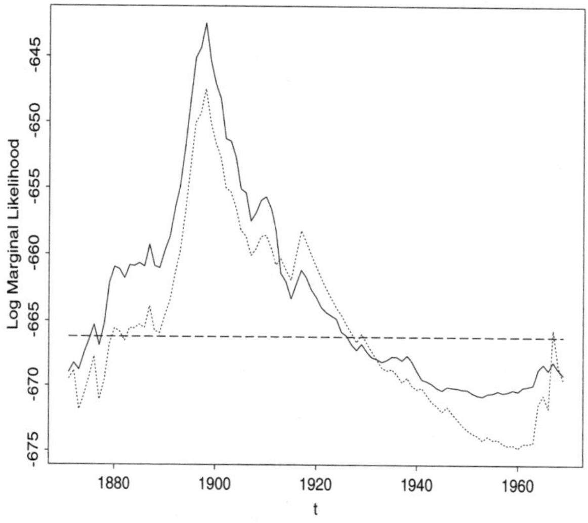

Figure 3.2 Comparison of the log marginal likelihoods for the Nile data under $\mathcal{M}_1$ (solid line) and $\mathcal{M}_2$ (dotted line) given the model parameters $\boldsymbol{\theta} = (t)$. The horizontal dashed line gives the corresponding result under $\mathcal{M}_0$.

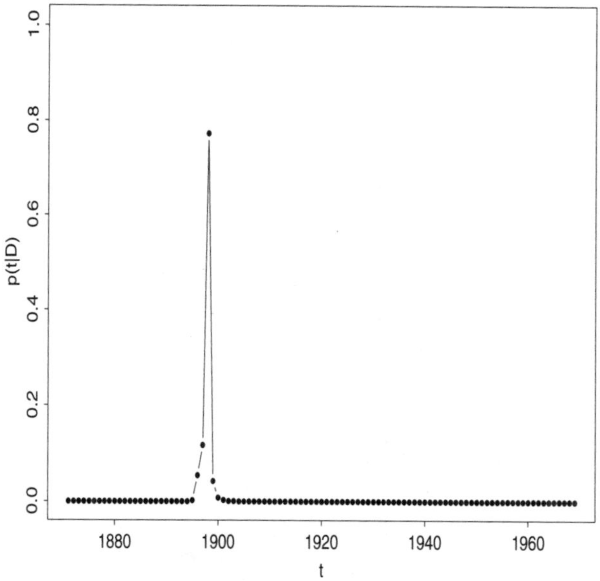

Figure 3.3 The posterior probability distribution of $\boldsymbol{\theta} = (t)$ for the Nile data under the mean-shift model, i.e. $p(\boldsymbol{\theta} \mid \mathcal{D}, \mathcal{M}_1)$.

at 1902, so we have found no evidence to support the idea that the change in the level occurred due to the initial functioning of the Aswan dam.

We can use Bayes' factors to determine which of the three models is most favoured by the data. That is, which of the three explanations most closely mimics the true process. To do this we must first determine the marginal likelihoods,

$$p(\mathcal{D} \mid \mathcal{M}_j) = \sum_{t=1871}^{1969} p(\mathcal{D} \mid \boldsymbol{\theta} = t, \mathcal{M}_j) p(\boldsymbol{\theta} = t),$$

for $j = 1, 2$. In this simple example, as we can sum over all the possible changepoints, this is straightforward but in general determination of such marginal likelihoods is not so easy.

Due to our prior choice for $\boldsymbol{\theta}$ (i.e. $p(\boldsymbol{\theta} = i) = 1/99$) this is just the average of the marginal likelihoods conditioned on the location of the changepoint. We find that the logarithms of $p(\mathcal{D} \mid \mathcal{M}_j)$ for $j = 0, 1, 2$ are -666.1, -646.7 and -651.5, respectively. The log Bayes factor, which is simply the difference in log marginal likelihoods of the models under consideration, is given by

$$\log \mathrm{BF}(\mathcal{M}_i, \mathcal{M}_j) = \log p(\mathcal{D} \mid \mathcal{M}_i) - \log p(\mathcal{D} \mid \mathcal{M}_j),$$

which, when each model is assumed equally likely before we see the data, is equal to the difference in log posterior probabilities. According to the Bayes factors, the models can be ordered in preference as

$$\mathcal{M}_1 \succ \mathcal{M}_2 \succ \mathcal{M}_0.$$

Again, looking on the log-scale can prove deceptive as the Bayes factors are

$$\begin{aligned}
\mathrm{BF}(\mathcal{M}_1, \mathcal{M}_0) &= 122;\\
\mathrm{BF}(\mathcal{M}_1, \mathcal{M}_2) &= 3 \times 10^8;\\
\mathrm{BF}(\mathcal{M}_2, \mathcal{M}_0) &= 2 \times 10^6.
\end{aligned}$$

If we wished to introduce a further model, $\mathcal{M}_3$, which allows for two mean-shifts at t_1 and t_2 ($> t_1$) and a constant error variance. We could compare this model with the previous ones, again using Bayes' factors, as this just involves determining the double sum,

$$p(\mathcal{D} \mid \mathcal{M}_3) = \sum_{t_1=1871}^{1968} \sum_{t_2=t_1+1}^{1969} p(\mathcal{D} \mid \mathcal{M}_3, \boldsymbol{\theta}_1 = (t_1, t_2)), \tag{3.1}$$

where $p(\mathcal{D} \mid \mathcal{M}_3, \boldsymbol{\theta} = (t_1, t_2))$ can be determined analytically using the results in Chapter 2. The summation above requires $\binom{99}{2} = 4851$ terms which can be undertaken quickly with modern computing power. In fact we find that it too is less favoured by the data than $\mathcal{M}_1$. This illustrates an extremely important feature in Bayesian modelling. Despite the fact that uniform priors were specified over the competing models and the

Table 3.1 Comparison of the posterior distribution of $\boldsymbol{\theta}$ for the mean-shift model, $\mathcal{M}_1$, using the Nile data. The true distribution is given, along with the approximations found using a sample generated by the first N iterations of a single run of an MCMC algorithm, denoted MCMC(N). Note that for illustration no burn-in period was used and each algorithm was started from $\mathcal{M}_0$.

$\boldsymbol{\theta}$	1895	1896	1897	1898	1899	1900	1901
$p(\boldsymbol{\theta} \mid \mathcal{D}, \mathcal{M}_1)$	0.002	0.055	0.118	0.773	0.042	0.008	0.002
MCMC(10^6)	0.002	0.058	0.118	0.771	0.041	0.008	0.002
MCMC(10^4)	0.002	0.064	0.072	0.836	0.023	0.004	0.000
MCMC(10^3)	0.000	0.155	0.000	0.810	0.000	0.030	0.000

changepoints, the model with the highest posterior density was not the most complex one.

In theory, we can generalise this method to determining the marginal likelihood under models with any number of changepoints. However, to compare the marginal likelihoods of the mean-shift models with up to k changepoints requires

$$\sum_{1}^{k} \binom{n}{i}$$

likelihood evaluations. So in this example and taking $k = 4$, that means approximately four million likelihood evaluations. It is because of this rapid explosion in the space of models that MCMC techniques have become popular for these types of situations. It is worth noting, however, that we are led to using MCMC methods here not because the summation involved is intractable, but just because it is computationally infeasible as it would take too long to perform. MCMC algorithms generate approximate random samples from the posterior distribution of the model. These samples can then be used to approximate the summations, or integrals, needed to evaluate the predictive distribution.

3.3 Curve Fitting with Splines

We now generalise the ideas in the previous section to situations where we do not have strong beliefs about the form of the underlying process (i.e. relationship between y and $\boldsymbol{x}$) that generated the data. This section illustrates how to use Markov chain Monte Carlo methods to draw samples from the posterior distribution of the model when analytic methods are impractical. As a starting point we assume that f, the true functional relationship between the response and the predictors, is continuous. Thus, we cannot use the mean-shift models seen earlier as they insist on jumps in the mean level at the changepoints. Hence, we choose to use what are known as *spline basis functions* to model f. These are similar to the models seen before except that at the changepoints the function remains continuous and instead jumps in one of its derivatives. However, the basic problem remains the same: to evaluate the predictive distribution we need to determine the posterior distribution over the number, and location, of the changepoints.

When we know little about the true underlying function, we want methods that will be able to 'automatically' capture the true nature of the curve. In these situations the aim is to determine ways to select default prior parameters which will work well in a wide variety of situations. To allow a very general class of models to be estimated well, we propose methods that can model smooth curves as well as those that might be rapidly varying in their mean level, or even in some of their derivatives.

We have already seen that for curves that require more than a very small number of basis functions to model their behaviour, analytic determination of the posterior probability of the model is computationally too expensive. In these cases an MCMC algorithm can be used to find an approximate sample from the target posterior distribution, i.e. a sample of models (curves). We can then base our posterior inference on this sample of models.

A widespread and general approach to modelling f is through regression splines. In particular, we focus on the truncated power series basis functions that lead to our estimate to f being given by

$$g(x) = \beta_0 + \sum_{i=1}^{k} \beta_i (x - t_i)_+^q, \qquad x \in \mathcal{X}, \quad t_1, \ldots, t_k \in \mathcal{T}, \tag{3.2}$$

where q is a positive integer and gives the order of the spline, $(x)_+$ is the positive part of x so equals x for $x > 0$ and 0 otherwise, and $\mathcal{T}$ is the set of candidate knot locations. The order q relates to the smoothness of the estimated function, as the highest derivative the model allows jumps in is the qth one. For now we shall take $q = 1$ so that (3.2) is just a sum of truncated linear spline terms (see, for example, de Boor 1978) and examples of this type of basis function are given later in Figure 3.18.

To be able to evaluate the predictive distribution $p(y \mid \boldsymbol{x}, \mathcal{D})$ we need to determine the posterior distribution of the basis functions to use in the model, i.e. $p(\boldsymbol{\theta} \mid \mathcal{D})$ where $\boldsymbol{\theta} = (t_1, \ldots, t_k)$. The estimating function in (3.2) with $q = 1$ has jumps in the first derivative at $t_1, \ldots, t_k$ so, ideally, $\boldsymbol{\theta}$ will be able to determine the t_i where these occur in the true function. More realistically we use (3.2) with $q = 1$ as a flexible function which is reasonably smooth but can accommodate sharp changes in the true function, even if these changes do not all take the form of jumps in the gradient of f.

To actually make inference about the target posterior $p(\boldsymbol{\theta} \mid \mathcal{D})$, we need to resort to Markov chain Monte Carlo methods, as exact analysis is not computationally feasible for this general model. We shall discuss two methods for sampling from this distribution. The first is based on the work of Denison *et al.* (1998a) and the second on that of Smith and Kohn (1996). These two approaches solve the same problem but the former uses the reversible jump method of Green (1995) while the latter utilises the Gibbs sampler (Gelfand and Smith 1990; Geman and Geman 1984). Both of these methods are outlined in Section 2.8. From further research it appears that while the ideas behind the papers are both well founded the methods can be further refined. Thus, in the expositions we shall give here we shall not follow either of the papers exactly. Instead, we motivate and suggest using slightly modified priors to those given by Smith and Kohn (1996) and Denison *et al.* (1998a).

3.3.1 Metropolis–Hastings sampler

When using the reversible jump Metropolis–Hastings technique of Green (1995) we consider the target posterior to be of varying dimension, dependent on the number of basis functions k. We use the prior specification for $(\boldsymbol{\beta}, \sigma^2)$ outlined in Section 2.4.1 and relate all the unknowns through a hierarchical prior specification taking

$$p(\boldsymbol{\beta}, \sigma^2, \boldsymbol{\theta}) = p(\boldsymbol{\beta} \mid \sigma^2, \boldsymbol{\theta}) p(\sigma^2) p(\boldsymbol{\theta}).$$

With the conjugate prior specification we know that the conditional priors on $\boldsymbol{\beta}$ and σ^2 are normal and inverse-gamma distributions, respectively. Section 3.5 will discuss the setting of these priors for a spline model in more detail but for now we take $p(\beta_0) \propto 1$, $p(\beta_1, \ldots, \beta_k \mid \sigma^2, \boldsymbol{\theta}) = N(\mathbf{0}, \sigma^2 v \boldsymbol{I})$ and $p(\sigma^2) = \mathrm{IG}(a, b)$.

We wish to use a prior specification for $\boldsymbol{\theta}$ which is, in some sense, as uninformative as possible. We do this with a discrete uniform specification that takes

$$p(\boldsymbol{\theta}) = \binom{T}{k}^{-1} \times \frac{1}{K+1}, \tag{3.3}$$

for $k = 0, 1, \ldots, K$, where $k = \dim(\boldsymbol{\theta})$, the number of elements in $\boldsymbol{\theta}$, $T = |\mathcal{T}|$, the size of the candidate set of knot locations, and K is the maximum number of changepoints allowed. The first term in (3.3) ensures that each model with dimension k has equal weight. Thus we assume that, given k, any set of changepoints is found by sampling k items from the candidate set $\mathcal{T}$ without replacement. This ensures that the elements of $\boldsymbol{\theta}$ are distinct, implying that K must be less than, or equal to, T. The second term assumes that each possible dimension k ($\in \{0, 1, \ldots, K\}$) is equally likely.

To sample from $p(\boldsymbol{\theta} \mid \mathcal{D})$ we can use the reversible jump algorithm and here we outline the move steps we employ to traverse this posterior probability surface. In this case we follow Denison *et al.* (1998a) and use the three move types *BIRTH*, *DEATH* and *MOVE*. The first two of these propose moves between different dimensions while the third one proposes moves within a dimension. Throughout we assume that the current model is of dimension k.

BIRTH. Propose, with probability b_k, to add a new basis function at a randomly chosen data point chosen from those that do not currently have a changepoint.

DEATH. Propose, with probability d_k, to remove a randomly chosen basis function from those present.

MOVE. Propose, with probability m_k, to alter a randomly chosen basis function, say $(x - t_i)_+$, by swapping t_i for a randomly chosen value in $\mathcal{T}$ which still ensures that all the basis functions are distinct (i.e. none of the t_i coincide).

We choose the proposal probabilities as $b_k = d_k = m_k = \frac{1}{3}$ for $k = 1, \ldots, K-1$ and $b_K = d_0 = m_0 = 0$, $b_0 = 1$ and $d_K = m_K = \frac{1}{2}$. These can be chosen in any way but, under this specification, the acceptance probability for a proposed move from model $\boldsymbol{\theta}$ (of dimension k) to $\boldsymbol{\theta}'$ (of dimension k') is

$$\min\left\{1, \frac{p(\mathcal{D} \mid \boldsymbol{\theta}')}{p(\mathcal{D} \mid \boldsymbol{\theta})} \times R\right\}, \tag{3.4}$$

where R is a ratio of probabilities given by $d_{k'}/b_k$ for a *BIRTH*, $b_{k'}/d_k$ for a *DEATH* and $m_{k'}/m_k$ for a *MOVE*. Note that, because of the simple way in which we assign the probabilities of proposing each move type, in most cases $R = 1$.

In general, the form of the acceptance probability for a reversible jump algorithm (Green 1995) is far more complex than (3.4) and is given by

$$\min\left\{1, \frac{p(\mathcal{D} \mid \boldsymbol{\theta}')}{p(\mathcal{D} \mid \boldsymbol{\theta})} \frac{p(\boldsymbol{\theta}')}{p(\boldsymbol{\theta})} \frac{q(\boldsymbol{\theta} \mid \boldsymbol{\theta}')}{q(\boldsymbol{\theta}' \mid \boldsymbol{\theta})} |J|\right\}, \tag{3.5}$$

the ratio of marginal likelihoods, prior and proposal distributions together with a Jacobian term $|J|$, which accounts for the change in scale when moving between models of potentially different dimensions.

We now describe how we obtain (3.4) for this model from the general acceptance probability for the reversible sampler in (3.5). Firstly, *MOVE* is just a standard update in an independence sampler so (3.4) follows immediately (see Section 2.8.2). Now, if we can show that (3.4) holds for *BIRTH* moves, it will also hold for *DEATH* moves by symmetry. So here we only concentrate on the acceptance probability when adding a basis function to $\boldsymbol{\theta}$, with dimension k, to obtain $\boldsymbol{\theta}'$, with dimension $k' = k + 1$. For this general *BIRTH* move the prior ratio is

$$\begin{aligned} \frac{p(\boldsymbol{\theta}')}{p(\boldsymbol{\theta})} &= \left\{\binom{T}{k+1}^{-1} (K+1)^{-1}\right\} \Big/ \left\{\binom{T}{k}^{-1} (K+1)^{-1}\right\} \\ &= \frac{k+1}{T-k}. \end{aligned} \tag{3.6}$$

The proposal ratio involves understanding both the *BIRTH* and the converse *DEATH* step. When adding a new basis function we use the proposal density $q(\boldsymbol{\theta}' \mid \boldsymbol{\theta}) = b_k/(T-k)$. This is made up of the probability of actually attempting the birth step together with that of choosing the particular new value $\theta_{k+1} = (t_{k+1})$, given $\boldsymbol{\theta}$. This can be done in $T - k$ ways as θ_{k+1} must be distinct to the k elements of $\boldsymbol{\theta}$ and there are only T possibilities in total. The probability of proposing the reverse move is $q(\boldsymbol{\theta} \mid \boldsymbol{\theta}') = d_{k+1}/(k+1)$. This is just the probability of proposing a *DEATH* step and then of choosing the proposed basis function as the one to remove.

As we saw in Section 2.8.3, after defining the model prior and proposal densities we can sample from the posterior using reversible jump simulation once the hyper-parameters have been set. Further, with the set-up we have advocated we find that

$$\frac{p(\boldsymbol{\theta}')}{p(\boldsymbol{\theta})} \frac{q(\boldsymbol{\theta} \mid \boldsymbol{\theta}')}{q(\boldsymbol{\theta}' \mid \boldsymbol{\theta})} = \frac{d_{k+1}}{b_k} = R,$$

so that the acceptance probability for a *BIRTH* step is exactly as that given in (2.43). We now give the algorithm for sampling from this model.

RJ-sampler:

```
Set θ^(0) = ();                          /*no basis functions*/
Set t = 0;
REPEAT
  /********************************/
  /*POSITION 1: If required */
  /*draw auxiliary variables here*/
  /********************************/
  Set u_1 to a draw from a U(0,1) distribution;
  IF u_1 ⩽ b_k
    θ' = Birth-proposal(θ^(t));
  ELSE IF b_k < u_1 ⩽ b_k + d_k
    θ' = Death-proposal(θ^(t));
  ELSE
    θ' = Move-proposal(θ^(t));
  ENDIF;
  Set u_2 to a draw from a U(0,1) distribution;
  IF u_2 < min{1, BF(θ', θ^(t)) × R}          /*acceptance?*/
    θ^(t+1) = θ';
  ELSE
    θ^(t+1) = θ^(t);
  ENDIF;
  /*************************************/
  /*POSITION 2: If required */
  /*sample other parameters (ψ_G, ψ_H) now*/
  /*************************************/
  t = t + 1;
  Store every mth value of θ^(t) after
    initial burn-in;
END REPEAT;
```

Birth-proposal($\boldsymbol{\theta}$):

```
Set z to a draw from a U{T\θ} distribution;
return(θ, z);
```

Death-proposal($\boldsymbol{\theta}$):

```
Set z to a draw from a U{1,...,k} distribution;
return(θ_{-z});
```

Move-proposal($\boldsymbol{\theta}$):

```
θ* = Death-proposal(θ);
θ** = Birth-proposal(θ*);
return(θ**);
```

For the time being you can ignore the references to Position 1 and Position 2. These comments have been added to help in describing some of the more elaborate reversible jump sampling algorithms we shall use later on.

This is the most straightforward version of the Metropolis–Hastings sampler for varying dimension problems (reversible jump). All the algorithms described later on in the book basically use the same code with small modifications depending on the problem at hand. In general, the **Birth-proposal** is highly dependent on the form of $\boldsymbol{\theta}$, whereas the **Death-proposal** and **Move-proposal** subroutines do not change. Also, we include a comment showing where we would add lines to update any other fixed-dimensional parameters that need sampling.

3.3.2 Gibbs sampling

We can write the spline model in (3.2) with $q = 1$ alternatively as

$$g(x) = \beta_0 + \sum_{i=1}^{T} \theta_i \beta_i (x - t_i)_+, \qquad x \in \mathcal{X}, \quad \theta_1, \ldots, \theta_k \in \{0, 1\}, \tag{3.7}$$

where t_i here denotes the ith member of $\mathcal{T}$ (as opposed to the ith member of $\boldsymbol{\theta}$ as before). We see that the model is described by a vector of indicators $\boldsymbol{\theta} = (\theta_1, \ldots, \theta_T) \in \boldsymbol{\Theta} = \{0, 1\}^T$. This vector of model parameters is of fixed dimension so the target posterior distribution, $p(\boldsymbol{\theta} \mid \mathcal{D})$, is also of fixed dimension. This allows us to use the Gibbs sampling algorithm outlined in Section 2.8.1. to generate our sample of models. We just iterate through the elements of $\boldsymbol{\theta}$ in turn, drawing samples from $p(\theta_i \mid \mathcal{D}, \boldsymbol{\theta}_{-i})$ at each stage. This is straightforward as this full conditional (posterior) distribution is Bernoulli with probability of success $1 - p(\boldsymbol{\theta}_i = 0 \mid \boldsymbol{\theta}_{-i}, \mathcal{D})$, which can be calculated using (2.37). Thinking about the curve fitting problem in this way simplifies the computational algorithm as no move types between models of different dimension need to be considered, even though the effective number of basis functions can vary. Software to run the model of Smith and Kohn (1996) is available from Statlib at `http://lib.stat.cmu.edu/`.

It should be stressed again that the choice of sampling algorithm to use (either Gibbs or reversible jump) is user-dependent. For simple problems (T small) the Gibbs sampler is attractive as it is more straightforward to program. This is an important issue as it also makes the program less susceptible to programming errors, and easier to debug. However, it should be remembered that for $T < 25$ analytic results can be determined which are preferable to ones found by sampling.

Throughout this book we shall place more emphasis on the reversible jump sampling algorithm. This is mainly because it generalises well for T large and works efficiently for T small, despite the slight increase in programming complexity. Once an algorithm of this sort has been established by the user it can be readily modified to handle a wide variety of problems. Also the reversible jump sampler is amenable to problems where the space of possible models, $\boldsymbol{\Theta}$, is large or even uncountable: this is in contrast the Gibbs sampler. For instance, if we use a continuous prior over some

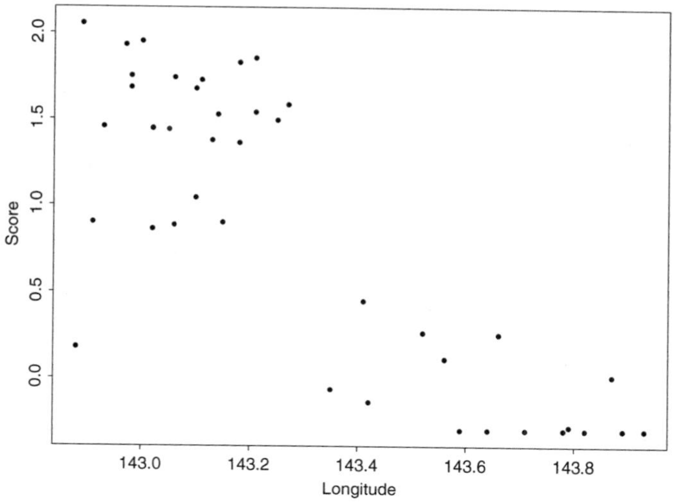

Figure 3.4 The Great Barrier Reef data.

interval for the location of each changepoint there is no concept of a full model so Gibbs sampling cannot be employed. An algorithm in Matlab to perform the reversible jump method for many of the methods in this book is available from the Web page `stats.ma.ic.ac.uk/~dgtd`.

3.3.3 Example: Great Barrier Reef Data

We now give a demonstration of curve fitting using the methodology set out above. However, it is worth recalling that throughout the previous work we have implicitly conditioned on the values a, b and v in the normal inverse-gamma prior for $(\boldsymbol{\beta}, \sigma^2)$. Hence, we shall also give an idea of how these parameters can affect the posterior inference.

Consider the data shown in Figure 3.4. This is the same subset of the complete dataset of Poiner *et al.* (1997) that was analysed in Chapter 5 of Bowman and Azzalini (1997). The complete dataset concerns the weight of fauna, denoted by a score on the log scale, captured at a series of locations near the Great Barrier Reef. The subset we look at is how this score relates to the longitude of the sampling location.

We can see from Figure 3.4 that the true nature of the underlying regression function is difficult to determine by eye. Although it appears that there may be a changepoint at about longitude 143.3, it is unclear whether to fit linear or constant terms around this changepoint.

To illustrate the methodology seen we fit a function of the form of (3.2) with $q = 1$ to the Great Barrier Reef data, using the Metropolis–Hastings reversible jump

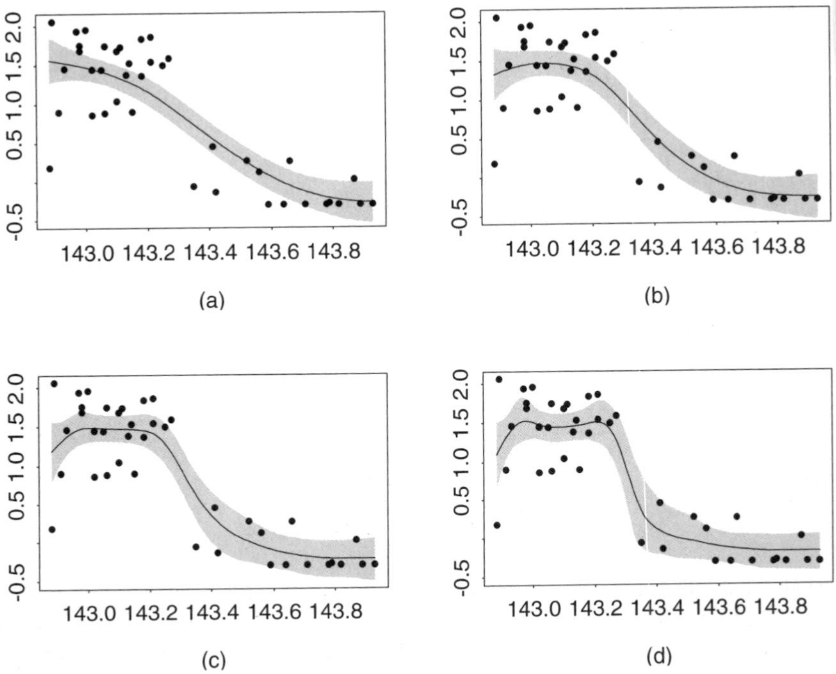

Figure 3.5 Posterior mean estimates $\tilde{g}(x)$ of $g(x)$, together with 95% credible intervals for the mean, for the Great Barrier Reef data: (a) $v = 1$; (b) $v = 10$; (c) $v = 100$ and (d) $v = 1000$. The horizontal axes give the longitude with score given on the vertical axes.

method to generate the sample of models. A uniform prior over 100 grid points from the minimum to the maximum longitude in the data is used. Another discrete uniform prior is taken over the number of linear spline basis functions k, from zero up to a maximum equal to the number of observed data points (i.e. we take $K = 42$).

The MCMC sampler was run for a total of 150 000 iterations with the first 50 000 discarded as burn-in iterations. From the last 100 000 iterations every fifth model visited was stored and taken to be in the generated sample. Thus the generated sample contained 20 000 models and the predictions in Figure 3.5 were found by approximating the true function by a pointwise average over each of the functions in the sample. The next subsection discusses the setting of the algorithm's parameters in more detail.

We now find the posterior fit to the data for various values of the prior variance of the coefficients. This allows us to see the effect this prior constant has on our posterior inference. In Figure 3.5 we display the posterior mean of g found for different values of v (recall that $p(\beta_i \mid \sigma^2) = N(0, \sigma^2 v)$). If we denote by $\boldsymbol{\theta}^{(i)}$ the ith model, or curve, in the posterior sample we can calculate the posterior mean curve at x in the same

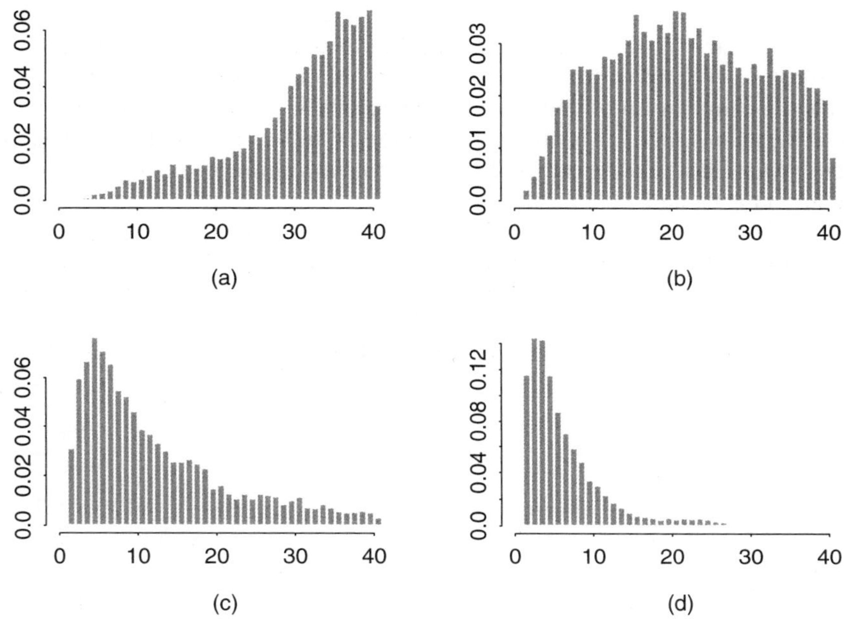

Figure 3.6 Histograms of the posterior distributions of k, the number of basis functions, for (a) $v = 1$; (b) $v = 10$; (c) $v = 100$ and (d) $v = 1000$.

way as (2.32), i.e. using

$$\tilde{g}(x) = E(y \mid x, \mathcal{D}) \approx \frac{1}{N} \sum_{i=1}^{N} E(y \mid x, \mathcal{D}, \boldsymbol{\theta}^{(i)}),$$

where N is the size of the posterior sample. To produce Figure 3.5 we calculated $\tilde{g}$ at 100 evenly spaced point over the range of x. We can also estimate credible intervals using the generated sample. The 95% ones shown in Figure 3.5 were found by discarding the lowest and highest 2.5% of the values of $\tilde{g}$ found from the sample, at each value of x. Note that these give credible intervals for the posterior mean of the regression function so many of the data points do not lie within them. Predictive intervals on the response y can be found by generating from the appropriate Student distribution in (2.30) for each model, and these are much wider.

From Figure 3.5 we see that as v increases $\tilde{g}$ becomes less smooth. When v is small we find that the prior constricts the function from large jumps so many basis functions are needed to adequately model the regression function (see Figure 3.6). Conversely, with v large the prior allows more flexibility in the posterior mean so big jumps in g can be accommodated. Also, fewer basis functions need to be used to model the function as each basis function has many more degrees of freedom (Figure 3.6).

Model complexity is often judged in terms of the number of basis functions, k, but these results suggest that this can be misleading. In Figure 3.5(a) $\tilde{g}$ is very smooth yet the posterior mean of k is relatively large and turns out to be 30.7. However, by eye we can see that in Figure 3.5(c) $\tilde{g}$ is more flexible than the estimate in Figure 3.5(a), in the sense that it follows that data closer, yet the average k is much smaller at only 7.9. It is the combination of prior choices that relates to the fidelity of the estimate to the observations. In fact, we find that the prior on the variance of the coefficients is more important in this respect than any reasonable prior on the number of basis functions in the model.

We find that as v increases beyond 1000 the posterior mean of k, $E(k \mid \mathcal{D})$ becomes increasingly small. This is due to the increasing flexibility each basis function coefficient is allowed when its prior becomes diffuse, but eventually Lindley's paradox (Section 2.5.3) becomes an overriding factor. In the limit as $v \to \infty$ the Bayesian curve fitting approach just fits a constant level and the posterior distribution of k is a point mass at zero.

3.3.4 Monitoring convergence of the sampler

In the last section posterior inference was made using a generated sample of 20 000 models. These were found by running the MCMC algorithm for 150 000 iterations in total. The first 50 000 iterations were discarded as burn-in iterations, only after which the sampling algorithm was judged to have converged. Then, every fifth model visited in the last 100 000 iterations was taken to be in the generated sample. We now explain why we made these choices, making particular reference to the linear spline model with v taken to be 100.

In the top plot in Figure 3.7 we display the logarithm of the marginal likelihood ($\log p(\mathcal{D} \mid \boldsymbol{\theta})$ found from (2.23)) at each of the 150 000 iterations for which we ran the MCMC sampler. We see that the initial model, from which the sampling algorithm started, had a particularly low log marginal likelihood (LML), but very quickly the LMLs of the models settled down to around −46. The variability around this value is due to the sampler visiting models which have both smaller and larger LMLs. We also see that, although we are using a sampling strategy that does not force the model to change at every iteration, the waiting times at a single model are not long. Thus the sampler quickly explores the space of possible models and the simple form of proposal distribution used is adequate.

The other plots in Figure 3.7 give the number of basis functions in the model at each iteration, as well as random draws from $p(\sigma^2 \mid \boldsymbol{\theta}^{(t)}, \mathcal{D})$. We see how the sampler moves between models with very different values on k even though proposed values of k only differ from the current ones by at most one. However, the dependence in k between successive values is clearly visible so convergence to $p(k \mid \mathcal{D})$ would be expected to be slow. In contrast, convergence to $p(\sigma^2 \mid \mathcal{D})$ should be much quicker due to the lack of dependence between successive draws of it based on the MCMC sampler. As in the top plot we see that after the poor initial value for it, it quickly settles down to a value centred around 0.4.

Figure 3.8 uses a qq-plot to highlight the differences between the estimated distribution of $p(\sigma^2 \mid \mathcal{D})$ using the first and last 10 000 models generated by the sampler.

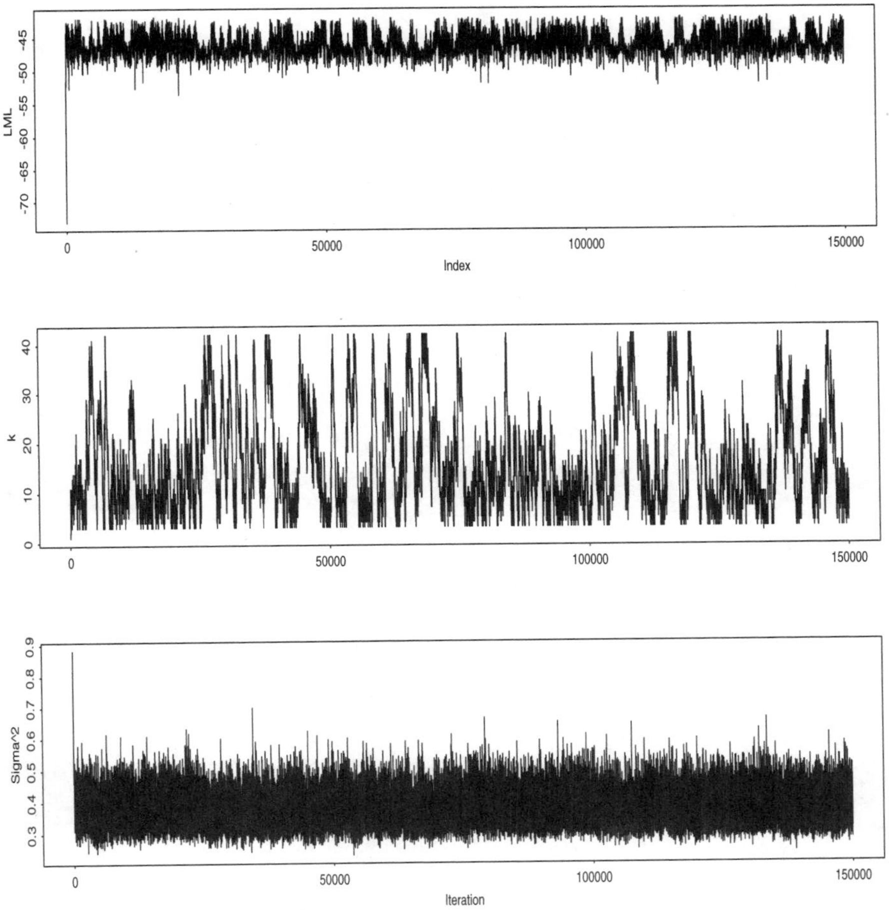

Figure 3.7 Plot of how the log marginal likelihood (LML) (top plot), number of basis functions k (middle), and the regression variance σ^2 (bottom) vary during the 150 000 iterations of the MCMC sampling algorithm for the linear spline model of Section 3.3.3 with $v = 100$.

We see that the densities are essentially the same except for a couple of points which relate to the initial iterations of the algorithm. This demonstrates that convergence to the posterior density of σ^2 is fast, and certainly much quicker than convergence to the posterior density of k. In a perfect world we require convergence to the posterior for all random variables for true convergence of the sampler to have taken place. However, in some instances we may, for practical reasons, be happy with convergence only for the quantities we are particularly interested in.

To diagnose (non-)convergence when we are interested in prediction it is most

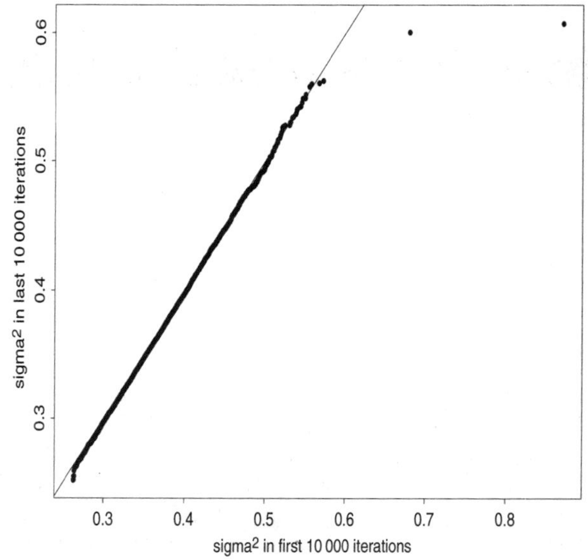

Figure 3.8 A qq-plot comparing the estimated posterior densities of σ^2 using the first and last 10 000 iterations of the sampler.

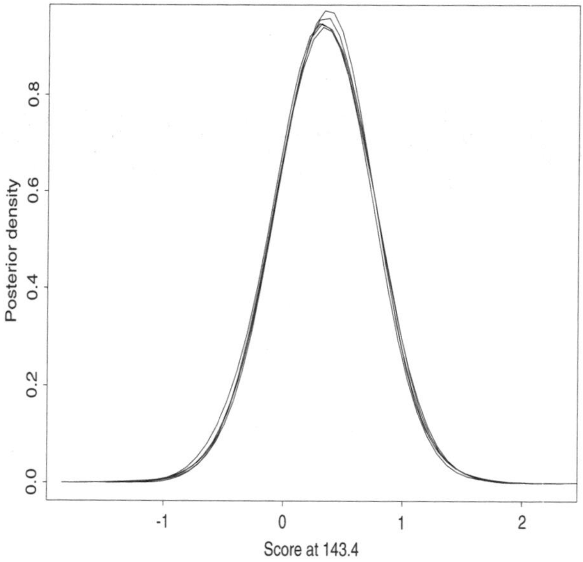

Figure 3.9 The predictive densities at $x = 143.4$ found by each group of 4000 models in the generated sample of 20 000 models.

appropriate to check that predictive densities have converged. Figure 3.9 shows that this is the case at the particular point, $x = 143.4$. Here we plot the estimates to the density $p(y \mid x = 143.4, \mathcal{D})$ found using each group of 4000 models in the generated sample (i.e. Group $i (\in \{1, \ldots, 5\})$ uses models $4000(i-1)+1, \ldots, 4000i$)). We see that the densities are all approximately equal suggesting that there is no reason to suspect non-convergence of the posterior predictive density at $x = 143.4$. To check that this is the case everywhere we could randomly select some other points and plot equivalent figures. In Section 7.5.6 we describe a quantitative method for testing for non-convergence, but here we just demonstrate the idea qualitatively.

We took every fifth model after the burn-in to be in the generated sample. This is to reduce the autocorrelation between some of the parameters in the models and so we need fewer models to provide us with accurate estimates to the posterior densities of interest. In practice, this subsampling of the models visited is also favoured as it ensures that the generated sample of reasonable size. In theory it would be better to make inference using all 100 000 models visited after the burn-in, rather than just the 20 000 found by 'thinning' the 100 000 by only including every fifth one. This is because the 100 000 models contain all the information in the thinned sample, plus the extra given by the further 80 000 models. However, storing the extra 80 000 models can be problematic and the additional information they provide is not that great due to their heavy dependence on the 20 000 models in the smaller sample. Taking every fifth model seems to have become a default choice but any other value, including no subsampling at all, is equally valid.

3.3.5 *Default curve fitting*

From Figure 3.5 we can see that the choice of the variance of the coefficients, v, has a significant effect on the fitted curve so setting it to a sensible value is crucial. Instead of setting v to a set of possible values and then judging which one we think gives the best results we should be able to use the data to determine appropriate settings of v. For instance, we could use an empirical Bayes choice of v found by maximising the marginal likelihood with respect to this parameter (see, for example, Efron 1996; George and Foster 2000; Samaniego and Neath 1996), or even by some sort of cross-validation procedure. However, the strategy advocated here is to allow for uncertainty in v and make inference about it at the same time as making inference about the model parameters. To do this we assign a prior distribution to the possible values of v, rather than a degenerate distribution on a single fixed value as before. We make our predictive inference over the new target posterior $p(\boldsymbol{\theta}, v \mid \mathcal{D})$, which includes the new random parameter v, so that

$$\begin{aligned}\tilde{g}(x) &= E(y \mid x, \mathcal{D}) \\ &\approx \frac{1}{N} \sum_{i=1}^{N} E(y \mid \mathcal{D}, \boldsymbol{\theta}^{(i)}, v^{(i)}).\end{aligned}$$

When we take v random we find that posterior inference is most straightforwardly made by assigning the conjugate gamma prior to v^{-1}, i.e.

$$\begin{aligned} p(v^{-1}) &= \text{Ga}(\delta_1, \delta_2) \\ &= \frac{(\delta_2)^{\delta_1}}{\Gamma(\delta_1)}(v^{-1})^{\delta_1 - 1}\exp(-\delta_2 v^{-1}), \qquad v > 0. \end{aligned}$$

Any prior assigned to a hyperparameter, as we have here, is called a *hyperprior*. Using this conjugate prior we can determine the full conditional posterior distribution of v^{-1} given the other random quantities as

$$\begin{aligned} p(v^{-1} \mid \mathcal{D}, \boldsymbol{\beta}, \sigma^2, \boldsymbol{\theta}) &= p(v^{-1} \mid \boldsymbol{\beta}, \sigma^2) \\ &\propto p(\boldsymbol{\beta} \mid v^{-1}, \sigma^2)p(v^{-1}) \\ &= \text{Ga}\left(\delta_1 + \tfrac{1}{2}k, \delta_2 + \frac{1}{2\sigma^2}\sum_1^k \beta_i^2\right) = \text{Ga}(\delta_1^*, \delta_2^*). \end{aligned} \tag{3.8}$$

Although this hyperprior is used to decrease the effect of our prior specification we now have to set two hyperparameters (δ_1 and δ_2) rather than just one (v), as before. Despite this, we find that the posterior inferences are more robust when we use a hyperprior. To illustrate why, compare the model with the precision of the coefficients fixed, so that $v^{-1} = 0.01$, with another model where $\delta_1/\delta_2 = 0.01$ (note that this leads to the mean of the hyperprior on v^{-1} also being 0.01). Now, if 0.01 is a poor choice of prior precision (say the 'true' precision was much lower), then when we have fixed it we have no way of compensating for our initial fixed setting of v^{-1}. However, with the hyperprior, for all finite values of δ_1 and δ_2 we always have some prior mass on values of v^{-1} less than 0.01 so, if the true posterior mean of the precision is less than 0.01 and the data bear this out, the posterior distribution of v^{-1} will be centred around values less than the prior mean, i.e. closer to the truth.

We see from (3.8) that the distribution of v^{-1} depends explicitly on both $\boldsymbol{\beta}$ and σ^2. The samplers that we have described previously have analytically integrated out these terms, but to update v^{-1} we need to know some current values for $\boldsymbol{\beta}$ and σ^2. This means that we must sample these values even though we (usually) do not want to make inferences about them. Fortunately, this is not a problem as we can use almost exactly the same code as given in Section 3.3.1 even though our target posterior distribution is now $p(\boldsymbol{\theta}, \boldsymbol{\beta}, \sigma^2, v \mid \mathcal{D})$.

First note that we can use a hybrid sampler (Section 2.8.4) to sample from the posterior of interest, taking $\boldsymbol{\psi}_{\text{G}} = (\boldsymbol{\beta}, \sigma^2, v)$ and $\boldsymbol{\phi} = \emptyset$ as we no longer wish to integrate out $\boldsymbol{\beta}$ and σ^2. However, we still know the full conditional posterior distributions of all the parameters in $\boldsymbol{\psi}_{\text{G}}$ so the easiest sampling strategy is to update each parameter conditional on the others. Hence, at iteration t, the sampling algorithm does something like this:

Draw $\boldsymbol{\theta}^{(t+1)}$ from $p(\boldsymbol{\theta} \mid \boldsymbol{\beta}^{(t)}, (\sigma^2)^{(t)}, v^{(t)})$,
Draw $(\boldsymbol{\beta}^{(t+1)}, (\sigma^2)^{(t)})$ from $p(\boldsymbol{\beta}, \sigma^2 \mid \boldsymbol{\theta}^{(t+1)}, v^{(t)})$,
Draw $v^{(t+1)}$ from $p(v \mid \boldsymbol{\theta}^{(t+1)}, \boldsymbol{\beta}^{(t+1)}, (\sigma^2)^{(t+1)})$.

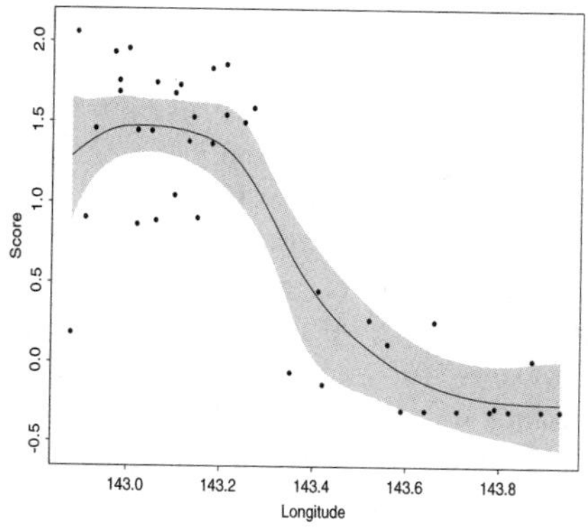

Figure 3.10 Posterior mean estimate for the Great Barrier Reef data when $p(v^{-1}) = \text{Ga}(0.001, 0.1)$.

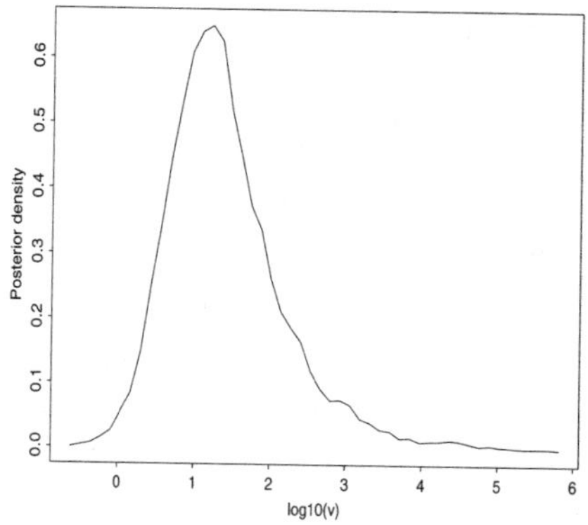

Figure 3.11 The posterior density of $\log_{10} v$.

However, we find that we can make the algorithm far more efficient by noting that

$$p(\boldsymbol{\theta} \mid v) = \int\int p(\boldsymbol{\theta} \mid \boldsymbol{\beta}, \sigma^2, v)p(\boldsymbol{\beta}, \sigma^2 \mid v)\,\mathrm{d}\boldsymbol{\beta}\,\mathrm{d}\sigma^2,$$

can be determined analytically. Then we can just sample $\boldsymbol{\theta}$ given v, draw $(\boldsymbol{\beta}, \sigma^2)$ given v and $\boldsymbol{\theta}$, and finally draw v given everything else. This allows us to use exactly the same algorithm as in Section 3.3.1 to sample $\boldsymbol{\theta}$ except that when sampling $\boldsymbol{\theta}$, everything is conditional on v. Remember that in the earlier work we assumed a fixed constant value of v so we did not need to highlight its dependence on the acceptance probability. Now we take v as a random variable that we need to make inference on. Anyway, this change in the algorithm is straightforward and just involves adding the following lines in the space indicated for sampling $\boldsymbol{\psi}_{\text{G}}$ (POSITION 2).

`Draw` $(\sigma^{-2})^{(t+1)}$ `from` $\text{Ga}(a^*, b^*)$`;`
`Draw` $\boldsymbol{\beta}^{(t+1)}$ `from` $N(\boldsymbol{m}^*, (\sigma^2)^{(t)}\boldsymbol{V}^*)$`;`
`Draw` $(v^{-1})^{(t+1)}$ `from` $\text{Ga}(\delta_1^*, \delta_2^*)$`.`

Note that the updated parameters of the sampling distributions are determined by the current settings of the model.

We now run this new algorithm on the Great Barrier Reef dataset. In the results we give in Figures 3.10 and 3.11 we set $\delta_1 = 0.001$ and $\delta_2 = 0.1$. These settings ensure that the prior mean is a reasonable value (0.01) and the variance is large. We see that the posterior mean estimate in Figure 3.10 is a compromise between the estimates with fixed v (Figure 3.5). Figure 3.11 displays the posterior density of v on the $\log_{10}$ scale. The posterior mean can be thought of as a mixture of the estimates we would get with fixed v, where the mixing weights are given by this posterior density. We see from Figure 3.10 that, not surprisingly, by allowing uncertainty in v we get wider credible intervals. This upholds the general principle that the less specific the priors the less certain you will be in the resulting posterior.

Empirical Bayes methods that 'plug-in' estimates to v found from the data approximate its distribution by a single value. These empirical estimates are usually taken as the mode of the distribution we might expect if we sampled the parameter, so in this case it would be close to the mode of the distribution given in Figure 3.11. The accuracy of this approximation will depend on both the concentration and skewness of the distribution for the parameter (v in this case) around its modal value. For this simple model, where sampling is still straightforward when we allow v to be random, empirical Bayes methods are not required. However, for more complex modelling tasks it may be useful to use such plug-in estimates for some parameters to simplify the sampling (see, for example, Clyde and George 2000; Kennedy and O'Hagan 2001)

3.4 Curve Fitting Using Wavelets

In this chapter, so far we have restricted our attention to a single basis set, namely the truncated splines. However, the methodology described in relation to that basis set is generic and can be applied using any basis set. Another such basis model that has received enormous attention within the statistics community are wavelet bases, the focus of this section.

Wavelet methods (see, for example, the books by Bruce and Gao (1996), Vidakovic (1999) and Percival and Walden (2000)) also estimate f with the linear in the parameters form of (2.9). However, we devote a whole section to wavelet regression as wavelet bases have some particularly attractive properties that makes them a good basis set to use in certain circumstances (e.g. when the underlying function is rapidly varying). Wavelets form a multiresolution decomposition of the curve so that the coefficients, $\boldsymbol{\beta}$, contain information about the nature of the curve at different scales. For instance, at the highest resolution a high value of one coefficient may tell us that the curve has a changepoint within the support of its corresponding basis function.

When using wavelet bases we think of the estimate to the true curve as being given by

$$g(x) = \beta_0 + \sum_{j=1}^{J-1} \sum_{i=1}^{n(j)} \beta_{ji} B_{ji}(x), \qquad x \in \mathcal{X}, \tag{3.9}$$

where $J = \log_2 n$ and $n(j) = n2^{j-J}$. We write this function using a double summation to highlight the multiresolution properties of wavelet bases. They depend on both scaling and translating what is known as a mother wavelet, $\psi(x)$. In particular

$$B_{ji}(x) = 2^{-j/2}\psi(2^{-j}x - i), \qquad j = 1, \ldots, J, \tag{3.10}$$

where j is known as the dilation number and i the translation number. As bases with the same value of j are just translations of each other we also refer to j as the resolution level. Although the number of samples, n, does not strictly have to be a power of two, in the majority of papers published on wavelets it has this form.

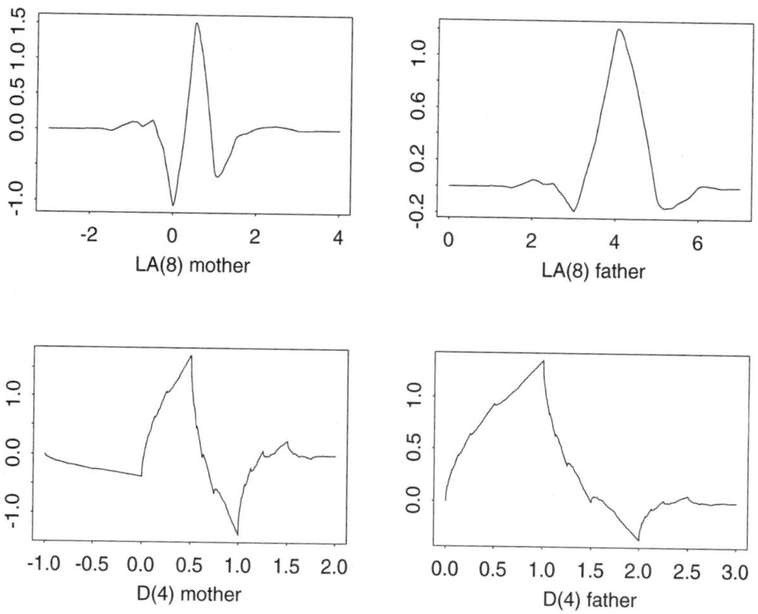

Figure 3.12 Examples of mother and father wavelets for LA(8) and D(4).

This is because wavelet regression is particularly computationally efficient, and hence attractive, when we can write $n = 2^J$ for some integer J.

Mother wavelets have a form that ensures that the B_{ji} form an orthonormal basis set, so

$$\int B_{ji}(x)B_{j'i'}(x)\,\mathrm{d}x = \begin{cases} 1, & i = i' \text{ and } j = j', \\ 0, & \text{otherwise.} \end{cases}$$

There are many different mother wavelets which give rise to basis functions that satisfy these conditions. In particular, in Figure 3.12 we display examples of mother, together with their related father, wavelets for Daubechies' least asymmetric (LA) and Daubechies' extremal phase (D) families (Bruce and Gao 1996; Daubechies 1992) of wavelet bases. Note that analytic expressions for these functions are not available as is usually the case for wavelets. In fact, of the most popularly used wavelets, only the very simple Haar wavelet (Haar 1910) admits an analytic form. Also, families of wavelets are typically indexed by their order (e.g. the least asymmetric wavelet of order 8, is written LA(8)) which relates to the number of vanishing moments of the bases. Hence, the higher the order of the wavelet the more smooth it is.

We have already mentioned how wavelets relate to different scales and translations of other functions. In Figure 3.13 we see how this relates to some LA(8) wavelets. These wavelets have support only on a finite range of values and we find that the bases

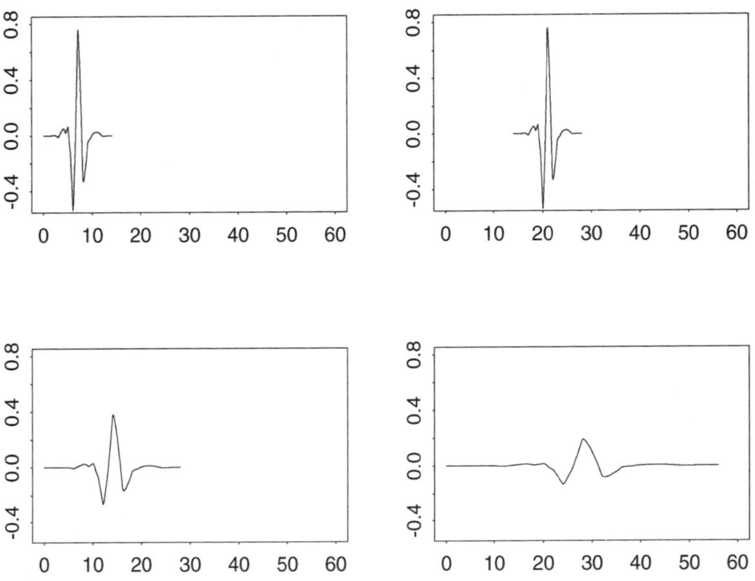

Figure 3.13 Examples of wavelet basis functions found by translating and scaling the LA(8) wavelet.

relating to low resolution levels (j low) are non-zero over a greater range than those with higher resolution. Thus low-level bases tend to pick up low frequency, or near global, structure, whereas those for which j is high pick up local changes in the data.

From the form of (3.9) and (3.10) we see the dependence of the wavelet basis on powers of 2. We find that when the size of dataset is an exact power of 2 ($n = 2^J$) and it is equally spaced a very efficient $O(n)$ algorithm (see, for example, Mallat 1989) can be used to calculate the wavelet basis matrix, $\boldsymbol{B}$. Although these restrictions can cause problems in traditional statistical settings when these conditions are not generally met, wavelets have proved particularly popular in time series and signal processing applications when datasets of this form are more common (see, for example, Abramovich *et al.* 2000; Nason and von Sachs 1999; Nowak and Baraniuk 1999; Percival 1995; Schiff *et al.* 1994; Serroukh *et al.* 2000).

By construction the basis set is orthonormal so we know that $\boldsymbol{B}'\boldsymbol{B} = \boldsymbol{I}$. Hence, to calculate the least squares estimates to the coefficients, $\hat{\boldsymbol{\beta}}$, is particularly simple as

$$\hat{\boldsymbol{\beta}} = (\boldsymbol{B}'\boldsymbol{B})^{-1}\boldsymbol{B}'Y = \boldsymbol{B}'Y.$$

Thus the coefficient vector is just a linear combination of the responses. We also find, by premultiplying the standard linear regression equation (2.11) by $\boldsymbol{B}'$, that

$$\boldsymbol{B}'Y = \boldsymbol{B}'\boldsymbol{B}\boldsymbol{\beta} + \boldsymbol{B}'\boldsymbol{\epsilon},$$
$$\hat{\boldsymbol{\beta}} = \boldsymbol{\beta} + \boldsymbol{\tau}. \tag{3.11}$$

This is the representation of the problem in the wavelet domain. We see that the errors in the wavelet domain, $\boldsymbol{\tau}$, are a linear combination of the errors in the standard representation in the data domain. Making the common assumption of normal errors in the data domain, $\boldsymbol{\epsilon} \sim N(\mathbf{0}, \sigma^2 \boldsymbol{I})$, we find that the errors in the wavelet domain $\boldsymbol{\tau} = \boldsymbol{B}'\boldsymbol{\epsilon}$ also follow an $N(\mathbf{0}, \sigma^2 \boldsymbol{I})$ distribution. Hence we can perform regression in the wavelet domain assuming the same error structure as we would have in the more usual data domain.

3.4.1 Wavelet shrinkage

Standard (non-Bayesian) nonlinear regression using wavelets is undertaken in three basic steps.

(i) Apply the discrete wavelet transform (DWT) to the standard matrix regression equation by premultiply it by $\boldsymbol{B}'$. In this way we obtain the representation of the problem in the wavelet domain, equation (3.11).

(ii) Shrink the empirical wavelet coefficients, $\hat{\boldsymbol{\beta}}$, according to some predetermined function, yielding the estimates to the true coefficients $\tilde{\boldsymbol{\beta}}$.

(iii) Invert the DWT, by premultiplying the shrunken coefficients $\tilde{\boldsymbol{\beta}}$ by $\boldsymbol{B}$ yielding the estimate to f, $\tilde{Y} = \boldsymbol{B}\tilde{\boldsymbol{\beta}}$.

Once the wavelet basis matrix has been calculated steps (i) and (iii) are trivial so it is step (ii) that is crucial in determining the properties of the estimates of the responses, $\tilde{Y}$. To carry out step (ii) we first need to choose a suitable shrinkage function. The most widely used functions of this type are the hard- and soft-thresholding schemes (Donoho and Johnstone 1994) given by

$$t_{\mathrm{H}}(x, \lambda) = xI(|x| > \lambda), \tag{3.12}$$
$$t_{\mathrm{S}}(x, \lambda) = \mathrm{sgn}(x)(|x| - \lambda)_+, \tag{3.13}$$

where $\lambda > 0$ is known as the threshold value and $\mathrm{sgn}(x)$ takes the value one when x is positive and -1 otherwise. From Figure 3.14 we see that the hard-thresholding rule just sets coefficients below λ to zero and leaves the others unaffected. In contrast soft-thresholding shrinks all the coefficients: those less than λ are set to zero and the others are reduced in magnitude by λ. However, in practice soft-thresholding is not used on the coefficients that correspond to basis functions that model low resolution features (i.e. low j) to prevent any shrinkage of the global features of the curve.

The threshold value λ has a significant effect on the fitted responses, $\tilde{Y}$. A simple choice is the so-called universal threshold (Donoho and Johnstone 1994) that takes

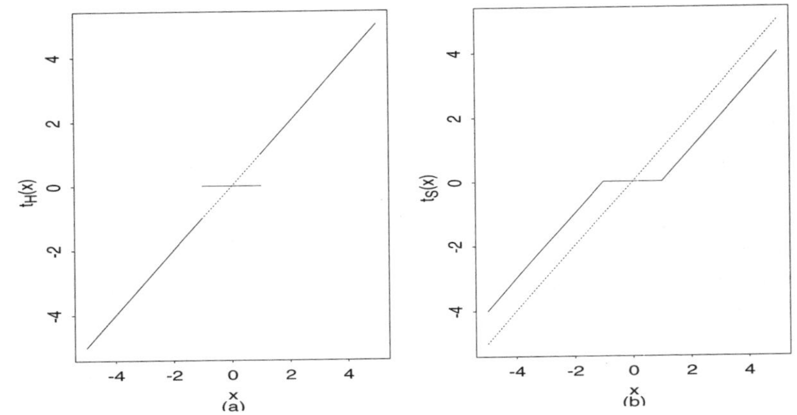

Figure 3.14 Thresholding schemes. The dotted lines gives the coefficient (or input) values to the thresholding functions given by the solid lines: (a) $t_H(x, 1)$; (b) $t_S(x, 1)$.

$\lambda = \hat{\sigma}\sqrt{2\log n}$, where $\hat{\sigma}$ is a robust estimate of the standard deviation of the noise. Donoho *et al.* (1995) propose the median absolute deviation estimator found with the coefficients of the highest level, that is

$$\hat{\sigma} = 1.483\,\text{median}(|\hat{\beta}_{J-1,1}|, \ldots, |\hat{\beta}_{J-1,2^{J-1}}|).$$

Universal thresholding is widely used due to its simplicity and as it is asymptotically optimal using both hard- and soft-thresholding schemes. However, for finite samples Donoho and Johnstone (1994) noted that other schemes may provide better results in terms of mean-squared error.

3.4.2 Bayesian wavelets

The hard and soft thresholding schemes of (3.12), (3.13) are particularly straightforward forms of shrinkage. Hard thresholding preserves large coefficients but can lead to visually unattractive reconstructions as noise 'spikes' can be present relating to coefficients of high resolution which are not set to zero. Soft thresholding gets around the problem of the noise spikes but leads to large coefficients being shrunk when they relate entirely to signal which can lead to large errors. For instance, the reconstruction of the Bumps example using wavelets given in Donoho and Johnstone (1994) had a larger mean-squared error than the actual simulated, noisy dataset. Although the noise had been removed the shrinkage of large coefficients by the soft thresholding method attenuated the high peaks severely leading to large squared differences between the truth and the reconstruction. We want a method that is in some way a combination of these two approaches and find that by using the Bayesian paradigm we can formulate models that have desirable shrinkage properties.

First we consider the method of Clyde *et al.* (1998), which models the terms in the wavelet domain (3.11). They assume that a single empirical wavelet coefficient, $\hat{\beta}_{ji}$, is related to the true value, β_{ji} via the expression,

$$\theta_{ji}\hat{\beta}_{ji} = \beta_{ji} + \tau_{ji},$$

where $p(\tau_{ji}) = N(0, \sigma^2)$ and $\theta_{ji} \in \{0, 1\}$ are unknown. This assumes that for $\theta_{ji} = 1$ the true coefficient, β_{ji} follows a $N(\hat{\beta}_{ji}, \sigma^2)$, where the least squares estimate is found using the discrete wavelet transform. Clyde *et al.* (1998) make use of a hierarchical prior specification taking

$$p(\beta_{ji} \mid \theta_{ji}) = N(0, \theta_{ji}\sigma^2 v_j), \qquad p(\theta_{ji}) = \mathrm{Br}(\gamma_j),$$

where γ_j and v_j are fixed hyperparameters that depend on the resolution level. Assuming $N(0, \sigma^2)$ errors we find (Clyde *et al.* 1998) that

$$E(\beta_{ji} \mid \mathcal{D}) = p(\theta_{ji} = 1 \mid \mathcal{D})\frac{v}{1+v}\hat{\beta}_{ji}. \tag{3.14}$$

This expectation minimises the posterior squared-error loss between the true coefficient value, β_{ji}, and our estimate, $\tilde{\beta}_{ji}$. Further, we can determine the predictive density as

$$p(y \mid x, \mathcal{D}) = \sum_{\boldsymbol{\theta} \in \boldsymbol{\Theta}} p(y \mid x, \mathcal{D}, \boldsymbol{\theta})p(\boldsymbol{\theta} \mid \mathcal{D}),$$

where the summation is over the 2^n possible models. Typically, n is large so sampling techniques need to be employed and the Gibbs sampler was suggested by Clyde *et al.* (1998) and this appears to be the favoured approach due to its simplicity (see Müller and Vidakovic (1999) for a collection of Bayesian wavelet papers). The setting of the hyperparameters v_j and γ_j is usually undertaken via empirical Bayes' methods (see, for example, Clyde and George 1999, 2000).

Chipman *et al.* (1997) proposed a similar method, known as adaptive Bayesian wavelet shrinkage, except that they suggested robustly estimating σ^2 rather taking it to be random. They then chose

$$p(\beta_{ji} \mid \theta_{ji}) = \theta_{ji}N(0, v_j) + (1 - \theta_{ji})N(0, c_j v_j),$$

where c_j is taken to be near zero. This follows the prior suggested by George and McCulloch (1993) which assumes that β_{ji} is a mixture of two normals, one with a small variance and one with a much larger variance. With this prior we find that

$$E(\beta_{ji} \mid \mathcal{D}) = \left\{p(\theta_{ji} = 0 \mid \mathcal{D})\frac{c_j v_j}{\sigma^2 + c_j v_j} + p(\theta_{ji} = 1 \mid \mathcal{D})\frac{v_j}{\sigma^2 + v_j}\right\}\hat{\beta}_{ji}.$$

This reduces to (3.14) when $c_j = 0$. Again empirical Bayes methods are used to estimate the hyperparameters.

Vidakovic (1998) proposed Bayesian wavelet shrinkage with heavy-tailed prior distributions, for example, student prior distributions were assigned to the wavelet coefficients. These worked particularly well as, compared to the model using normal priors, large coefficients were shrunk less and those near zero shrunk more. In a similar spirit, Clyde and George (2000) extend this work by investigating the shrinkage estimates produced when the priors on the β_{ji}s are taken to be scale mixtures of normal distributions. They provide extensive empirical comparisons between a variety of possible models as well as providing useful analytic results, where available.

A quite different method of wavelet shrinkage was proposed by Holmes and Denison (1999). Here, the prior was again taken to be multivariate normal but a separate variance was assumed for each wavelet coefficient, so $p(\beta_{ji} \mid v_{ji}) = N(0, v_{ji})$, and each coefficient was taken independent *a priori*. They suggested using independent priors on the v_{ji} that penalised the degrees of freedom associated with that basis function, rather than the conjugate choice, taking

$$p(v_{ji} \mid \sigma^2) \propto \exp\left\{-\frac{cv_{ji}}{\sigma^2 + v_{ji}}\right\},$$

for all i, j, and assigning a standard inverse-gamma prior to σ^2. Note that here c is not a level-dependent hyperparameter.

The definition of the degrees of freedom is not unique but they used the accepted definition of Hastie and Tibshirani (1990), which takes it as the trace (i.e. the sum of the diagonal terms) of the smoothing matrix $\boldsymbol{S} = \boldsymbol{B}(\boldsymbol{B}'\boldsymbol{B} + \boldsymbol{V}^{-1})^{-1}\boldsymbol{B}'$, where $\boldsymbol{V}$ is diagonal with entries v_{ji} and represents the prior variance matrix for the coefficients $\boldsymbol{\beta}$. The authors do not suggest a specific choice of c but they show how certain values of it correspond to the maximum *a posteriori* model being the same one that would be chosen under a variety of classical model choice criteria. They find that $c = \log n$ is a particularly good value for wavelet regression, and this setting relates to the risk inflation criterion of Foster and George (1994).

All of the Bayesian wavelet shrinkage methods described lead to shrinkage similar to that displayed in Figure 3.15. This particular figure was produced from results using the method of Holmes and Denison (1999) but a similar shrinkage function is found using the other methods. Bayesian wavelet shrinkage tends to lead to a compromise between the hard and soft thresholding rules of (3.12), (3.13). It can shrink coefficients values to very small values but not to exactly zero, which is why they are known as *shrinkage* models. This is fine for making predictions but as the signal can usually be represented well by only a few coefficients formal thresholding methods are sometimes more appropriate (e.g. for data compression). Throughout this book we are focusing on prediction so we do not describe Bayesian thresholding models here but the interested reader is referred to the work of Abramovich *et al.* (1998), Müller and Rios Insua (1998b), Moulin and Liu (1999) and Ruggeri and Vidakovic (1999).

3.5 Prior Elicitation

In this section we concentrate on how the prior parameters and the model assumptions that we make affect our prior distribution of the curves. Although poor prior parameter

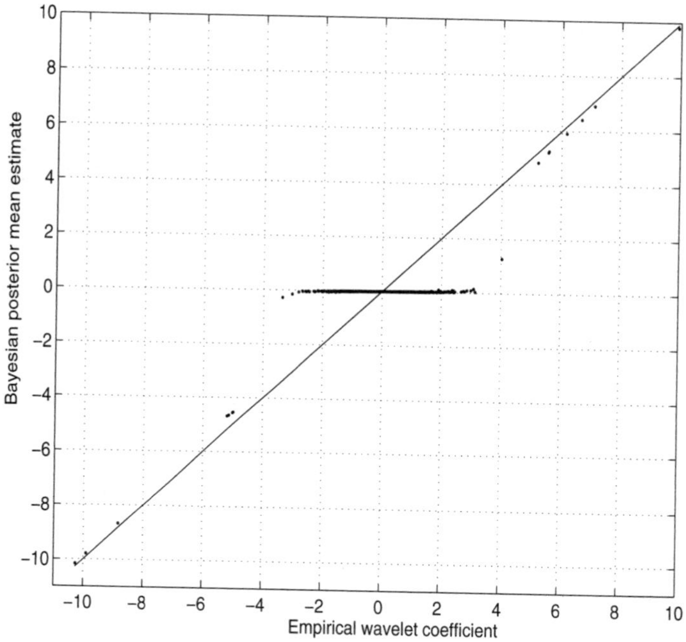

Figure 3.15 An example of Bayesian wavelet shrinkage. The horizontal axis gives the empirical wavelet coefficient values, $\hat{\beta}_{ji}$, and the vertical axis the shrunken Bayesian posterior mean estimates, $E(\beta_{ji} \mid \mathcal{D})$.

specification only significantly affects the posterior inference when there is a small amount of data, poor model assumptions will generally produce poor results in any statistical analysis. We do not embark on a wide-ranging study into the effect of poor prior choice on the posterior inference, instead we concentrate on choosing priors that reflect sensible beliefs. The ideas presented here relate to all types of basis functions that we might use for curve fitting, although the majority of the exposition focuses on truncated spline bases.

3.5.1 The model prior

We see first how the prior on the form of the basis functions can affect the inference. We illustrate one particular problem with the widely recommended cubic spline basis functions (see, for example, de Boor 1978; Durrleman and Simon 1989; Knott 1999), that also applies to the linear splines we have already encountered. We demonstrate the importance of sampling from the prior models we use to ensure that they have the properties we would expect of them.

We shall now look at samples from the prior of the Bayesian cubic spline model,

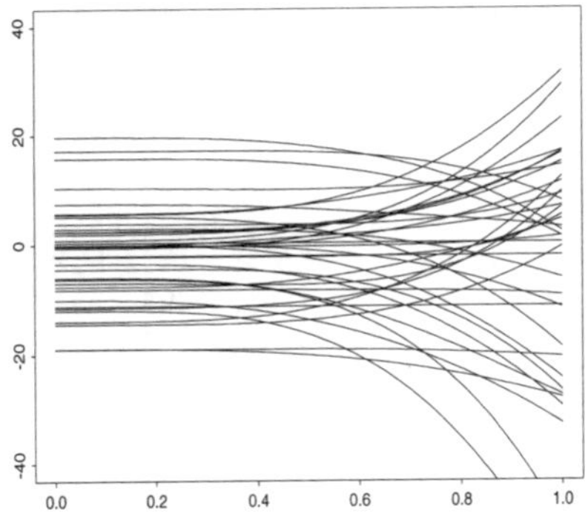

Figure 3.16 Fifty realisations from prior of the one-sided cubic spline model.

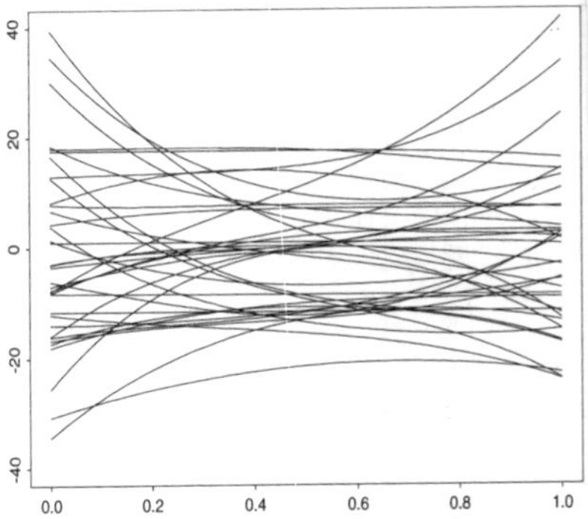

Figure 3.17 Fifty realisations from prior for the two-sided cubic spline model.

which we write using the truncated power series representation, i.e.

$$g(x) = \beta_0 + \sum_{i=1}^{k} \beta_i (x - t_i)_+^3, \tag{3.15}$$

and again write $\boldsymbol{\theta} = (t_1, \ldots, t_k)$ (this is just (3.2) with $q = 3$). For illustrative purposes we fix the prior regression variance and prior variance of the coefficients so that $\sigma^2 = 1$ and $v = 100$. Further, we assume the model priors are given by (3.3) with $\mathcal{T} = \{0, 0.01, \ldots, 0.99, 1\}$ so $T = 101$ and choose $K = 50$, the maximum number of knots allowable.

In Figure 3.16 we display 50 draws from the prior for this model. It is obvious that the prior curves are distributed differently according to x, most notably at the endpoints $x = 0$ and $x = 1$. This is because from the form of g, given in (3.15), $g(0) = \beta_0$ so $\text{var}\{g_1(0)\} = \text{var}(\beta_0) = 100$. However, at $x = 1$ we find that

$$\begin{aligned}
\text{var}\{g(1) \mid \boldsymbol{\theta}\} &= \text{var}\left\{\beta_0 + \sum_{i=1}^{k} \beta_i (1 - t_i)^3\right\} \\
&= \text{var}(\beta_0) + \sum_{i=1}^{k} (1 - t_i)^6 \,\text{var}(\beta_i) \\
&= \text{var}\{g(0)\}\left\{1 + \sum_{i=1}^{k} (1 - t_i)^6\right\} \\
&> \text{var}\{g(0)\}.
\end{aligned}$$

In the absence of further information, we would tend to believe that the variance of the estimating function should be the same at both endpoints. Any other prior cannot be justified as it inputs prior information which we have no knowledge about, hence g of the form in (3.15) is inappropriate. The reason that the early papers in Bayesian curve fitting (Denison *et al.* 1998a; Smith and Kohn 1996) used similar forms to (3.15) is probably due to the widespread use of such models in the non-Bayesian literature. It is only by drawing from the model prior that we can see the problems with (3.15) so we feel that it is always important to check that draws from the prior reflect the knowledge assumed.

Fortunately, it is straightforward to modify g so that the behaviour at both endpoints is identical. The reason that this is not the case with (3.15) is that the candidate basis functions are all *one sided*, that is they are all non-zero only when x is greater than some value. However, we can enlarge the set of candidate bases to also include opposite bases which are non-zero only when x is less than some value. Hence, choosing the candidate set of bases as

$$\{(x)_+^3, (x)_-^3, (x-0.01)_+^3, (x-0.01)_-^3, \ldots, (x-1)_+^3, (x-1)_-^3\},$$

where $(x-t)_- = (t-x)_+$, we obtain a two-sided cubic spline model. We plot 50 realisations from the prior for this two-sided spline model in Figure 3.17 and find that the curve behaves identically at both ends. Further, we see that the variance of the prior decreases towards the midpoint of the range because here none of the basis functions that are non-zero have a particularly high value of $(x-t_i)^6$.

The increase in variance of cubic splines near the endpoints is well-known. Natural splines (see, for example, de Boor 1978) combat this by assuming that the function is linear outside the boundary knots. This tends to decrease test error outside the range of the data. We can also use other spline representations to provide more stable prior curves (i.e. ones with lower variance) such as the B-spline (de Boor 1978) suggested by Biller (2000). B-splines are defined via the Cox–de Boor recursion formula (de Boor 1978)

$$\begin{aligned} B_{i0}(x) &= I(t_i \leqslant x < t_{i+1}), \\ B_{iq}(x) &= \frac{x-t_i}{t_{i+q}-t_i} B_{i,q-1}(x) + \frac{t_{i+q+1}-x}{t_{i+q+1}-t_{i+1}} B_{i+1,q-1}(x). \end{aligned}$$

Note that we use double subscripts to index the basis functions here to illustrate the recursive nature of the construction of the B-spline bases.

An advantage of the cubic B-spline basis functions, for which $q=3$, is that they are non-zero only at most five interior knot locations (see Figure 3.19). Thus a design matrix produced with B-spline bases ensures that $\boldsymbol{B}'\boldsymbol{B}$ is banded. The condition number of such banded matrices is often relatively low (compared to those found using truncated linear splines) giving more numerical stability when calculating of $(\boldsymbol{B}'\boldsymbol{B}+\boldsymbol{V}^{-1})^{-1}$. This is especially noticeable when the number of knots in the model is high. Nevertheless, we chose to illustrate Bayesian curve fitting using the truncated power series spline representation for two reasons: they are simple to write down and it is straightforward to understand the effect of each basis function (i.e. they give the points at which there are jumps in the derivative of the estimating function, g).

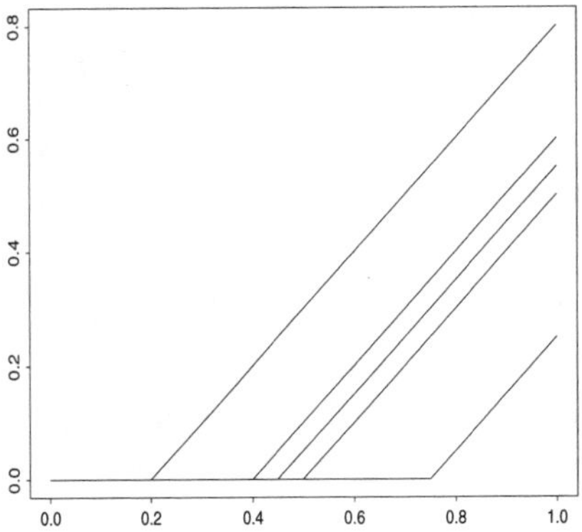

Figure 3.18 One-sided piecewise linear spline basis functions with knot points $t_1 = 0.2$, $t_2 = 0.4$, $t_3 = 0.45$, $t_4 = 0.5$ and $t_5 = 0.75$.

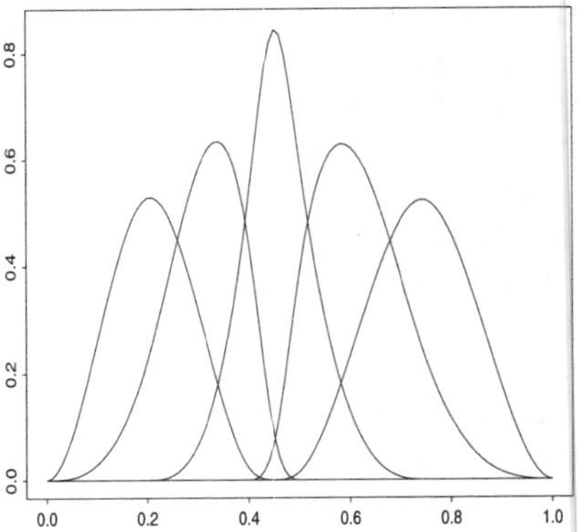

Figure 3.19 *B*-spline basis functions with the same knot points as those used in Figure 3.18.

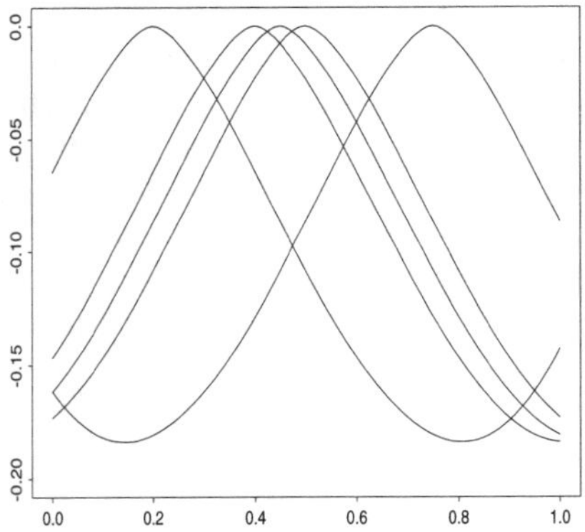

Figure 3.20 Thin-plate spline basis functions with the same knot points as those used in Figure 3.18.

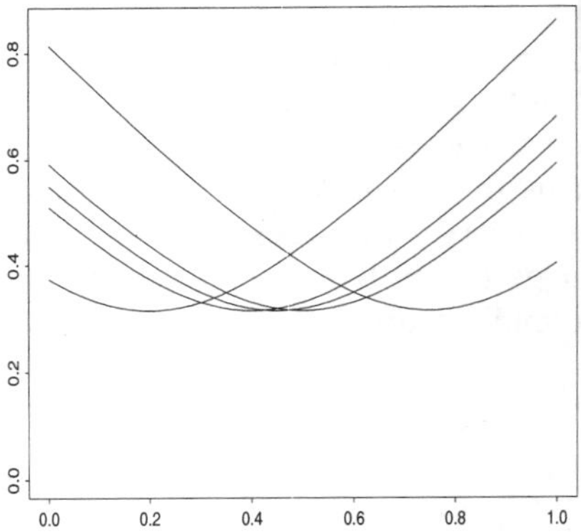

Figure 3.21 Hardy's multiquadric basis functions with the same knot points as those used in Figure 3.18 and with $\kappa^2 = 0.1$.

In Figures 3.20 and 3.21 we plot examples of other standard basis functions we could use for curve fitting. Namely the thin-plate spline bases (Green and Silverman 1994), for which

$$B_i(x) = (x - t_i)^2 \log |x - t_i|,$$

and Hardy's multiquadric bases (Franke 1982), where

$$B_i(x) = |\kappa^2 + (x - t_i)^2|^{1/2},$$

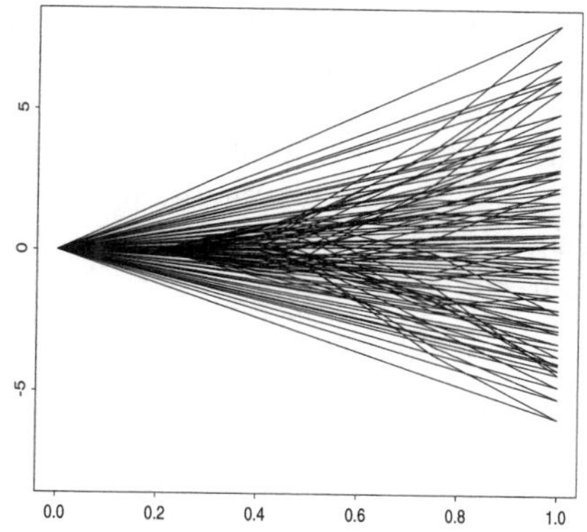

Figure 3.22 Fifty random curves found using the basis functions in Figure 3.18. The coefficient of each basis function was drawn from a uniform distribution on $[-5, 5]$.

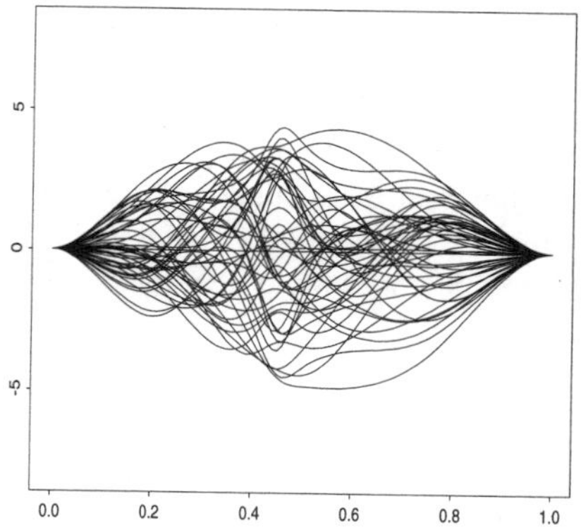

Figure 3.23 Same as Figure 3.22 but using the basis functions in Figure 3.19.

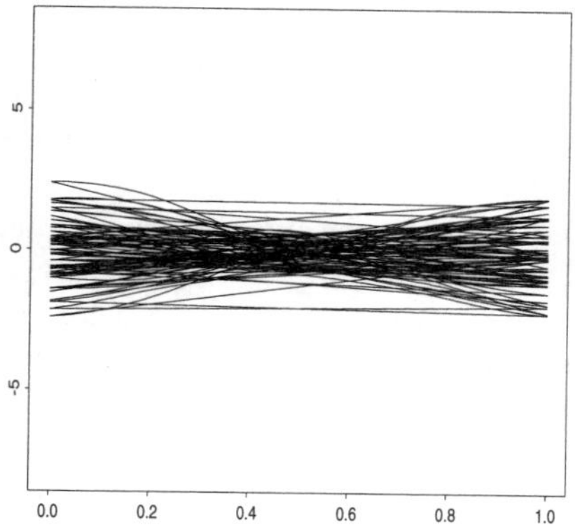

Figure 3.24 Same as Figure 3.22 but using the basis functions in Figure 3.20.

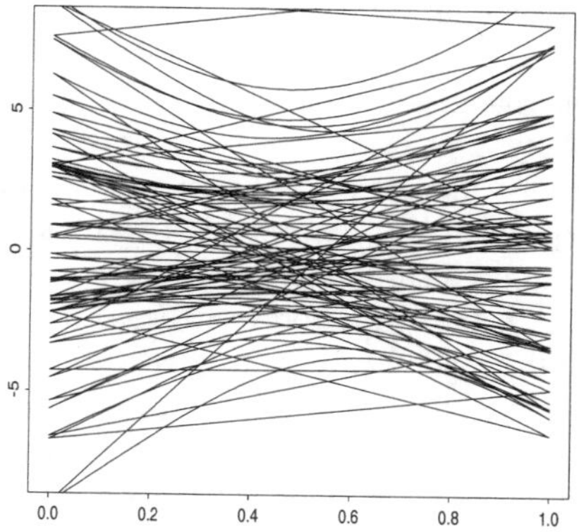

Figure 3.25 Same as Figure 3.22 but using the basis functions in Figure 3.21.

with κ^2 a parameter of the model that controls the smoothness of the basis functions. These are some of the most common examples of possible bases but, of course, others exist.

In Figures 3.22–3.25 we show how the different basis functions relate to curves with different properties. We plot 50 curves found using the basis functions of Figures 3.18–3.21 with the coefficients of each curve drawn uniformly from $[-5, 5]$, i.e. we plot 50 realisations of $\sum_1^5 \beta_i B_i(x)$, where $\beta_i \sim U[-5, 5]$. Note that we use the same random values of the coefficients for each plot and do not include an intercept term.

An interesting features to note is how the B-spline basis leads to a function that is zero at both ends so, in general, we should include global basis functions in the regression, such as $1, x, x^2$ and x^3, to allow the model to accurately estimate curves for which the responses at each endpoint are not the same. Nevertheless, having low variance at the end points is an attractive feature of this basis set. We also see how the choice of basis function affects the scale of the curves, as demonstrated by the difference in Figures 3.24 and 3.25. This might suggest that the scale of the priors on the coefficients should be related to the actual basis set used.

3.5.2 Prior on the model parameters

Here we look at how to assign $p(\boldsymbol{\theta})$ given that we have already decided which type of basis functions to use. So far we have suggested that, in the absence of any additional information, to use the specification,

$$p(\boldsymbol{\theta}) = \binom{T}{k}^{-1} \times \frac{1}{K+1}, \tag{3.16}$$

for $k = 0, 1, \ldots, K$. This assumes that each value of k is equally likely *a priori* up to some maximum value K which is typically chosen so large that its effect on the posterior is negligible. The first term ensures that if there are T candidate locations for the changepoints then, given k changepoints, each collection of $(t_1, \ldots, t_k)$ is equally likely. This prior for discrete models can easily be adapted when the changepoints can be chosen continuously on some interval $\mathcal{T} = [t_{\min}, t_{\max}]$. In these cases we just replace the first term in (3.16) with $k!(t_{\max} - t_{\min})^{-k}$, the density of the appropriate uniform distribution.

A prior of the form (3.16) does not place an explicit penalty on the dimension of the model. However, as we have seen earlier, the marginal likelihood contains a penalty on the dimension which depends on the prior variance of the coefficients. For this reason the uniform prior is adequate and it will continue to be the default used throughout the rest of the book.

Smith and Kohn (1996) use a uniform prior on the set of all possible models. This means that the probability that a given basis function is included in the model (3.7) is $\frac{1}{2}$, hence $k \sim \text{Bi}(T, \frac{1}{2})$ and

$$p(k) = \binom{T}{k}\left(\frac{1}{2}\right)^k\left(\frac{1}{2}\right)^{T-k} = \binom{T}{k}2^{-T}, \qquad k = 0, 1, \ldots, T.$$

For realistic values of T this means that there is very little prior weight on models with few basis functions and most of the weight is on models with about half the basis functions in. This goes against our natural wishes of wanting to estimate the model in as parsimonious a fashion as possible. For this reason in further work Kohn *et al.* (2001) suggested refining this prior by assigning a hierarchical beta prior to the probability of a single basis function being included. In the literature this type of prior has tended to be favoured when using Gibbs sampling even though it is also easy to adopt (3.16).

Another popular prior is taking $k \mid \lambda \sim \text{Poi}(\lambda)$ so that

$$p(k \mid \lambda) = \frac{\lambda^k \exp(-\lambda)}{k!}, \qquad k = 0, 1, 2, \ldots.$$

Here λ is a hyperparameter to be set, or elicited, through the use of a hyperprior on it. As the maximum number of bases, K, is usually finite, it is more usual to use the truncated Poisson distribution for k, hence

$$p(k \mid \lambda) = \frac{1}{\sum_{i=0}^{K}(\lambda^i/i!)} \frac{\lambda^k}{k!}, \qquad k = 0, 1, \ldots, K.$$

In the reversible jump literature this choice was originally adopted by Green (1995) and continues to be popular (Andrieu *et al.* 2001a; Denison *et al.* 1998a; Heikkinen and Arjas 1998). This suffers from the same problem as the binomial prior in that, in general, the prior favours model with around λ basis functions rather than those with only a few bases.

If a nonuniform prior on k is to be adopted, a natural one that captures the concept of parsimony is the truncated geometric one, say

$$p(k \mid \lambda) = \frac{\lambda(1-\lambda)^k}{1-(1-\lambda)^{K+1}}, \qquad k = 0, 1, \ldots, K.$$

This has its prior mode on the model with no basis functions and as the number of basis functions increases the model prior for that number necessarily decreases.

Whenever attempting to set priors on $\boldsymbol{\theta}$, it is worth remembering that penalties on k control the complexity of the model only in terms of how many random variables it contains. As can be seen by Figures 3.5 and 3.6 this does not necessarily give a good idea of how flexible the model. For example, the posterior mean curve in Figure 3.5(d) is more flexible than that in Figure 3.5(b), even though fewer basis functions were used in its reconstruction.

3.5.3 The prior on the coefficients

As we have already seen, using a normal prior for $p(\boldsymbol{\beta} \mid \sigma^2, \boldsymbol{\theta})$ together with an inverse-gamma prior on the regression variance, σ^2, ensures that $p(\mathcal{D} \mid \boldsymbol{\theta})$ can be calculated analytically. In Section 2.4.1 we took $p(\boldsymbol{\beta} \mid \sigma^2, \boldsymbol{\theta}) = N(\mathbf{0}, \boldsymbol{V})$ but did not specify $\boldsymbol{V}$, which can be taken to be either a random quantity or constant matrix.

In Section 3.3.3 we chose $\boldsymbol{V} = \text{diag}(\infty, v, \ldots, v)$, where v was a fixed constant. This assumes that the constant coefficient has a prior which is constant over the whole real line while the other coefficients are independent and identically distributed zero-mean normal random variables. Setting v is not trivial due to Lindley's paradox and no definitive advice can be given. In Section 3.3.5 we took v to be random and assigned a hierarchical prior to it to robustify our method against poor choices of it.

In the absence of information it makes sense to set the prior means of the basis coefficients $\beta_1, \ldots, \beta_k$ to be zero as this reflects ignorance about the sign of the coefficients *a priori*. It is also reasonable to take the prior on the constant term, β_0 to be uniform on the real line (this is an improper normal prior with infinite variance) as there is rarely evidence to suggest that the overall mean response of a dataset is zero. Lindley's paradox for vague priors is only a problem when we use improper priors of varying dimension, so using it only for β_0, which is always included in the model, does not cause difficulties.

The assumption of prior independence between the coefficients is often made but makes more sense when the basis functions in the model lead to $\boldsymbol{B}'\boldsymbol{B}$ being diagonal. This is only true for the special case when the complete set of candidate bases are orthogonal. However, as we do not generally use orthogonal bases the assumption of prior independence is made only in the absence of any further information which could guide us to a better prior. We shall refer to this prior as a ridge prior for reasons that will become clear at the end of this subsection.

Another choice of prior for the coefficients is the g-prior of Zellner (1986). This particular specification introduces prior dependence between the coefficient values, unlike the ridge prior already described. The g-prior for $\boldsymbol{\beta}$ is given by

$$p(\boldsymbol{\beta} \mid \sigma^2, \boldsymbol{\theta}) = N\{\boldsymbol{0}, c\sigma^2(\boldsymbol{B}'\boldsymbol{B})^{-1}\},$$

where c is a positive scale factor that needs to be chosen. Note that the prior depends only on the predictor locations through the basis design matrix $\boldsymbol{B}$ and not on the response values.

The g-prior has some nice properties. Firstly, basis functions along similar projections have corresponding coefficients that are highly correlated *a priori*. Note that basis functions B_i and B_j describe data along a similar projection in data-space when

$$b_{ij} = \frac{\int_{\mathcal{X}} B_i(x)B_j(x)\,\mathrm{d}x}{\int_{\mathcal{X}} B_i^2(x)\,\mathrm{d}x \int_{\mathcal{X}} B_j^2(x)\,\mathrm{d}x} \tag{3.17}$$

is close to one. In the case of truncated linear splines this occurs when two knot points are very close together. Further, this prior correlation decreases with b_{ij} and, in the limiting case when $b_{ij} = 0$, the coefficients β_i and β_j are independent. This makes sense as in this case the basis functions are orthogonal to each other.

The g-prior leads to the conditional posterior distribution of the coefficients being given by

$$p(\boldsymbol{\beta} \mid \sigma^2, \boldsymbol{\theta}) = N\left\{\frac{c}{1+c}\hat{\boldsymbol{\beta}}, \frac{c}{1+c}\sigma^2(\boldsymbol{B}'\boldsymbol{B})^{-1}\right\}, \tag{3.18}$$

where $\hat{\boldsymbol{\beta}}$ again represents the least squares estimate to $\boldsymbol{\beta}$. Hence the conditional posterior mean of $\boldsymbol{\beta}$ is just a shrunken version of the least squares estimate with equality as $c \to \infty$.

One conceptual problem with using the g-prior is that it is impossible to generate from the prior for $\boldsymbol{\beta}$ unless we know beforehand the locations at which the responses will be drawn. In some contexts this is not a problem, for example, economic data will often be available in monthly or quarterly intervals over a known time span, but in general the basis matrix $\boldsymbol{B}$ is dependent on the observed predictor locations.

Taking $c = n$ has proved a popular choice among practitioners (see, for example, Gustafson 2000; Smith and Kohn 1996) as then the prior scales naturally with the size of the data. Kass and Wasserman (1996) call this the unit information prior.

Choosing between the ridge and g-priors has tended to be a matter of personal taste. In the literature the g-prior appears to have been favoured more yet, personally, we prefer the ridge prior as it leads to the same estimates of the coefficients as the ridge regression model of Hoerl and Kennard (1970a,b). Such models have been shown empirically to have good predictive power (see, for example, Dempster *et al.* 1977b; Gunst and Mason 1977), outperforming more complex ones on a wide variety of test examples. Ridge estimators perform particularly well when some of the basis functions are highly correlated. In this situation the inverse of the prior variance (in the Bayesian viewpoint of ridge models) is added to each diagonal element of $\boldsymbol{B}'\boldsymbol{B}$ before it needs to be inverted. This conditions the matrix by ensuring that the smallest eigenvalue of $(\boldsymbol{B}'\boldsymbol{B} + \boldsymbol{V})$ is further from zero than the smallest eigenvalue of $\boldsymbol{B}'\boldsymbol{B}$.

In the case when $\boldsymbol{B}'\boldsymbol{B}$ is diagonal the basis functions are orthogonal and the ridge and g-priors both give rise to shrinkage estimators of the true coefficients $\boldsymbol{\beta}$. These have become popular since the remarkable result of James and Stein (1961), sometimes known as Stein's paradox. This paper demonstrates that the maximum likelihood estimator is inadmissible, in that there exists a class of biased estimators for $\boldsymbol{\beta}$ that have uniformly smaller risk. We can think of the risk as just the mean-squared error between the true coefficients and their estimates. Thus, both prior choices are well founded.

We find that the g-prior is of the form of shrinkage estimator (Zellner and Vandaele 1974), where the contraction factor is $n/(n+1)$. From p. 175 of Gruber (1998) and (3.18) we see that the actual James–Stein estimator assumes that we take c according to

$$\frac{c}{c+1} = 1 - \frac{p-2}{\hat{\boldsymbol{\beta}}\boldsymbol{B}'\boldsymbol{B}\hat{\boldsymbol{\beta}}},$$

in situations where $p > 2$, i.e.

$$c = \frac{\|\boldsymbol{B}\hat{\boldsymbol{\beta}}\|^2}{(p-2)} - 1.$$

However, to calculate $\hat{\boldsymbol{\beta}}$ we need to know the response values. Thus this choice of c depends on the data, so this formulation for the 'prior' would also depend on the data. Obviously, priors, by their name, are assumed to be taken without knowledge of the data, so choices of c in this way are to be discouraged. Priors that depend only on the structure of the data, such as the g-prior, are widely accepted; it is the dependence of the prior on the observed responses that leads to priors that cannot be considered strictly Bayesian. Nevertheless, priors that do depend on the responses are still known as *empirical Bayes'* priors.

3.5.4 *The prior on the regression variance*

In this book we place our prior on the regression variance through the inverse-gamma distribution so that $p(\sigma^2) = \text{IG}(a, b)$ for $a, b > 0$ (see (2.13)). We shall now look at a few properties of this distribution to see how the choice of a and b affect these.

We find that the inverse-gamma distribution always has a mode at $b/(a+1)$ and that it tends to zero as σ^2 tends to zero and infinity. We also find that the ℓth moment of the distribution is given by

$$E\{(\sigma^2)^\ell\} = \frac{b^\ell \Gamma(a-\ell)}{\Gamma(a)} = b^\ell \prod_{i=1}^{\ell} (a-i)^{-1}, \qquad \text{for } a > \ell, \tag{3.19}$$

and is undefined for $a \leqslant \ell$.

The reference prior for σ^2 is the inverse-gamma one with $a = b = 0$, i.e. $p(\sigma^2) \propto \sigma^{-2}$. This is an improper prior and so cannot be normalised, but it minimises the effect of the prior choice on the posterior. Obviously, one immediate difficulty with taking $a = b = 0$ is that it is not possible to draw from the prior so we cannot determine how the IG(0, 0) prior represents our beliefs.

Authors who have wanted to ensure a proper prior specification tend to take a and b as equal to some small value, say 0.01. This leads to essentially identical posterior inference when compared with the reference prior, but now draws from the prior model are possible. This is the approach we shall take throughout this book, but we shall still highlight an undesirable consequence of this choice.

We see from (3.19) that the ℓth moment of the inverted-gamma distribution is finite only if $a > \ell$. So by taking $a = 0.01$ we have specified a prior on σ^2 which has no finite moments, in particular, the mean and variance of the prior are undefined. We can only fix $E(\sigma^2)$ if we take $a > 1$ and $\text{var}(\sigma^2)$ if $a > 2$. Unfortunately, for small to medium sized datasets, taking a and b much more than zero does affect the posterior inference to a significant degree so cannot be adopted generally. Further, it can be difficult to know sensible values of $E(\sigma^2)$ and $\text{var}(\sigma^2)$ with which a and b can be fixed.

3.6 Robust Curve Fitting

Assuming that the errors between the responses and the estimating function are normally distributed can lead to the resulting curves being overly affected by any outliers in the dataset. This is due to the small amount of probability associated with points far from the mean of a normal density. For this reason the normal density is said to have 'thin tails'. So, modelling with the normal distribution can lead to curve estimates whose shape at certain locations is highly dependent on only a few unrepresentative points. This property is, in general, not desirable so in this section we present two methods that can be used to *robustify* the general methodology against such problems.

In the context of this section we are presenting models that are robust with respect to the data, not to the priors as in standard Bayesian robustness (see, for example,

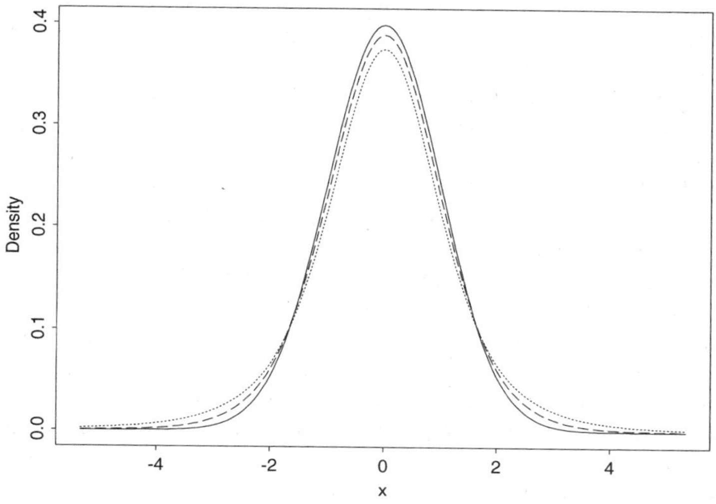

Figure 3.26 A comparison of the normal density with mean zero, scale parameter (i.e. variance) one (solid line) with similar Student distributions with degrees of freedom 4 (dotted line) and 10 (dashed line).

Berger 1982; Berger and Moreno 1994; Lavine *et al.* 1991). Bayesian robustness is concerned with how the posterior is affected by prior distributional assumptions and parameter settings; a topic we have covered to some extent in Section 3.5.

3.6.1 Modelling with a heavy-tailed error distribution

We now present a simple way to counter the difficulties associated with the small amount of probability in the tails of the normal distribution by using a more dispersed distribution for the errors. The distribution we suggest is the Student distribution with unknown scale and degrees of freedom, d (see the PDF given in (2.31)). The Student distribution is similar to the normal distribution for large values of d but, when d is small, it has much thicker tails than the normal (see Figure 3.26). Thus, large residuals between the fitted curve and the observed response are not penalised to such an extent. This has the effect of producing posterior predictive distributions that are more robust to the observed response values. However, the main problem with allowing for Student errors is that conjugate priors for $\boldsymbol{\beta}$ and σ^2 no longer exist so we need to use a different method for sampling from the posterior of interest.

First let us write down an important identity. If $p(u \mid z) = N(\mu, \sigma^2 z)$ and $p(z) = \mathrm{Ga}(d/2, d/2)$ then marginally (see Problem 3.9) $p(u) = \mathrm{St}(\mu, \sigma^2, d)$, i.e.

$$\mathrm{St}(u \mid \mu, \sigma^2, d) = \int_0^\infty N(u \mid \mu, \sigma^2 z)\, \mathrm{Ga}(z \mid d/2, d/2)\, \mathrm{d}z. \tag{3.20}$$

Thus, the Student distribution can be thought of as being given by a mixture of normal distributions with variance $\sigma^2 z$, scaled by gamma distributions. We sometimes refer to distributions with this property as being *scale mixtures*, in this case, of normal distributions (see, for example, Andrews and Mallows 1974; Haro-Lopez and Smith 1999; O'Hagan 1979, 1988; Pericchi and Smith 1992; West 1987).

The relationship in (3.20) gives us an easy way to sample random draws from Student distributions with d degrees of freedom. This just involves drawing z from a $\mathrm{Ga}(d/2, d/2)$ distribution and then, given this value of z as well as σ^2, drawing u from an $N(\mu, \sigma^2 z)$ distribution. This sampled value of u is then a draw from the required Student distribution.

As already mentioned, the problem with allowing for Student errors is that the likelihood, $p(\mathcal{D} \mid \boldsymbol{\beta}, \sigma^2, \boldsymbol{\theta})$, is now multivariate Student rather than multivariate normal. As no conjugate prior distributions exist for $\boldsymbol{\beta}$ and σ^2 exist, we find that the original efficiency of our sampling algorithms is lost. However, (3.20) gives us a relationship between the Student and normal distributions which we can use to allow us to still integrate out the coefficients and the regression variance. The strategy we employ uses *auxiliary* variables which, when added to the unknowns in the model, reduce the likelihood to a standard form (in this case a normal distribution) for which a conjugate prior specification exists for some of the parameters. In this example this allows us to assign the same conjugate priors over $\boldsymbol{\beta}$ and σ^2 as before, except that this time the priors are given conditionally on the auxiliary variables. Now let us look at the proposed auxiliary variable method in more detail.

The standard regression model can be written

$$y_i = g(x_i) + \epsilon_i,$$

for $i = 1, \ldots, n$, and where $g(x_i)$ is a linear combination of coefficients and basis functions. In the past we have assumed that each ϵ_i is independent normal with zero-mean and variance σ^2. However, we wish to model the errors ϵ_i as independent $\mathrm{St}(0, \sigma^2, d)$ distributions. This leads to each $(y_i - g(x_i))$ also being a Student distribution so the likelihood itself, $p(\mathcal{D} \mid \boldsymbol{\beta}, \sigma^2, \boldsymbol{\theta})$ is also Student. The scale mixture representation of a Student distribution tells us that, if $z_i \sim \mathrm{Ga}(d/2, d/2)$ and $e_i \sim N(0, \sigma^2)$, then

$$p(\epsilon_i) = \int p(\epsilon_i \mid z_i) p(z_i)\, \mathrm{d}z_i = \mathrm{St}(0, \sigma^2, d),$$

but $p(\epsilon_i \mid z_i) = N(0, z_i \sigma^2)$. Hence, if we let $z = (z_1, \ldots, z_n)$ be an n-vector of independent $\mathrm{Ga}(d/2, d/2)$ random variates, then

$$y_i = g(x_i) + z_i e_i$$

defines a normal linear model, with the only difference from earlier being that the error terms are now $N(0, z_i \sigma^2)$. Writing $\boldsymbol{\Sigma} = \mathrm{diag}(z_1, \ldots, z_n)$ we see that the likelihood of the model is

$$p(\mathcal{D} \mid \boldsymbol{\beta}, \sigma^2, \boldsymbol{\theta}, z) = (2\pi)^{-n/2} |\boldsymbol{\Sigma}|^{-1/2} \exp\left\{ -\frac{1}{2\sigma^2} (Y - \boldsymbol{B\beta})' \boldsymbol{\Sigma}^{-1} (Y - \boldsymbol{B\beta}) \right\}.$$

We see that the introduction of the auxiliary variables z has led to the likelihood reducing to a multivariate normal distribution with mean $\boldsymbol{B\beta}$ and error variance $\sigma^2\boldsymbol{\Sigma}$. This is the same as the likelihood for the normal linear model used previously to analyse the Great Barrier Reef data except that now the variance is $\sigma^2\boldsymbol{\Sigma}$ rather than just $\sigma^2\boldsymbol{I}$. We find that this produces little extra trouble as, conditional on z, we can again determine the marginal likelihood, $p(\mathcal{D} \mid \boldsymbol{\theta}, z)$. This is of the same form as (2.23) but with the updated parameters now given by

$$\boldsymbol{V}^* = (\boldsymbol{B}'\boldsymbol{\Sigma}^{-1}\boldsymbol{B} + \boldsymbol{V}^{-1})^{-1}, \tag{3.21}$$

$$\boldsymbol{m}^* = \boldsymbol{V}^*(\boldsymbol{V}^{-1}\boldsymbol{m} + \boldsymbol{B}'\boldsymbol{\Sigma}^{-1}Y), \tag{3.22}$$

$$b^* = b + (\boldsymbol{m}'\boldsymbol{V}^{-1}\boldsymbol{m} + Y'\boldsymbol{\Sigma}^{-1}Y - (\boldsymbol{m}^*)'(\boldsymbol{V}^*)^{-1}\boldsymbol{m}^*)/2, \tag{3.23}$$

with $a^* = a + n/2$ as before (Problem 3.9). Obviously, by taking $\boldsymbol{\Sigma} = \boldsymbol{I}$ these results reduce to (2.17), (2.18) and (2.20).

The auxiliary variables have allowed us to set up the model in the same way as for the normal error case, conditional of z. So, given z, we can use the same sampling strategies to update the model parameters $\boldsymbol{\theta}$. However, we are interested in the posterior of the model parameters unconditionally on z so we must marginalise out over these auxiliary variables. We know that

$$\begin{aligned} p(\boldsymbol{\theta} \mid \mathcal{D}) &= \int p(\boldsymbol{\theta} \mid \mathcal{D}, z)p(z)\,\mathrm{d}z \\ &\approx \frac{1}{N}\sum_{i=1}^{N} p(\boldsymbol{\theta} \mid \mathcal{D}, z^{(i)}), \end{aligned} \tag{3.24}$$

where each $z^{(i)}$ is an n-vector of independent $\mathrm{Ga}(d/2, d/2)$ random variates. Hence, to draw samples from the required distribution we just need to be able to draw samples from $p(\boldsymbol{\theta} \mid \mathcal{D}, z^{(i)})$. As this posterior can be handled in an identical way to that when the errors are assumed normal, we can use the same algorithm as in Section 3.3.3 to approximate it. We run the sampler so that at the ith iteration we first draw a realisation $z^{(i)}$ and then update $\boldsymbol{\theta}^{(i-1)}$ conditional on $z^{(i)}$ to obtain $\boldsymbol{\theta}^{(i)}$. Thus the predictive distribution is given by

$$\begin{aligned} p(y \mid x, \mathcal{D}) &= \int \left\{ \int p(y \mid x, \mathcal{D}, \boldsymbol{\theta}, z)p(\boldsymbol{\theta} \mid \mathcal{D}, z)\,\mathrm{d}\boldsymbol{\theta} \right\} p(z)\,\mathrm{d}z \\ &\approx \frac{1}{N}\sum_{i=1}^{N} p(y \mid x, \mathcal{D}, \boldsymbol{\theta}^{(i)}, z^{(i)}). \end{aligned}$$

All we have to do to implement the auxiliary variables method is draw a realisation from the distribution of z and then update $\boldsymbol{\theta}$ conditional on this value. However, this leads to different expressions for $\boldsymbol{m}^*$, $\boldsymbol{V}^*$ and b^* in the marginal likelihood so requires alteration of the code to perform this calculation, as well as changing the definition of the function which calculates the marginal likelihood to allow z to be sent to it.

However, we can minimise the changes to the algorithm required for this method by noting that $p(\mathcal{D} \mid \boldsymbol{\theta}, z)$ is the same as the marginal likelihood if we perform a linear regression assuming $p(\boldsymbol{\epsilon}) = N(\mathbf{0}, \sigma^2 \boldsymbol{\Sigma})$. By using a standard linear transformation of this normal distribution we know that $p(\boldsymbol{\Gamma}\boldsymbol{\epsilon}) = N(\mathbf{0}, \sigma^2 \boldsymbol{I})$, where $\boldsymbol{\Gamma}'\boldsymbol{\Gamma} = \boldsymbol{\Sigma}^{-1}$. In this simple case $\boldsymbol{\Gamma} = \text{diag}(\sqrt{z_1^{-1}}, \dots, \sqrt{z_n^{-1}})$. Hence, the linear model,

$$\boldsymbol{\Gamma} Y = \boldsymbol{\Gamma} \boldsymbol{B} \boldsymbol{\beta} + \boldsymbol{\Gamma} \boldsymbol{\epsilon}, \tag{3.25}$$

is of the same form as that given in (2.11). This means that to incorporate Student errors, instead of having to rewrite the parts of the code that calculate the marginal likelihood, we just need to send to the marginal likelihood function $\boldsymbol{\Gamma} Y$ and $\boldsymbol{\Gamma} \boldsymbol{B}$ rather than Y and $\boldsymbol{B}$ as before.

So far we have seen how to perform curve fitting when we have Student errors with some known degrees of freedom, d, but we have not said anything about sensible values of d. A popular choice is to take $d = 4$ as this has significantly thicker tails than the normal. However, we could allow the data to determine d and draw it along with the model parameters. We would need to assign a sensible prior to d and then update it, conditional on the current values of $\boldsymbol{\theta}$ and z, most probably by a Metropolis–Hastings step with a suitably chosen proposal distribution.

Remember that it is only because of the natural relationship between the Student and normal distributions that we can utilise conjugacy by the introduction of the auxiliary variables. This gives us a particularly simple way to allow for Student errors to provide a more robust curve fit. Other error distributions do not give us such neat results which is why this one has been highlighted.

3.6.2 *Outlier detection models*

Modelling with a heavy-tailed distribution assumes that all the errors come from a single distribution which happens to have heavier tails than a normal. However, we may think that the errors come from two separate distributions, the majority from an $N(0, \sigma^2)$ but some outlying points from a more dispersed $N(0, \kappa^2\sigma^2)$ distribution ($\kappa^2 > 1$). The points relating to the overdispersed distribution could then be considered outlying. A method that allows us to determine which points relate to which of these two competing error distributions does two things. Firstly, it robustifies the curve fit as large residuals have little effect on the likelihood as they are associated with the overdispersed normal distribution. Secondly, the method automatically classifies each point as either an outlier or not, even attaching probabilities to these two events. This method dates back, at least, to the work of Box and Tiao (1968) and is described first in connection with modern computational methods by Verdinelli and Wasserman (1991).

We can formulate this model mathematically by adding a layer of hierarchy to the error distribution. So, instead of the usual model for which each ϵ_i is assumed to follow an $N(0, \sigma^2)$, we assign it a normal mixture prior where

$$\begin{aligned} p(\epsilon_i) &= p(\epsilon_i \mid \omega_i = 0)p(\omega_i = 0) + p(\epsilon_i \mid \omega_i = 1)p(\omega_i = 1) \\ &= N(0, \sigma^2)p(\omega_i = 0) + N(0, \kappa^2\sigma^2)p(\omega_i = 1), \end{aligned}$$

for $i = 1, \ldots, n$ and where ω_i is a binary indicator variable which takes the value 1 if the ith point is considered an outlier and 0 otherwise. The probabilities $p(\omega_i = 0)$ and $p(\omega_i = 1)$ can either be chosen beforehand or taken as random with the restriction that $p(\omega_i = 0) + p(\omega_i = 1) = 1$. Smith and Kohn (1996) suggested taking $p(\omega_i = 1)$ as fixed and equal to 0.05, expecting 1 in 20 points to be outliers. Wong and Kohn (1996) instead propose assigning a beta prior with known parameters to $p(\omega_i = 1)$. The choice of the scaling by which the normal distributions differ is also important. Smith and Kohn (1996) report good results when using $\kappa^2 = 100$.

This model has substantial similarities with the Student error distribution described in the last subsection. As we see from (3.20), the Student distribution is made up of an infinite mixture of normal distributions with the mixing distribution controlled by the degrees of freedom parameter, d. In contrast the error distribution in this model is a mixture of two normal distributions.

Sampling from this model proceeds by using Gibbs steps to update the ω_is at each iteration. We can simulate the ω_is in a similar way to simulating the θ_is in Section 2.8.1 as we can determine exactly

$$p(\omega_i \mid \mathcal{D}, \boldsymbol{\theta}, \boldsymbol{\omega}_{-i}) \propto p(\mathcal{D} \mid \boldsymbol{\theta}, \omega_i, \boldsymbol{\omega}_{-i})p(\omega_i), \tag{3.26}$$

for $\omega_i = 0, 1$. This is because the marginal likelihood, $p(\mathcal{D} \mid \boldsymbol{\theta}, \boldsymbol{\omega})$ for this model can be calculated in the same way as for the Student errors model but with $\boldsymbol{\Sigma}$ now being a diagonal matrix with ith entry along the diagonal being given by 1 if $\omega_i = 0$ and κ^2 otherwise.

The two methods for robust curve fitting both allow the variance of the error distributions to be related to the diagonal matrix $\boldsymbol{\Sigma}$. When modelling with a Student the ith diagonal entry of $\boldsymbol{\Sigma}$ is just a draw from a $\text{Ga}(d/2, d/2)$ distribution, whereas for this model the entry is either 1 or κ^2 depending on the distribution in (3.26).

Having described robust procedures in this section it is worth remarking that there may be situations where we do not want to use such methods. Consider Figure 3.27(a). If we were to use a robust method to estimate the true curve, we would end up with function that stays close to zero and is not overly influenced by the supposed 'outlying' points for which $y > 10$. However, the true curve is given in Figure 3.27(b) so any robust method would not get close to this truth. In this situation we want a method with completely different properties as we wish it to be very flexible so that it can capture the large jumps seen in the truth. This means that we allow the estimating function to actually follow the 'outliers' rather than attempt to ignore them. Before we analyse a dataset we must determine whether we are in a situation where 'outliers' give important information or should be ignored. Firstly, this information should determine which sort of basis function to use as this will have the greatest effect on the posterior estimate to the curve. In the case of Figure 3.27, because of the local properties of

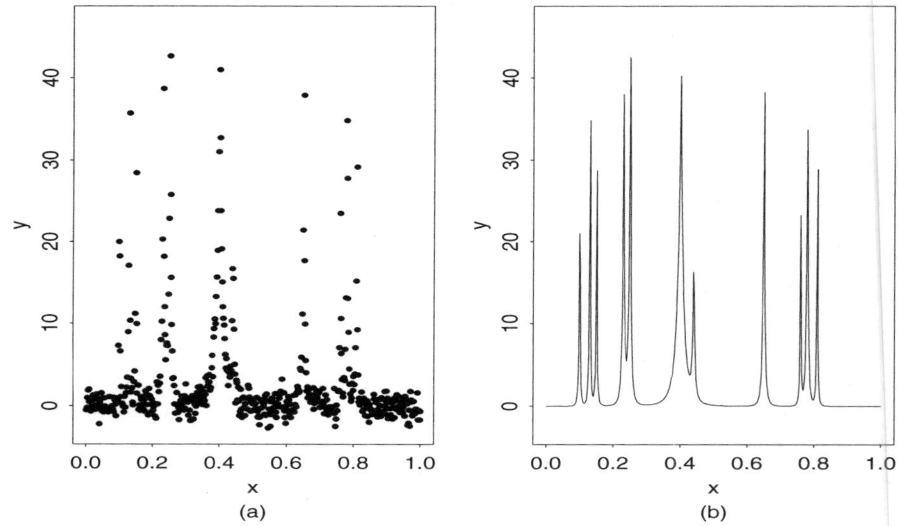

Figure 3.27 (a) A realisation and (b) the true curve for the simulated Bumps data given in Donoho *et al.* (1995) with $n = 512$.

wavelet bases these would be more appropriate than spline bases here. This choice will be even more important than the actual error distribution used. However, for smoother functions, spline bases often produce more appealing posterior estimates to f.

3.7 Discussion

This chapter is the most important in the book. It introduces the ideas that will be expanded on throughout the other chapters. It reviews the basic ideas of fitting basis function models in a Bayesian framework, utilising the results of Chapter 2. Further, it demonstrates why sampling strategies such as MCMC are required for these models as the normalising constants for the model posteriors quickly become infeasible to calculate exactly. We shall see in the next chapter how these basic ideas form the foundation of the analysis of basis functions for regression irrespective of the dimension of the predictor space. Moreover, we find that the priors on the model parameters for curve fitting generalise easily to cases where there are more than one predictor.

We highlighted two methods to perform robust regression. These methods can also be applied to the other regression models when overdispersed errors need to be considered. Also, the auxiliary variable method will be shown to be useful in other contexts, most notably when the responses are non-normal, as we shall see in Chapter 5, when we first encounter generalised linear models.

The models outlined in this chapter are intended to be basic templates that can be extended to include known prior information. For instance, one simple extension to the standard normal model in this chapter is to allow the errors to be correlated, rather than just taking them to be independent, i.e. we can take $\boldsymbol{\epsilon} \sim N(\mathbf{0}, \sigma^2 \boldsymbol{\Sigma})$ for some non-diagonal, positive-definite, symmetric matrix $\boldsymbol{\Sigma}$. When $\boldsymbol{\Sigma}$ is known we find that the linear regression of $\boldsymbol{\Gamma B \beta}$ on $\boldsymbol{\Gamma} Y$, where $\boldsymbol{\Gamma}'\boldsymbol{\Gamma} = \boldsymbol{\Sigma}^{-1}$, has independent $N(0, \sigma^2)$ errors allowing standard methods to be performed. The calculation of the $\boldsymbol{\Gamma}$ is not as straightforward as in the case when it is diagonal in Section 3.6.1. However, a standard function of nearly all numerical software packages performs the Cholesky decomposition of a positive-definite symmetric matrix yielding the Cholesky factor $\boldsymbol{\Gamma}$. Examples of curve fitting in the presence of such correlated errors is given by Smith and Kohn (1996) and Denison (1997).

Bayesian methods also allow us to easily extend the methods given. This is most often done through the adoption of hierarchical prior distributions. For example, the assumption of a known error variance structure may be inappropriate. By adopting a prior on this matrix we can try and infer the correlation structure from the data. This can either be done by adopting priors on parameters that define the elements of the matrix, or even by adopting a prior on the matrix elements themselves.

3.8 Further Reading

The first example presented in this chapter was the Nile discharge dataset. This is a well-analysed dataset and a particular favourite of the community involved in changepoint modelling. Detecting changepoints is a well-researched field with many papers specifically dedicated to it (see, for example, Broemeling 1974; Carlin *et al.* 1992; Chernoff and Zacks 1964; Fan and Brooks 2000; Lee 1998; Rekaya *et al.* 2000; Smith 1975; Smith and Spiegelhalter 1980; Stephens 1994; Wu and Fitzgerald 1995; Zacks 1982). These papers do not, in general, want to approximate the underlying regression function but instead make inference about the location of the changepoint, or changepoints. We return to changepoint modelling in Chapter 7, where we present a method that can be used when the responses are non-normal.

Curve fitting with splines with an unknown number of knot points at unknown locations appears to have been suggested initially in a Bayesian paper by Halpern (1973). Here Halpern discussed a general approach to fitting a piecewise linear function (which is equivalent to a linear spline) to a scatterplot using a conjugate prior specification which is very similar to that described in this chapter. Halpern's excellent paper was written well before the computational power existed that could be harnessed to provide the results given here but nevertheless it is an article of great importance in the general area of flexible spline fitting. In a non-Bayesian context the credit for introducing splines with unknown locations and numbers of knots, sometimes known as free-knot splines, is often given to Smith (1982), who uses a stepwise selection algorithm to determine the form of the estimating function. Further work on finding the location and number of knot locations to use when curve fitting includes the TURBO and BRUTO methods (Hastie and Tibshirani 1990), which perform a fast stepwise searches for the basis functions. One big difference between the Bayesian and non-Bayesian methods in this field is that the Bayesian paradigm gives us a good way of combining the models we visit during the sampling algorithm. However, non-Bayesian work has typically suggested different criteria for choosing the 'best' set of knot points.

There is a considerable literature relating to Bayesian splines, where, instead of viewing splines as specific types of basis function models, they are taken as the solution of the constrained minimisation problem,

$$g(x) = \arg\max_h \left[\sum_{i=1}^{n} \{y_i - h(x_i)\}^2 + \mu \int_{\mathcal{X}} \{h^{(m)}(u)\}^2 \, du \right], \qquad (3.27)$$

where the second term can be thought of as a penalty on the 'roughness' of the function h. The early papers by Wahba (1978, 1983) show how the spline estimate found when estimating μ by generalised cross-validation (Craven and Wahba 1979) is equivalent to a Bayes estimate for g when taking it as a sample function from a certain zero-mean Gaussian prior. Further work by Abramovich and Steinberg (1996) generalises the work of Wahba (1983).

Other important work on curve fitting in a Bayesian framework includes O'Hagan (1978), who uses Gaussian processes to perform curve fitting. An advantage of this approach is that the pointwise credible intervals that it produces are dependent on the distribution of the data as the correlation between responses is nearly always taken as a function of the distance between them. Other important work on Gaussian process includes Neal (1996), Rasmussen (1996), Williams and Rasmussen (1996), Williams (1998) and Neal (1999). Upsdell (1996) also suggests using Gaussian process priors but places these on the derivatives of the function in question. The advantage of this is that it admits nonstationary covariance functions on the mean level. Also of interest is Angers and Delampady (1992), who assign priors to the derivatives of the true regression function g at some point $x_0 \in \mathcal{X}$ and then reconstruct g over the whole interval of interest $\mathcal{X}$ using the Taylor expansion.

In contrast to using splines of known order with an unknown number of knot points Mallick (1998) provides a curve fitting method which approximates g by a known number of piecewise polynomials, where each polynomial is of unknown order. More recently, Punska *et al.* (1999) segment time series into an unknown number of autoregressive series, each one being of unknown order. This model could also be used to combine the ideas of Denison *et al.* (1998a) and Mallick (1998) to fit curves with an unknown number of piecewise polynomials, each of unknown order, and with unknown breakpoints.

The choice of priors on the coefficients of the basis functions is particularly important in determining the curve fit. The two priors we considered were

$$p(\boldsymbol{\beta} \mid v) = N\{\mathbf{0}, \sigma^2 \operatorname{diag}(\infty, v, \dots, v)\} \qquad \text{(ridge)},$$
$$p(\boldsymbol{\beta} \mid c) = N\{\mathbf{0}, c\sigma^2 (\boldsymbol{B}'\boldsymbol{B})^{-1}\} \qquad (g\text{-prior}).$$

We suggest adopting a robust approach to the regression by assigning hyperpriors to v or c depending on the prior choice. Examples of this are given by Andrieu (1999), Andrieu *et al.* (2001b), Kohn *et al.* (2001), amongst others.

Generalisation of the ridge prior is possible by assigning a different prior variance for each coefficient in the model, i.e. take $V = \text{diag}(\infty, v_1, \ldots, v_p)$. This is the Bayesian equivalent of the generalised ridge regression model briefly introduced in Hoerl and Kennard (1970b). A Bayesian perspective on the method is given by Goldstein and Smith (1974) but it appears that the data do not contain enough information to allow accurate estimation of all the v_i ($i = 1, \ldots, p$) except for wavelet regression (Clyde and George 2000; Clyde *et al.* 1998). Hence, comparative prediction results have, in general, proved disappointing (Lawless 1981). Copas (1983) expounds the idea that shrinkage estimators are particularly well suited to prediction. Also Lindley (1995) points out that multiple shrinkage of the coefficients in a linear model is a generalisation of models that select basis functions, so should be preferable. This has led to more recent research on determining good multiple shrinkage estimators (Denison and George 2001; George 1986a,b; George and Oman 1996) and this appears a promising area of future research.

Recently, George (1999) has suggested using priors on the set of model basis functions that depend on the correlation structure of the basis set. The idea proposes identifying priors that dilute the prior weight over different models, rather than those with different basis functions. This is worthwhile as, in the extreme case when two basis functions are identical, a standard flat prior over which bases to include assigns disproportionate weight to models that are essentially the same. However, a dilution prior would aim to spread this prior weight over the model space, taking into account the fact that the basis functions are the same automatically. Unfortunately, eliciting such priors is not easy and is the subject of current research.

3.9 Problems

3.1 Suppose that we wish to fit just a constant level to a dataset, so we take $y_i = \beta + \epsilon_i$ for $i = 1, \ldots, n$. Assuming that the errors ϵ_i are independent normally distributed random variables with unknown variance σ^2, and taking the prior on β as $p(\beta) = N(0, v)$, show that the marginal likelihood of the model is given by

$$p(\mathcal{D}) = \frac{(2b)^a \Gamma(a + n/2)}{\pi^{n/2}\Gamma(a)\sqrt{1 + vn}} \left(2b + s_{yy} - \frac{vn\overline{y}^2}{1 + vn}\right)^{-a-n/2},$$

where

$$s_{yy} = \sum_i y_i^2, \quad \overline{y} = \frac{1}{n}\sum_i y_i \quad \text{and} \quad \mathcal{D} = \{y_i, x_i\}_1^n.$$

For the Nile discharge dataset we find that $s_{yy} = 2\,835\,157$ and $\overline{y} = 0$ (by construction). By taking $a = b = 0.01$ and $v = 1000$ confirm that this expression gives the same log marginal likelihood value for the model $\mathcal{M}_0$ of Section 3.2.1 given in the text, that is -666.1.

Now use the given expression for $p(\mathcal{D})$ to determine the marginal likelihood of model $\mathcal{M}_2$ of Section 3.2.1, which assumes that there is a change in both mean and variance at some point $t \in \{x_1, \ldots, x_{n-1}\}$.

3.2 Prove the relationship given in (3.20).

Suppose instead that $p(y \mid v) = \text{Poi}(v)$, so that

$$p(y \mid v) = \frac{v^y \exp(-v)}{y!} \qquad \text{for } y = 0, 1, \ldots, \tag{3.28}$$

and take $p(v) = \text{Ga}(\alpha, \beta)$. Determine the marginal distribution of y for this scale mixture of Poisson distributions.

3.3 Suppose that a linear model is described by the parameter vectors $\boldsymbol{\theta}$ and $\boldsymbol{\phi}$, where $\boldsymbol{\theta}$ are random variables indicating which basis functions are present, and $\boldsymbol{\phi} = (\boldsymbol{\beta}, \sigma^2)$ are parameters for which we know the conditional posterior $p(\boldsymbol{\phi} \mid \boldsymbol{\theta}, \mathcal{D})$. Often we run a sampler to make inference about $p(\boldsymbol{\theta} \mid \mathcal{D})$, and analytically integrate out $\boldsymbol{\phi}$. Suppose instead that we wish to make inference about $p(\boldsymbol{\theta}, \boldsymbol{\phi} \mid \mathcal{D})$, so that we wish to sample $\boldsymbol{\phi}$ instead of integrating it out.

Show that if we use the conditional priors to make proposals, for general k, the acceptance probability for a death step is given by

$$\min\left\{1, \frac{p(\mathcal{D} \mid \boldsymbol{\theta}')}{p(\mathcal{D} \mid \boldsymbol{\theta})}\right\}.$$

Note that this is the same as if we did not sample $\boldsymbol{\phi}$. This result is given in the appendix of Holmes and Mallick (2000a).

3.4 Derive equations (3.21)–(3.23).

3.5 Suppose that we have a model such that

$$d = z + \epsilon,$$

where we observe d and want to estimate z. We assume that the error ϵ follows an $N(0, \sigma^2)$ with σ^2 known. If we assign a double exponential prior to z of the form

$$p(z) = \frac{\lambda}{2\sigma^2} \exp\left\{-\frac{\lambda}{\sigma^2}|z|\right\},$$

where λ is a fixed constant, show that the posterior $p(z \mid d)$ is maximised when

$$\tfrac{1}{2}(d - z)^2 + \lambda|z|$$

is minimised. Use the fact that d and z are of the same sign, so that $(d - z)^2 = (|d| - |z|)^2$, to show that the maximum *a posteriori* value of z is given by

$$\tilde{z} = \text{sgn}(d) \cdot (|d| - \lambda)_+.$$

Explain how this result relates to the soft thresholding rule given in (3.12).

Similar methods can be used to show that if we had used

$$p(z) \propto \exp\{-\lambda|z|I(|z| \leqslant \lambda) + \tfrac{1}{2}\lambda^2 I(|z| > \lambda)\}$$

instead the MAP value is given by the hard threshold rule (Vidakovic 1999, p. 177).

3.6 Consider the hierarchical model for Bayesian wavelet analysis given by Clyde *et al.* (1998) and in Section 3.4.2. Show that the marginal prior distribution, $p(\beta_{ji} \mid \gamma_j)$ is a mixture of normal distributions given by

$$p(\beta_{ji} \mid \gamma_j) = (1 - \gamma_j)N(0, 0) + \gamma_j N(0, v_j\sigma^2).$$

Show that the marginal posterior distribution of θ_{ji} is again Bernoulli with $p(\theta_{ji} = 1 \mid Y) = O_{ji}/(1 + O_{ji})$, where O_{ji} is the posterior odds that $\theta_{ji} = 1$, i.e.

$$O_{ji} = (1 + v_j)^{-1/2} \frac{\gamma_j}{1 - \gamma_j} \exp\left(- \frac{v_j \hat{\beta}_{ji}^2}{2\sigma^2(1 + v_j)} \right).$$

4

Surface Fitting

4.1 Introduction

We have seen in the previous chapter how Bayesian regression splines provide a convenient and flexible method of fitting curves to data. The curve fitting models described gave rise to posterior predictive densities, $p(y \mid x, \mathcal{D})$, which can then be used to make posterior inference. This chapter uses essentially the same methodology to make inference in the more general case when there is more than one predictor variable. Hence the input, $\boldsymbol{x}$, is now vector valued rather than just a scalar as before, and we are led to approximating a regression surface rather than just a one-dimensional function.

Although the basic approach is the same, the choice of basis functions to use is more critical for surface fitting. We find that it is sensible to let the number of predictors influence this choice and demonstrate the capabilities of some well-known sets of basis functions (e.g. linear planes, spline bases and neural networks).

In Section 4.2 we introduce the additive model, which extends univariate curve fitting by approximating multidimensional functions as sums of univariate curves. Section 4.3 goes on to describe multidimensional extensions to univariate splines. After this, Section 4.4 gives examples of basis functions that are appropriate when there are many predictor variables and then, in Section 4.5, we extend the methodology to handle univariate time series data. Finally, Section 4.6 contains references for further reading on the topics raised by the chapter.

4.2 Additive Models

4.2.1 *Introduction to additive modelling*

Curve fitting provides an explicit representation of the dependence between the response variable y and the predictor x, via the posterior predictive $p(y \mid x, \mathcal{D})$ found by integrating (or summing) over the model space. We assume that the true relationship between y and x is such that $E(y \mid x) = f(x)$ for some unknown function f that we wish to estimate. The parameters of the model are those that define the function $g(x)$ which is used to approximate f, as well those that define the noise process. For example, we looked at the linear spline model, where we assumed that the truth can be approximated by

$$g(x) = \beta_0 + \sum_{i=1}^{k} \beta_i (x - t_i)_+. \tag{4.1}$$

With the standard Gaussian noise assumptions the model parameters are $(\beta_1, t_1, \ldots, \beta_k, t_k, \sigma^2)$. However, by adopting appropriate priors we find that to determine $p(y \mid x, \mathcal{D})$ we only need to sum over the set of possible knot points, $\boldsymbol{\theta} = (t_1, \ldots, t_k)$, as the coefficients and noise variance can be integrated out analytically.

One way to extend this model when the number of predictors, p, is greater than one is by using a sum of one-dimensional linear spline functions, like that given in (4.1). Hence we could take

$$g(\boldsymbol{x}) = \beta_0 + g_1(x_1) + g_2(x_2) + \cdots + g_p(x_p), \tag{4.2}$$

where $\boldsymbol{x} = (x_1, \ldots, x_p)$. Thus the overall regression function is a sum of p univariate functions, or curve fits. The univariate functions can be modelled with univariate splines, as we shall assume here, but any other set of basis functions which can be used for scatterplot smoothing are equally valid.

A model of the general type in (4.2) is known as an additive model. Hastie and Tibshirani (1990) provides an extensive account of these following on from their initial papers on the subject (Hastie and Tibshirani 1986, 1987); see also the parallel developments in Stone (1985a,b). The additive model (4.2) manages to retain interpretability by restricting nonlinear effects in the predictors to enter into the model independently of one another.

We find that inference for Bayesian additive models is straightforward when we use a basis function representation (such as the regression splines) for the individual curves. In this case we notice that (4.2) can be written similarly to the models we used for scatterplot smoothing. That is, we can still write $g(\boldsymbol{x})$ in (4.2) as a linear combination of basis functions and coefficients,

$$g(\boldsymbol{x}) = \beta_0 + \sum_{j=1}^{p} \sum_{i=1}^{k_j} \beta_{ji} (x_j - t_{ji})_+, \tag{4.3}$$

where x_j is the jth predictor in $\boldsymbol{x}$ and k_j is the number of knots points used to approximate the jth curve. Each one-dimensional function is again described by the parameters β_{ji} (the coefficients) and t_{ji} (the knot points).

As (4.3) is just a linear model we can use exactly the same Bayesian linear model results, seen in Chapters 2 and 3, to make posterior inference for additive models. Thus the fact that a general set of predictors is now a vector, rather than just scalar, is of little consequence. In matrix notation we can again write

$$Y = \boldsymbol{B}\boldsymbol{\beta} + \boldsymbol{\epsilon}, \tag{4.4}$$

with $\boldsymbol{\epsilon} \sim N(\mathbf{0}, \sigma^2 \boldsymbol{I})$, $\boldsymbol{\beta} = (\beta_0, \boldsymbol{\beta}_1, \ldots, \boldsymbol{\beta}_p)$ with $\boldsymbol{\beta}_j = (\beta_{j,1}, \ldots, \beta_{j,k_j})$ and

$$\boldsymbol{B} = \begin{pmatrix} 1 & B_{1,1}(\boldsymbol{x}_1) & \cdots & B_{1,k_1}(\boldsymbol{x}_1) & B_{2,1}(\boldsymbol{x}_1) & \cdots & B_{p,k_p}(\boldsymbol{x}_1) \\ \vdots & \vdots & \ddots & \vdots & \vdots & \ddots & \vdots \\ 1 & B_{1,1}(\boldsymbol{x}_n) & \cdots & B_{1,k_1}(\boldsymbol{x}_n) & B_{2,1}(\boldsymbol{x}_n) & \cdots & B_{p,k_p}(\boldsymbol{x}_n) \end{pmatrix}$$
$$= (1 \;\; \boldsymbol{B}_1 \;\; \cdots \;\; \boldsymbol{B}_p),$$

where

$$\boldsymbol{B}_j = \begin{pmatrix} (x_{j1} - t_{j,1})_+ & \cdots & (x_{j1} - t_{j,k_j})_+ \\ \vdots & \ddots & \vdots \\ (x_{n1} - t_{j,1})_+ & \cdots & (x_{n1} - t_{j,k_j})_+ \end{pmatrix}.$$

Now that the data are multivariate we need a different prior on the model space to that given in Chapter 3. However, we can easily generalise this work and take the priors on the knot locations in each estimating function g_j as being independent, so that

$$p(\boldsymbol{\theta}) = p(\boldsymbol{\theta}_1) \times \cdots \times p(\boldsymbol{\theta}_p),$$

where $\boldsymbol{\theta}_j = (t_{j,1}, \ldots, t_{j,k_j})$. Further, we can adopt identical univariate priors, as in (3.3), to each of the terms above so that

$$p(\boldsymbol{\theta}_j) = \binom{T_j}{k_j}^{-1} \times \frac{1}{K_j + 1},$$

where T_j and K_j are the number of candidate knot locations, and the maximum number of allowable knots, respectively, for the jth predictor.

We can again assign a conjugate prior for the coefficients and regression variance, taking $p(\boldsymbol{\beta}, \sigma^2) = \text{NIG}(\boldsymbol{m}, \boldsymbol{V}, a, b)$. In this way we can follow the standard reversible jump sampling algorithm seen in the previous chapter, except that the *BIRTH* step is now given by the following.

Birth-proposal($\boldsymbol{\theta}$):

```
Set u to a draw from a U{1,...,p} distribution;
Set z to a draw from a U{T_u\θ_u} distribution;
return(θ,z);
```

Here $\mathcal{T}_j$ is just the set of candidate knot locations for the jth spline. This *BIRTH* step first randomly chooses to add a term to the uth function, g_u, and then randomly picks a new knot location from those not already included in g_u. This step proposes the new model using the conditional priors as proposal distributions and the same is done in the *MOVE* step. The *DEATH* step just chooses one spline term to delete so is identical to that seen before. As before, because we still use the conditional priors as proposal distributions, we find that the probability of accepting the proposed model is the minimum of one and the Bayes factor.

The main difference between the model for Bayesian additive modelling, compared to that used for scatterplot smoothing, is that the number of columns in the basis matrix,

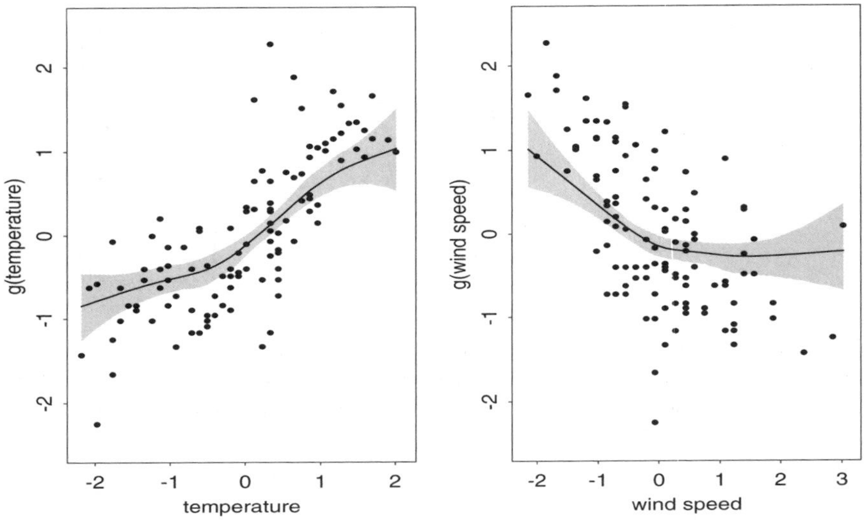

Figure 4.1 The posterior means of the fitted functions using a Bayesian additive model for the ozone data together with 95% credible intervals for the mean.

$\boldsymbol{B}$, is typically greater. Even though in both cases we can assign conjugate priors to the coefficient vector and regression variance, to determine the conditional posterior when additive modelling we need to invert a square matrix of dimension $\sum_1^p k_j$. As the inversion of any $k \times k$ matrix is an $\mathcal{O}(k^3)$ calculation this is typically far more computationally expensive than the inversion required when scatterplot smoothing (for which $p = 1$).

The extra computation effort required is not surprising. We are now fitting a complete regression surface in p dimensions rather than a curve. However, it does demonstrate a point that recurs throughout this chapter. In a basis function representation, surface fitting is conceptually a straightforward generalisation of one-dimensional curve fitting. However, we find that there are many competing approaches to the surface fitting problem because of the subtler complexities introduced in higher dimensions. First though, we consider an example of additive modelling.

4.2.2 Ozone data example

To illustrate the Bayesian additive model approach we consider a regression problem which involves predicting the ozone level from measurements of wind speed (m s^{-1}) and temperature (Fahrenheit). The data are taken from a study by Bruntz *et al.* (1974) (see also Cleveland *et al.* (1988) and Hastie and Tibshirani (1990)). As in previous approaches we take the response to be the cube root of ozone so that the errors in the model are better approximated by a normal distribution.

We display the results of fitting an additive model using linear splines in Figure 4.1. Here we plot both the posterior mean of the estimated regression functions and the 95% credible intervals for this mean. The plot is produced by determining the predictions for y, using each model in the generated sample, firstly with only those basis functions involving wind speed and then with only those involving temperature. For sensible plotting we accounted for the non-identifiability of the resulting estimates by removing the mean level of each of the individual curve found from the sample before averaging across them.

Additive models by definition are constructed using univariate functions and hence do not lend themselves readily to model interactions between predictors. For instance in this ozone example, it might be expected that high temperatures combined with low wind speeds would together result in an increased level of ozone. The additive model which we have outlined cannot model such interactions between predictor variables. However, if the form of interactions that might be present are known beforehand, we can include additional additive functions that depend on some scalar quantity dependent on two or more of the predictors, e.g. we could add to (4.2) functions like $g_{p+1}(x_1x_2)$ and $g_{p+2}(x_1 + x_2x_3)$. Obviously, the form of these interactions must be specified in advance, and this is typically difficult. Also, with many predictors it might be unclear which ones to use, and including too many leads to less precise predictions.

Another potential problem with additive models is that they are not rotationally invariant, meaning that an arbitrary linear transformation of the data $\boldsymbol{XA}$, for a positive-definite $p \times p$ matrix $\boldsymbol{A}$, changes the resulting predictions. In many situations rotational invariance is not of concern but when, for example, the $\boldsymbol{x}$ coordinates define a spatial location this difficulty cannot be overlooked. In order to accommodate interactions and/or rotational invariance we must consider basis functions that are based on the complete vector of predictors, $\boldsymbol{x}$, rather than just one-dimensional simplifications of it.

4.2.3 Further reading on Bayesian additive models

When interactions between predictors are known not to be present, or when interpretability of the model is a key factor, additive models are a powerful modelling tool. Further, they provide a natural first approach to relaxing strong linear assumptions in standard regression models.

Smith and Kohn (1996) give an outline to a Bayesian approach to additive modelling using the Gibbs sampler. They highlight the fact that additive modelling requires the same approach as univariate regression and even show how to robustify the fit against possible outliers. However, they do note a problem with using the g-prior. They find that with the large design matrices needed to fit additive models, this prior specification can lead to difficulties in inverting the $\boldsymbol{B}'\boldsymbol{B}$ matrix as it becomes almost singular. This problem can be overcome by adopting a ridge prior instead of the g-prior as the addition of the prior precision to every diagonal element of the $\boldsymbol{B}'\boldsymbol{B}$ matrix necessarily reduces its condition number. Similar work on Bayesian additive models by Kohn and co-workers is also of interest, notably Wong and Kohn (1996), Smith *et al.* (1998) and Shively *et al.* (1999).

Hastie and Tibshirani (2000) describe a 'Bayesian backfitting' algorithm to Gibbs sample the additive curves. In their approach, each curve is updated in turn, conditional on all the other curves being fixed. This approach is derived from the method of fitting additive models in a non-Bayesian setting via the backfitting algorithm (Hastie and Tibshirani 1990). However, this conditional updating appears unnecessary as all the coefficients can be updated together in a single step once we see that the additive model with regression splines is no more than a single linear regression in a nonlinear base.

By updating the coefficients of each spline conditionally on the others each sweep of the backfitting sampler requires inverting p, $k_j \times k_j$ matrices. In contrast, the method we suggested involves inverting a matrix of dimension $\sum_1^p k_j$. Thus, for most problems the 'backfitting' approach of Hastie and Tibshirani (2000) requires less computational effort per iteration. However, it seems intuitive to update as many parameters jointly as possible as the joint posterior takes into account posterior correlation, which may be high if the true underlying structure contains interactions, and should lead to the algorithm converging quicker.

4.3 Higher-Order Splines

4.3.1 *Truncated linear splines*

An obvious way allow us to be able to capture interaction terms is to generalise the additive model to be able to include spline terms that possibly depend on more than one variable, e.g. by allowing basis functions of the form,

$$(x_i - t_1)_+(x_j - t_2)_-,$$

for $i \neq j$ and where t_1 and t_2 are the knot locations of the basis. This is precisely what Smith and Kohn (1997) suggested for regression when there are only two predictors, who model such a function with

$$g(\boldsymbol{x}) = \beta_0 + \sum_{i=1}^{T_1} \theta_i \beta_i (x_1 - t_{1i})_+ + \sum_{j=1}^{T_2} \theta_{T_1+j} \beta_{T_1+j} (x_2 - t_{2j})_+ \\ + \sum_{i=1}^{T_1} \sum_{j=1}^{T_2} \theta_{ij} \beta_{ij} (x_1 - t_{1i})_+ (x_2 - t_{2j})_+,$$

where the candidate knots for predictor i are $t_{i,1}, \ldots, t_{i,T_i}$ and θ_i are indicators for which bases are present. We can then sample from the model space $\boldsymbol{\theta}$, given by the concatenation of all the $\theta_\bullet$ above, using the Gibbs sampler.

The number of indicator functions that we have to sample over is $K = T_1 + T_2 + T_1 T_2$, so the number of candidate basis functions allowed in each dimension is crucial to the computational effort required to approximate the true surface. Smith and Kohn (1997) suggest that the allowable knot locations follow the marginal density of the predictors. Firstly, they suggest taking the knots to be every 'mth' value of each sorted predictor where m is a function of the dataset and is chosen so that $T_1 = T_2 = 9$. Hence, the simulation requires updating 99 indicator variables, one for each candidate basis. Smith and Kohn (1997) also describe a way to generalise this method to take into account the multivariate density of the predictors, rather than just the individual marginals. Thus, the knot location chosen in one dimension affects the probability of choosing a particular location in any of the others. Smith and Kohn (1997) go on to demonstrate the effectiveness of this approach for bivariate surface estimation with both simulated and real data.

In a further paper, Kohn *et al.* (2001) extend this work for other basis functions noting that a bivariate surface can be estimated in an analogous way for any set of bivariate bases $B_1, \ldots, B_K$ such that

$$g(x_1, x_2) = \beta_0 + \sum_{i=1}^{K} \theta_i \beta_i B_i(x_1, x_2).$$

Possible choices of such basis functions are the thin-plate spline, multiquadric and cubic basis functions given, respectively, by

$$B_i(\boldsymbol{x}) = \|\boldsymbol{x} - \boldsymbol{t}_i\| \log \|\boldsymbol{x} - \boldsymbol{t}_i\|,$$
$$B_i(\boldsymbol{x}) = (\kappa^2 + \|\boldsymbol{x} - \boldsymbol{t}_i\|^2)^{1/2},$$
$$B_i(\boldsymbol{x}) = \|\boldsymbol{x} - \boldsymbol{t}_i\|^3,$$

where $\|\cdot\|$ is any sensible distance metric, nearly always taken to represent the Euclidean norm. These are all examples of radial basis functions; Franke (1982) provides a discussion on these and other basis functions. The name comes from the fact that the outputs of the bases depend solely on the distance between the knot, $\boldsymbol{t}_i$, and the input (predictor) vector, $\boldsymbol{x}$.

Kohn *et al.* (2001) suggest taking the possible knot locations $\boldsymbol{t}_1, \ldots, \boldsymbol{t}_K$ at some prespecified number of cluster centres (Kohn *et al.* 2001). The cluster centres can found by any standard clustering algorithm such as K-means (Everitt 1993; Hartigan 1975; Hartigan and Wong 1979). In this way the complexity of the problem can be controlled by the number of cluster centres to use (Kohn *et al.* (2001) take $K = 100$). Note that this method does not take into account the observed responses, only the observed density of the predictors.

When using reversible jump sampling there is no need to specify the full model *a priori*. So, when Holmes and Mallick (1998) tackled the regression problem and suggested using radial basis functions they took the candidate knot locations to be the observed data points. This approach ensures that the knot locations exactly mimic the observed marginals of the predictors. Also, as the number of random parameters which must be sampled does not necessarily increase with n, the computational complexity of the algorithm is not made infeasible when we use so many candidate knot sites.

4.4 High-Dimensional Regression

4.4.1 Extending to higher dimension

We now consider how the models introduced in the previous section might perform in higher dimensions. The main difficulty as the dimension increases is that datasets of equal size become more and more spread out. This is known as the 'curse of dimensionality', a term most often attributed to Bellman (1961). Consider the following example of it involving a randomly generated set of n points in the unit hypercube in p dimensions, $[0, 1]^p$. If $p = 1$ it is obvious that an interval of length 0.25 which is entirely contained within [0,1] will contain, on average, one-quarter of the randomly generated points. However, to contain on average a quarter of the points in p dimensions the interval length must scale like $(0.25)^{1/p}$. So, for example, with $p = 10$ we find that the hypercube must have side length $(0.25)^{1/10} = 0.871$ in order to expect it to contain $n/4$ points. This is hardly a local region and because of this we find that it is difficult to estimate parameters of 'local' basis functions when they are of a high dimension, as there are no local clusters.

Unfortunately, because of the curse of dimensionality, basis functions that rely on distances between vectors in predictor space can perform poorly in more than a few dimensions. This is because, in high dimensions, standard distance metrics do not give a reliable idea of 'closeness' between two vectors as they are not resistant to outliers. Further, as points are so spread out there are very few data points close to each knot so estimation of its coefficient can be poor. As a rule-of-thumb we find that basis functions based on distance (e.g. radial basis functions) perform poorly in situations where p is much greater than 3.

Another approach to overcoming the difficulties associated with the curse of dimensionality is to approximate the true function with a sum of lower-dimensional functions. The simplest way to do this is the additive modelling method seen earlier. When we project the data-space into lower-dimensional subspaces, the data are less spread out so more points can be used to estimate coefficients relating to such basis functions. For example, suppose that we wish to estimate a five-dimensional regression function, f. We could take the approximate the truth, g, as a sum of three functions, each with dimension less than or equal to two, i.e. we could assume

$$g(X_1, \ldots, X_5) = g_1(X_2) + g_2(X_1, X_5) + g_3(X_3, X_4).$$

Thus, the difficulty in modelling data for which p is high is in determining a method that can both approximate the unknown functions g_1, g_2 and g_3 as well as the predictors to include in each function. In practice, we also need to know how many low-dimensional functions we need to approximate the truth.

The simplest form of a high-dimensional estimate is the additive model (4.2). However, this makes the restrictive assumption that there are no interactions between parameters. As we pointed out earlier, if we know the form of possible interactions beforehand, these can be included as one-dimensional functions which are dependent on more than one variable. However, to use such a method we need to know the form of the interactions *a priori* and this is often unrealistic when p is large and there are $\frac{1}{2}p(p+1)$ possible two-way interaction terms. Uncovering unknown interactions between variables is a major focus of data analysis, so it is extremely useful to design methods that can automatically detect and model interaction effects. In the remaining sections of this chapter we examine such models.

Once more we see that the basic regression framework introduced in Chapter 3 is still present. Again, we use a linear combination of coefficients and basis functions, where now the form of the basis functions has been chosen to be able to both detect, and flexibly model, low-order interactions in the high-dimensional predictor space.

4.4.2 The BWISE model

Gustafson (2000) introduced the Bayesian regression model with interactions and smooth effects and coined the name BWISE for it. The BWISE model approximates the true function with $g(\boldsymbol{x})$ of the form,

$$g(\boldsymbol{x}) = \beta_0 + \sum_{i \in I_1} g_i(x_i) + \sum_{(j,k) \in I_2} g_{jk}(x_j, x_k),$$

where I_1 is the set of all variables included only as main effects and I_2 is the set of pairs of predictors that are assumed to have a bivariate effect. As we suggested earlier, this model approximates the high-dimensional regression function as a sum of low-dimensional functions. The main effect terms, g_i, are taken to be either linear terms or cubic splines with the method automatically selecting whichever is most appropriate. To ensure continuity of the overall function Gustafson (2000) gives appropriate conditions on the bivariate interaction terms, g_{jk}. These depend on whether the main effect terms relating to the interactions are included linearly or as splines.

As the model can be written as a linear combination of coefficients and basis functions, it reduces to a standard Bayesian linear model. Gustafson (2000) chooses to exploit conjugacy to make inference and adopts a prior motivated by the roughness penalty approach to spline fitting (Green and Silverman 1994). Inference is either carried out by analytically evaluating the marginal likelihoods of all the models, if p is small, or approximated via a search scheme based on the Metropolis–Hastings algorithm (Hastings 1970).

As demonstrated by Gustafson (2000), the model is particularly well suited to modelling data when the deviations from linearity are not great. This is because of the constraints needed on the interactions to ensure continuity and because the knots in the spline terms are fixed before the analysis. Demonstration of this point is provided in Gustafson (2000) and code to perform the BWISE model is available from `http://www.stat.ubc.ca/people/gustaf/`.

4.4.3 The BMARS model

One of the most popular multiple nonlinear regression methods is the multivariate adaptive regression spline (MARS) model which was introduced by Friedman (1991).

This formed the foundation of the Bayesian MARS method of Denison *et al.* (1998c) which we shall outline here. Again, the differences between this model and others introduced is mainly in the form of the basis functions. For Bayesian MARS with k non-constant basis functions the model takes the standard linear in the basis set form,

$$g(\boldsymbol{x}) = \beta_0 + \sum_{i=1}^{k} \beta_i B_i(\boldsymbol{x}),$$

where $B_i(\boldsymbol{x})$ is now designed to model interactions and is given by

$$B_i(\boldsymbol{x}) = \prod_{j=1}^{J_i} [s_{ij}(x_{w_{ij}} - t_{ij})]_+, \qquad i = 1, 2, \ldots, k,$$

where $[\cdot]_+ = \max[0, \cdot]$, J_i is the degree of interaction of basis B_i, the s_{ij} are sign indicators taking values ± 1, the w_{ij} index which predictor variables have an associated knot and the t_{ij} are knot points. It is common to ensure that each basis function is at most linear in any variable, so for each i we make the restriction that all the w_{ij} are distinct. Hence, a MARS basis is just a tensor product of univariate linear spline terms.

To understand the initially confusing notation we present a small example. Suppose $p = 6$ and that the fourth basis function in the model is given by

$$B_4(\boldsymbol{x}) = B_4(x_1, \ldots, x_6) = [x_2 - 1]_+[-(x_5 + 3)]_+.$$

This basis function is composed of two linear spline terms so the interaction level is $J_4 = 2$. The sign indicators are $s_{14} = 1$ and $s_{24} = -1$, and note that, by allowing them to be either ± 1 we are ensuring that the spline terms are two sided (see Section 3.5.1). The knot points, or thresholds, are $t_{14} = 1$ and $t_{24} = -3$, and the labels for the predictor splits are $w_{14} = 2$, $w_{24} = 5$, showing that x_2 and x_5 are the first and second terms in this basis function. The response of this basis function is shown in Figure 4.2.

As in the curve fitting models of Section 3.3.5 we suggest default priors for the variance of the noise process and the regression coefficients. Hence in the Bayesian MARS model we assume a uniform prior over the number of basis functions, and take

$$p(\boldsymbol{\beta}, \sigma^2) = \mathrm{NIG}(\boldsymbol{0}, v\boldsymbol{I}, 0.01, 0.01).$$

Discrete uniform priors are also adopted over the knot locations, $\{t_{ij}\}$, which are restricted to be at the observed predictor values, and the sign indicators, $\{s_{ij}\}$. In almost all situations we would restrict the interaction level to be at most 2, with $p(J_j) = U\{1, 2\}$. This is chosen on two counts; first by limiting the splines to have only low-order interactions it is easier to interpret the resulting model. Secondly, it would require a large amount of data to uncover higher-order interactions with any certainty due to the curse of dimensionality. However, these are just our recommendations and if higher-order interactions are suspected then the prior should be adjusted accordingly. The remaining prior is over the variable indices, w_{ij}. Denison *et al.* (1998c) suggested

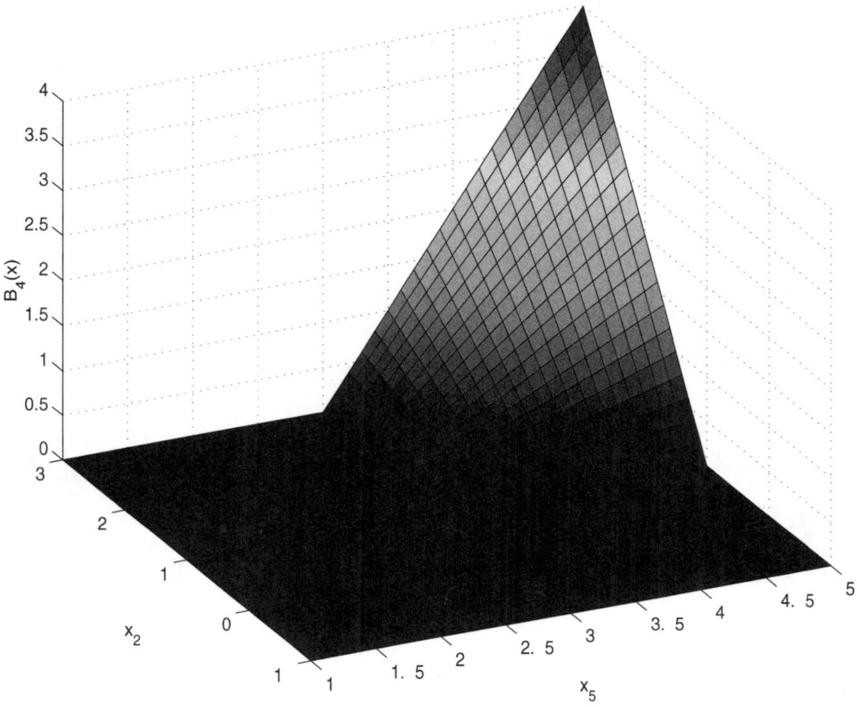

Figure 4.2 The response of the MARS basis $B_4(\boldsymbol{x}) = [x_2 - 1]_+[-(x_5 + 3)]_+$ given in equation (4.4.3). Note the change in direction on the two axes.

only allowing one and two-way interaction terms in the model and adopting a uniform prior over all these possible types of bases, i.e. over the p possible *types* of main effect basis and the $\frac{1}{2}p(p-1)$ possible types of interaction basis. Here basis are defined as being of the same type if they contain the same variables. However, when interpretation is the main aim and we would prefer the model to use simple main effect terms rather than interactions, a prior that puts weight $1/(2p)$ for each type of main effect, and $1/\{p(p-1)\}$ for each type of interaction, may be more appropriate. This was suggested by Mallick *et al.* (1999) and corresponds to putting half the prior mass on main effect terms and the other half on interactions.

When using a MARS model we can make inference with the reversible jump sampling algorithm outlined in Section 3.3.1. That is, we can generate models with different numbers and locations of splines. This mixes over MARS models with different interaction and main effect terms and, in essence, forms a guided search over the model space which uncovers important basis functions from the data itself. If a strong interaction exists, then we would expect it to be retained by the model as removing it would significantly decrease the marginal likelihood.

When sampling with MARS bases we use the following *BIRTH* step when we start with model $\boldsymbol{\theta} = (\boldsymbol{t}, \boldsymbol{s}, \boldsymbol{w})$ (where these vectors include all the t_{ij}, s_{ij} and w_{ij}).

Birth-proposal($\boldsymbol{\theta}$):

```
Set u_1 to a draw from a U{1,2} distribution;
FOR j = 1 : u_1
  Set s_j to a draw from a U{-1,+1} distribution;
  IF u_1 == 1
    Set w_j to a draw from a U{1,...,p} distribution;
  ELSE
    Set w_j to a draw from a U{1,...,p}\{w_1}
      distribution;
  Set t_j to a draw from a U{T_{w_j}};
ENDFOR;
return(θ, {s_j, w_j, t_j}_1^{u_1});
```

This just adds the basis function

$$B(\boldsymbol{x}) = \prod_{j=1}^{u_1} [s_j(x_{w_j} - t_j)]_+,$$

to the model. The death step is the same as before and we could also incorporate the same move step if we wished. However, Denison *et al.* (1998c) find that alternative move steps which do not make such drastic jumps around the model space also work well. Two such steps they suggest are just randomly picking a knot point t_{ij} in a basis function by drawing it from the prior, or completely redrawing one of the factors in a basis function (i.e. sample the variable and knot point together). This second step is identical to the standard move step if you randomly choose to alter a main effect term but not if an interaction term is chosen. The reason that we may wish to incorporate both of these move proposals is to allow the sampler to make both bold and local moves in the model space. Local moves help the sampler to slowly explore local modes in the posterior probability space, whereas bold moves allow the sampler to escape these local maxima. Both types of move have merit in a random sampler.

To demonstrate the capabilities of the Bayesian MARS model we shall analyse the Boston housing dataset of Harrison and Rubenfeld (1978). This has been the subject of many statistical analyses involving nonlinear regression including the work of Belsey *et al.* (1980), Quinlan (1993) and Neal (1996). The regression problem is to relate the 1970 median price (MEDV) of owner-occupied homes in 506 regions (or tracts) around Boston, to 13 predictor variables, listed in Table 4.1. After 50 000 iterations we judged that the sampling algorithm had converged so we then ran it for a further 500 000 iterations from which every fifth model was taken to be in the generated sample.

Table 4.1 The description of the variables in the Boston housing data.

Variable	Description
MEDV	median value of owner occupied homes in thousands of dollars
CRIM	per capita crime rate by town
ZN	proportion of residential land zoned for lots over 25 000 square feet
INDUS	proportion of non-retail business acres per town
CHAS	Charles River variable (= 1 if tract bounds river; 0 otherwise)
NOX	nitric oxides concentration (parts per 10 million)
RM	average number of rooms per dwelling
AGE	proportion of owner-occupied units built before 1940
DIS	weighted distances to five Boston employment centres
RAD	index of accessibility to radial highways
TAX	full-value property-tax rate per $10 000
PTRATIO	pupil-teacher ratio by town
BK	$1000(\mathrm{Bk} - 0.63)^2$, where Bk is the proportion of blacks by town
LSTAT	% lower status of the population

In Figure 4.3 we display the posterior mean main effect terms for each of the predictors in the dataset. These curves are found by determining

$$E\{g_j(x)\} = \frac{1}{T}\sum_{t=1}^{T}\sum_{\substack{i:J_i=1,\\ w_{1i}=j}} \beta_i^{(t)} B_i^{(t)}(x),$$

where T is the number of models in the generated sample, indexed with the superscript. The second summation ensures that the curves are found by only considering main effect basis functions involving the jth predictor, so we average the predictions over the basis functions relating to the desired main effect. Now, similarly to the univariate curves found with additive models, estimating the g_j in this way leads to each of them being unidentifiable (as $g_j(x) + g_i(x) = (g_j(x) + c) + (g_i(x) - c)$). As such the mean level of each effect is unidentifiable so we can only use Figure 4.3 to guide us in the shape of the main effect terms. We see evidence that the number of rooms (RM) has no effect on the median value of the house (MEDV) when there are four or five rooms (RM), but linearly increases for houses with six or more rooms. The other obvious feature is the linear decrease in value as the percentage of lower status in the population (LSTAT) increases. The other variables' main effects were not observed frequently in the generated sample so did not have a great influence on the response. However, this figure does demonstrate how threshold basis functions, like those in the MARS model, allow for insightful interpretation of the relationship between the response and predictors. It also demonstrates the ability of the MARS model to automatically ignore variables that have little effect on the response, aiding the overall interpretability of the model.

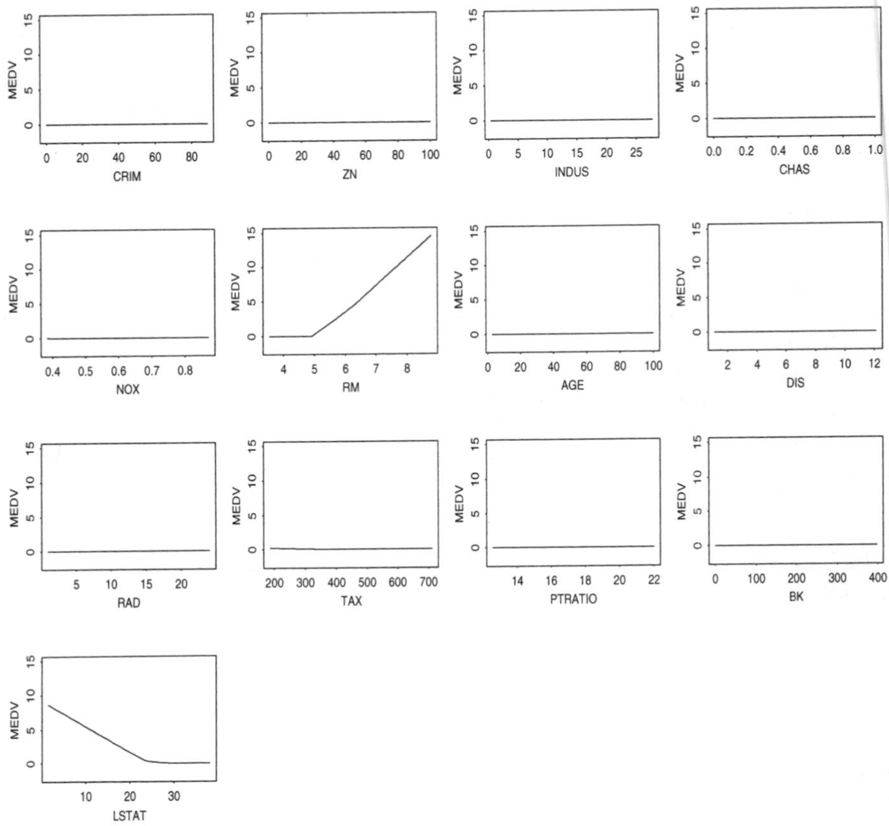

Figure 4.3 The posterior mean main effects for each of the predictor variables in Table 4.1.

The most frequently observed interaction term in the generated sample was the one between DIS and LSTAT. We plot the posterior mean of this interaction in Figure 4.4. This highlights both the advantages and disadvantages of the MARS basis functions. The flexibility of the basis set allows the model to capture the effect that a low proportion of lower status population, combined with a small distance to an employment centre, leads to high value of properties (probably expensive inner-city apartments). Also, the interaction surface picks up the decreasing linear trend in the relationship between MEDV and LSTAT for values of LSTAT less than 25%.

The flexibility of the model can be both a benefit and a disadvantage. In this data analysis the large peak in the interaction surface between DIS and LSTAT when both predictors are near zero gives us important information about the dataset. However, outside the convex hull of the data the basis function, or functions, that capture this peak would generalise very badly as they have such a high coefficient, estimated

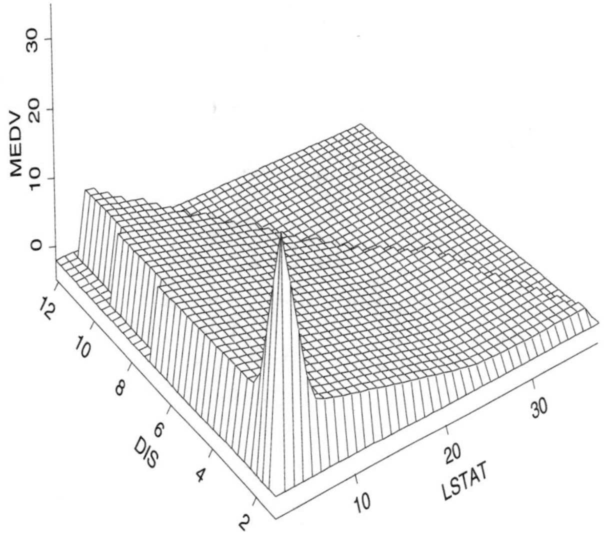

Figure 4.4 The posterior mean interaction between DIS and LSTAT. Points outside the convex hull of the dataset are represented by a MEDV value of −2.

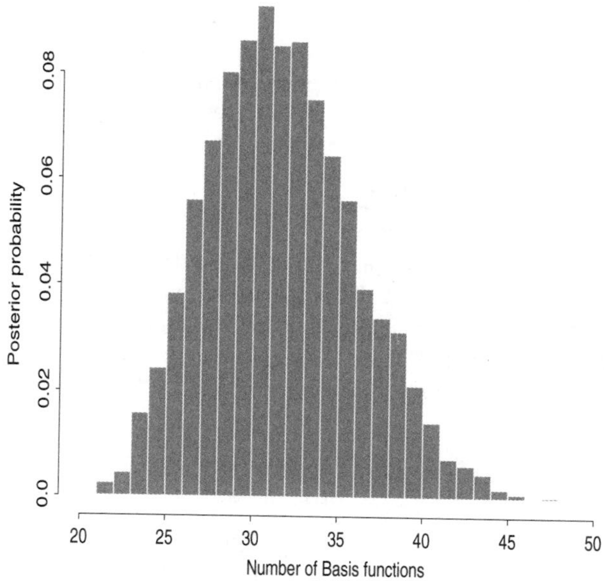

Figure 4.5 The posterior distribution of the number of basis functions in the generated sample for the BMARS model.

from only a very small proportion of the overall dataset. In this example this is not a great problem as we will never have values of DIS or LSTAT less than zero but in other applications this might not be the case. In these situations this can lead to wild predictions for the response at points outside the convex hull of the data, though it should be noted that we would expect to see an associated large increase in the variance of the predictive distribution associated with extrapolated forecasts.

4.4.4 Piecewise linear models

An alternative to the BMARS model for high-dimensional regression is the piecewise linear model first described in Breiman (1993) and subsequently analysed in a Bayesian framework by Holmes and Mallick (2001). We can motivate this model by first noting that the one-dimensional linear spline term, $[x - t_0]_+$, can be written alternatively as

$$[x - t_0]_+ \equiv [(1 \;\; x) \cdot (-t_0 \;\; 1)]_+, \tag{4.5}$$

where $a \cdot b$ indicates the dot (or inner) product of a and b. In two dimensions, a possible generalisation of this truncated spline basis function is given by

$$B_i(x_1, x_2) = [t_{i1}x_1 + t_{i2}x_2 - t_{i0}]_+ = [(1 \;\; x_1 \;\; x_2) \cdot (-t_{i0} \;\; t_{i1} \;\; t_{i2})]_+.$$

In this way the t_{i1} and t_{i2} terms act like coefficients of a linear plane in two dimensions, whereas t_{i0} can be thought of as an intercept. This basis functions known as a piecewise linear spline as it consists of the linear plane $t_{i1}x_1 + t_{i2}x_2 = t_{i0}$ for values of (x_1, x_2) which satisfy $t_{i1}x_1 + t_{i2}x_2 > t_{i0}$, and zero everywhere else (see Figure 4.6).

This idea can easily be extended to higher dimensions by thinking of piecewise linear terms such as

$$[(1 \;\; \boldsymbol{x}) \cdot \boldsymbol{t}_i]_+ = \left[t_{i0} + \sum_{j=1}^{p} t_{ij}x_1\right]_+,$$

where $\boldsymbol{x} = (x_1, \ldots, x_p)$ and $\boldsymbol{t}_i = (t_{i0}, t_{i1}, \ldots, t_{ip})$, and we have dropped the negative sign before t_{i0} as it serves no particular purpose.

We can use these piecewise linear planes as basis functions as suggested by Breiman (1993) and Holmes and Mallick (2001). This piecewise linear model can be written in the usual linear regression in the basis set form as

$$g(\boldsymbol{x}) = \beta_0 + \sum_{i=1}^{k} \beta_i[(1 \;\; \boldsymbol{x}) \cdot \boldsymbol{t}_i]_+. \tag{4.6}$$

One problem with this model is that the coefficient β_i and the elements of vector $\boldsymbol{t}_i$ are unidentifiable, i.e. multiplying β_i by a constant and dividing $\boldsymbol{t}_i$ by the same constant leaves the model unchanged. To compensate for this Holmes and Mallick (2001) suggest fixing the scale of the parameters $\boldsymbol{t}_i$ that determine the orientation of each plane: they take $\|(t_{i1}, \ldots, t_{ip})\| = 1$ for $i = 1, \ldots, k$. This allows us to think

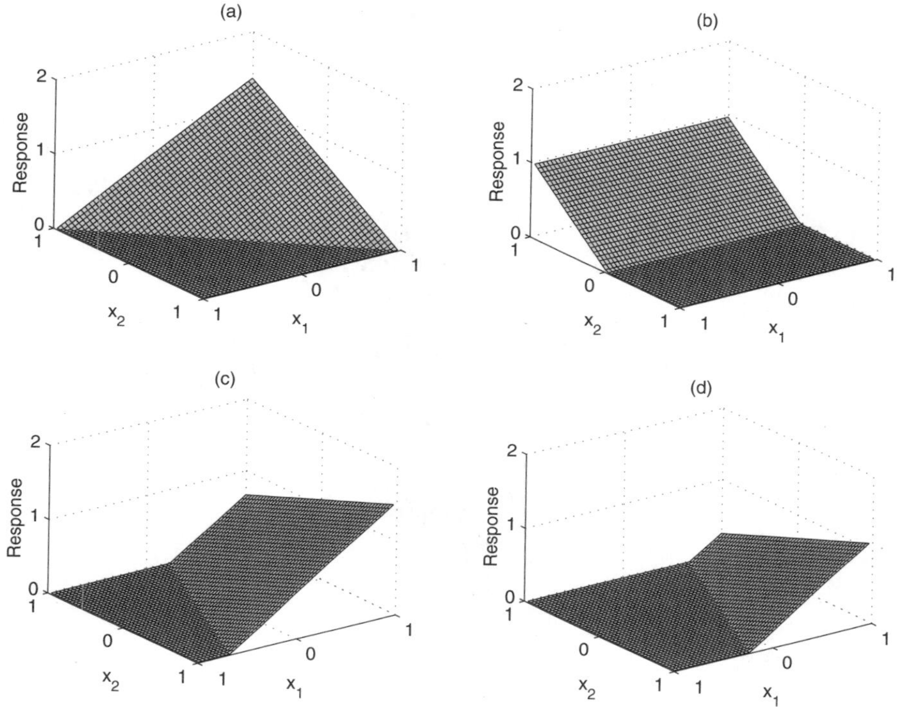

Figure 4.6 Piecewise linear basis functions $(\{1 \ \ \boldsymbol{x}\} \cdot \boldsymbol{t})_+$ in two dimensions. In (a) $\boldsymbol{t} = \{0, \sqrt{\frac{2}{3}}, \sqrt{\frac{1}{3}}\}$, (b) $\boldsymbol{t} = \{0, 0, 1\}$, (c) $\boldsymbol{t} = \{0, \sqrt{\frac{2}{3}}, -\sqrt{\frac{1}{3}}\}$ and (d) $\boldsymbol{t} = \{-5, \sqrt{\frac{2}{3}}, -\sqrt{\frac{1}{3}}\}$. Note that the first component of $\boldsymbol{t}$ relates to the location of the plane. Further, we find that (b) is identical to the BMARS basis function $[x_2 - 0]_+$. (Reproduced by permission of the Royal Statistical Society.)

of the regression coefficients, $\boldsymbol{\beta}$, as giving us information about the gradients of the planes that make up the basis set.

The piecewise linear model is particularly attractive as it automatically imposes mean level continuity on the regression surface with discontinuities in the first derivatives along the boundaries defined by the hyperplane $[(1 \ \ \boldsymbol{x}) \cdot \boldsymbol{t}_i] = 0$ of each spline basis. This is similar to the BMARS basis functions seen before but, in addition, the piecewise linear model is a sum of piecewise linear planes so is locally linear everywhere (see Figure 4.7). Thus, at its boundaries the regression function increases linearly at most, and also, we find that this model is rotationally invariant as the basis functions themselves are defined by rotations. This is unlike the BMARS model which imposes the restriction that the linear planes added are all parallel to the axes of the predictors.

In order to accommodate variable selection of relevant predictors and interactions we can allow each basis function to be of lower dimension than p, similarly to the BMARS bases which we chose earlier to have maximum dimension two. Then, to

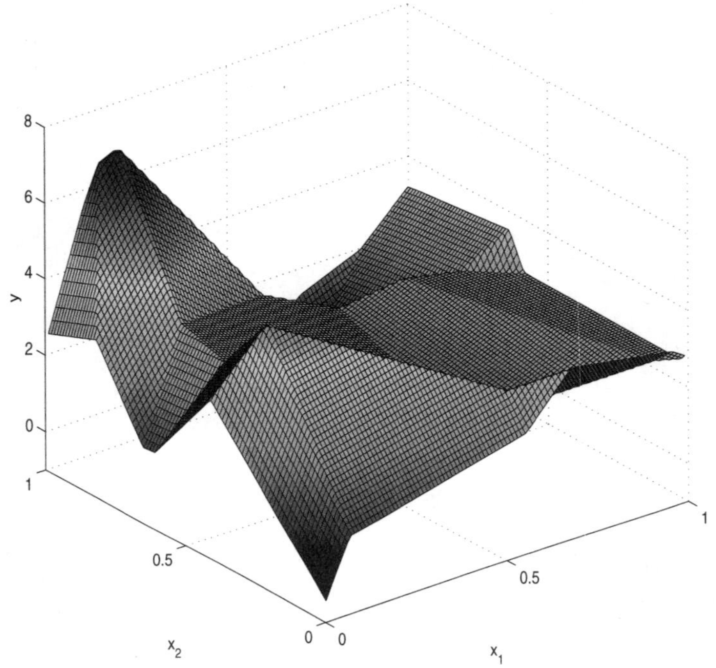

Figure 4.7 An example of a single realisation from a piecewise linear surface in two dimensions. (Reproduced by permission of the Royal Statistical Society.)

compare the two basis functions we can adopt similar notation to the BMARS model description of Section 4.4.3 and write the piecewise linear basis functions as

$$B_i(\boldsymbol{x}) = \left[t_{i0} + \sum_{j=1}^{J_i} x_{w_{ij}} t_{ij} \right]_+ ,$$

where the w_{ij} $(j = 1, \ldots, J_i)$ give the index of the predictor variables used by the ith basis function and now $\boldsymbol{t}_i = \{t_{i0}, \ldots, t_{iJ_i}\}$ again defines the orientation and location of basis B_i. By excluding a predictor from B_i we align the linear planes to be perpendicular to the excluded predictors. For example, Figure 4.6(b) shows a basis function independent of x_1. As for the BMARS model we typically recommend a prior on the interaction level such that $p(J_i) = U\{1, 2\}$ so three-way, or higher, interactions are not allowed.

We now outline the priors Holmes and Mallick (2001) recommend using over the parameters that define each basis functions, as all the others are the same as for BMARS. Firstly, they place a prior over $\{t_{i1}, \ldots, t_{ip}\}$ such that they lie on the unit circle. To do this Holmes and Mallick (2001) suggest drawing each t_{ij} $(j = 1, \ldots, J_i)$ from standard normal distributions and then, once they have all been drawn, dividing

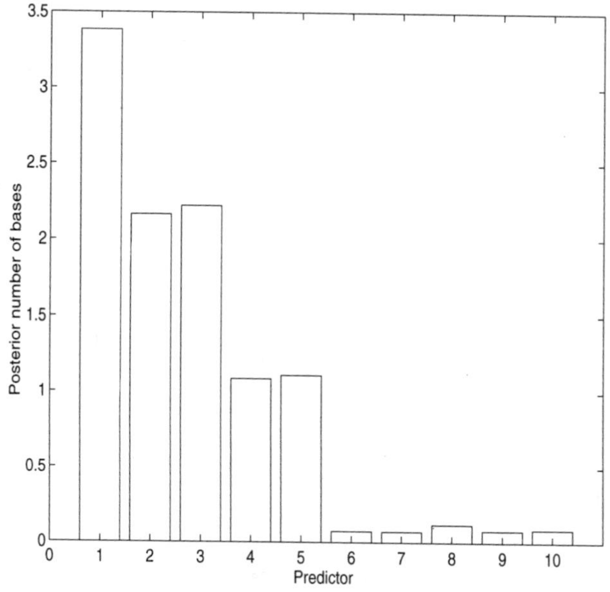

Figure 4.8 The posterior number of basis functions which include each of the 10 predictors for the dataset simulated according to (4.7). (Reproduced by permission of the Royal Statistical Society.)

each t_{ij} by $t_{i1}^2 + \cdots + t_{i,J_i}^2$. To ensure that the ridges of each plane pass through regions of the predictor space, $\mathcal{X}$, that contain data points we then assign a conditional prior on t_{i0} such that

$$p(t_{i0} \mid t_{i1}, \ldots, t_{iJ_i}) = U\left\{\sum_{j=1}^{J_i} x_{1,w_{ij}} t_{ij}, \sum_{j=1}^{J_i} x_{2,w_{ij}} t_{ij}, \ldots, \sum_{j=1}^{J_i} x_{n,w_{ij}} t_{ij}\right\}.$$

This constrains the ridge of the plane associated with the ith basis to pass through at least one of the observed predictor locations as

$$t_{i0} + \sum_{j=1}^{J_i} x_{k,w_{ij}} t_{ij} = 0,$$

for at least one value j ($\in 1, \ldots, n$). As for the BMARS model the prior on the indices w_{ij} is taken to be uniform without replacement on the p predictors.

Posterior inference proceeds using the same basic reversible jump sampling algorithm as that outlined in Section 3.3.1. We draw the generated sample of models,

$\boldsymbol{\theta} = (\boldsymbol{t}, \boldsymbol{w})$ (where these vectors include all the t_{ij} and w_{ij}), with the *BIRTH* step for the for a new basis function given by the following algorithm.

Birth-proposal($\boldsymbol{\theta}$):

```
Set u_1 to a draw from a U{1,2} distribution;
FOR j = 1 : u_1
  IF u_1 == 1
    Set w_j to a draw from a U{1,...,p} distribution;
  ELSE
    Set w_j to a draw from a U{1,...,p}\{w_1}
        distribution;
  Set t_j to a draw from an N(0,1);
ENDFOR;
Normalise t_1,...,t_{u_1} so that ||t_1,...,t_{u_1}|| = 1;
Set u_2 to a draw from a U{1,...,n} distribution;
Set t_0 = -sum_{j=1}^{u_1} x_{u_2,w_j} t_j
return(θ, {w_j, t_j}_1^{u_1});
```

This just adds the basis function

$$B(\boldsymbol{x}) = \left[t_0 + \sum_{j=1}^{u_1} x_{w_j} t_j \right]_+,$$

to the model and the ridge of the plane passes through predictor location $\boldsymbol{x}_{u_2}$. The *DEATH* step just randomly removes a basis function so is the same as that outlined before. The within-dimension *MOVE* step can be made up of a *BIRTH* followed by *DEATH* step. More local move steps could be undertaken by selecting a basis function at random, say B_i, and then either just updating the first coefficient t_{i0} from the prior, or we could perform a Metropolis–Hastings update to $t_{i1}, \ldots, t_{iJ_i}$.

An attractive feature of using the piecewise-linear basis functions is that the prediction at any point can be written as a local linear model. This is because all of the basis functions are themselves locally linear so we can write the estimating function at a general point as

$$g(\boldsymbol{x}) = \boldsymbol{\gamma}_{\boldsymbol{x}} \cdot \boldsymbol{x},$$

where the local coefficients $\boldsymbol{\gamma}_{\boldsymbol{x}}$ are made up of the sum of the output of the basis functions evaluated at $\boldsymbol{x}$.

The expected regression surface is recovered by averaging over the model space of different piecewise linear surfaces with different numbers and locations of bases. This tends to produce a smooth surface which can be viewed as a local linear model with distributions on the local linear coefficients, $\boldsymbol{\gamma}_{\boldsymbol{x}}$ for $\boldsymbol{x} \in \mathcal{X}$. The posterior distributions of the local linear coefficients at any design point show the contribution of each predictor to the uncertainty in the posterior predictive density at that point.

We now present an example of using the piecewise linear spline model to demonstrate its efficiency in determining which predictors are important. The example is taken from Friedman (1991) and has 10 predictor variables (i.e. $p = 10$) and 100 data points (i.e. $n = 100$). Each data point is drawn from the 10-dimensional uniform distribution on the unit simplex in 10 dimensions, $[0, 1]^{10}$. The true regression function is taken to be

$$f(\boldsymbol{x}) = 10\sin(\pi x_1 x_2) + 20(x_3 - \tfrac{1}{2})^2 + 10x_4 + 5x_5. \tag{4.7}$$

This function only involves the first 5 of the 10 predictors so half the predictors are irrelevant and we would like our method to uncover this fact.

Holmes and Mallick (2001) demonstrated how the piecewise linear model outperformed many other methods in terms of prediction, when applied to this dataset. However, here we just display in Figure 4.8 the posterior number of basis functions which contained each predictor, when running the algorithm in the same manner as the BMARS model described before. Here we see how the irrelevant predictors $X_6, \ldots, X_{10}$ are hardly included in the model. Further, X_4 and X_5, which linearly affect the true regression function (4.7) linearly appear in, on average, around only one basis function. This is because a good single piecewise linear basis function has been determined for each of these predictors. For the remaining predictors with more complex relationships between themselves and the output of f, more basis functions involving these predictors are needed in the model.

4.4.5 *Neural network models*

Neural networks are another example of basis function models. They have received enormous attention particularly within the computer science and engineering fields. One reason for this is that they are known to be universal approximators, which means that they can approximate any continuous function to within any prespecified degree of accuracy (for a proof see Ripley 1996, p. 174). Another reason for the interest in neural networks is the biological interpretation of artificial neural networks as simulating the firing of neurons (see, for example, McCulloch and Pitts 1943). It could also be argued that there has been a lot of hype associated to the neural network field as they can be viewed as 'just another basis set'. However, they have undeniably shown that, when used carefully, they can provide impressive empirical results in terms of predictive accuracy. This is why, at the time of writing, there are at least six international journals dedicated to this single form of basis function modelling alone.

The neural network (NN) models originated within computer science with the founding papers of McCulloch and Pitts (1943), Rosenblatt (1958) and Widrow and Hoff (1960). The original NN bases suggested were piecewise step functions, equivalent to those presented in Section 4.4.4 but using piecewise constants rather than piecewise linear terms. After an initial flurry of interest in the subject, problems of parameter estimation and functional specificity effectively diminished interest in the field. However, during the mid 1980s there was a well-documented resurgence of interest, often attributed to the publication of the book by Rumelhart *et al.* (1986). In particular the book by Rumelhart *et al.* (1986) described a gradient descent learning algorithm for the estimation of the model parameters $\boldsymbol{\theta} = \{\boldsymbol{\beta}, \boldsymbol{t}\}$.

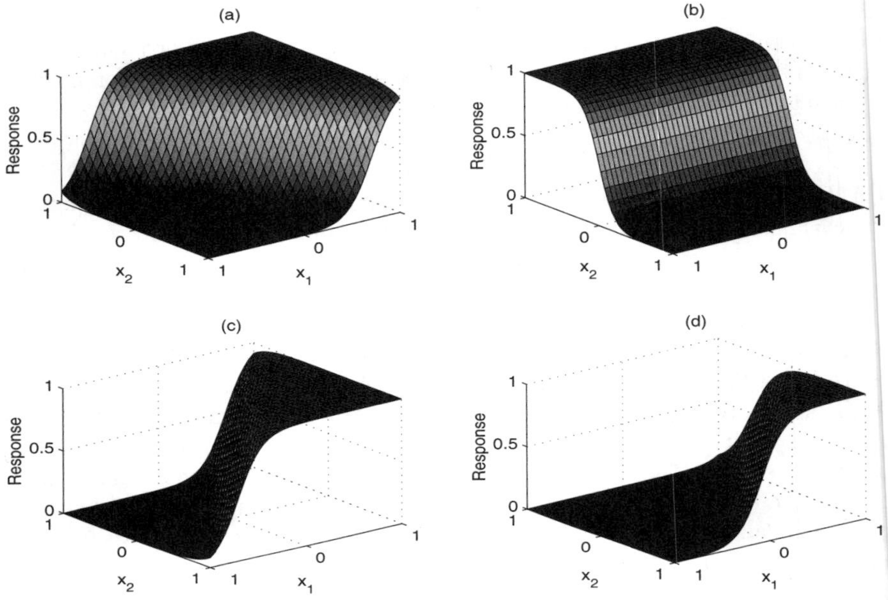

Figure 4.9 Neural network logistic basis functions $1/(1 + \exp\{(1\ \ \boldsymbol{x}) \cdot \boldsymbol{t}))$ in two dimensions. In (a) $\boldsymbol{t} = 10 \times \{0, \sqrt{\frac{2}{3}}, \sqrt{\frac{1}{3}}\}$, (b) $\boldsymbol{t} = 10 \times \{0, 0, 1\}$, (c) $\boldsymbol{t} = 10 \times \{0, \sqrt{\frac{2}{3}}, -\sqrt{\frac{1}{3}}\}$ and (d) $\boldsymbol{t} = 10 \times \{-\frac{1}{2}, \sqrt{\frac{2}{3}}, -\sqrt{\frac{1}{3}}\}$. In addition to the three parameters which fix the location and orientation, a further one, set at 10 in the above plot, determines the steepness of the slope. Compare this plot with Figure 4.10 which has the slope parameter set to 50.

Neural networks come in many form, however, we shall concentrate on the most popular one known as feed-forward neural networks which use 'sigmoidal' basis functions to estimate the true regression function f. An example of such a basis function is the logistic function for which

$$B(\boldsymbol{x}) = \frac{1}{1 + \exp(\boldsymbol{x} \cdot \boldsymbol{t})}, \tag{4.8}$$

which maps $(-\infty, \infty)$ onto $(0, 1)$ and where $\boldsymbol{t} = (t_0, t_1 \ldots, t_p)$ is a set of basis function parameters. Here, we have implicitly included an intercept term in $\boldsymbol{x}$ so that $\boldsymbol{x} = (1, x_1, \ldots, x_p)$. Another common basis function is the hyperbolic tangent one

$$B(\boldsymbol{x}) = \frac{\exp(\boldsymbol{x} \cdot \boldsymbol{t}) - \exp(-\boldsymbol{x} \cdot \boldsymbol{t})}{\exp(\boldsymbol{x} \cdot \boldsymbol{t}) + \exp(-\boldsymbol{x} \cdot \boldsymbol{t})},$$

which maps $(-\infty, \infty)$ onto $(-1, 1)$. The logistic is the more common of the two and, furthermore, the hyperbolic tangent function can be recovered from the logistic one after a linear transformation and mean shift, which is implicitly incorporated within most models. The neural network basis functions are similar to piecewise linear bases in that they are based on linear projections through the data such that each basis

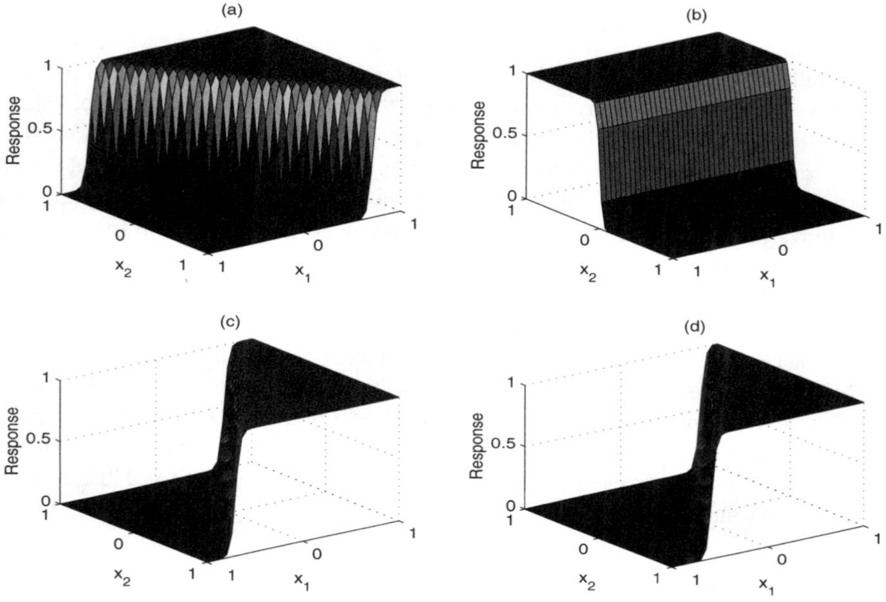

Figure 4.10 Same as Figure 4.9 but with the slope parameter set to 50 rather than 10.

has an orientation and position in the predictor space (compare Figure 4.6 with Figure 4.9). However, the sigmoidal bases have an additional degree of freedom relating to their slope which, unlike the piecewise linear bases, is not linearly dependent on $\boldsymbol{\beta}$. Figure 4.10 illustrates the effect of the slope parameter.

Methodologically, there is a major separation in neural network approaches to the basis function models considered up to this point. In neural networks hierarchies of 'hidden layers' are often constructed using the sigmoidal bases. That is, the original predictor space $\mathcal{X}$ is first mapped onto a k-dimensional space spanned by k basis functions. Then, rather than fit a linear model in this new space, as we have done for the other basis function models, the space is again passed through another set of NN basis functions. This operation can be repeated until finally a linear model is fit to the final layer of transformation. The number of transformations (hidden layers) is an unknown parameter that must be estimated, though typically just one or two layers are adopted.

The Bayesian treatment of neural networks was pioneered by Buntine and Weigend (1991), MacKay (1992a,b) and Neal (1996), and reviewed in Bishop (1995), Ripley (1996), MacKay (1995), Thodberg (1995) and Lampinen and Vehtari (2001). It is typical to adopt normal priors on the basis coefficients $\boldsymbol{t}$, which are known as 'input weights' in the neural network literature. Then normal priors are put on the regression coefficients $\boldsymbol{\beta}$, commonly referred to as 'output weights'.

Most of the literature on Bayesian neural networks considers models with a fixed number of basis functions. Having specified the priors, common forms of inference for these models are the evidence framework discussed in MacKay (1992a,b) and a hybrid Markov chain Monte Carlo scheme reported in Neal (1996). The evidence approach is essentially a Laplace approximation to the posterior distribution of $\{\boldsymbol{t}, \boldsymbol{\beta}\}$, so $p(\boldsymbol{t}, \boldsymbol{\beta} \mid \mathcal{D})$ is assumed normal. The hybrid MCMC method is based on the algorithm of Duane *et al.* (1987) which uses gradient information in the update proposals of the MCMC sampler. The use of gradient information allows large updates to be proposed with a reasonable acceptance rate. This is extremely useful as the posterior distribution of $\{\boldsymbol{t}, \boldsymbol{\beta}\}$ invariably has extreme correlation that can make standard MCMC approaches very slow to converge. Husmeier *et al.* (1999) and Penny and Roberts (1999) compare and contrast the two methods concluding that the MCMC sampling scheme appeared to provide better performance in terms of error rates on out-of-sample data. The extension to include a prior distribution on the dimension of the model was considered by Müller and Rios Insua (1998a,b), who utilise a reversible jump MCMC algorithm, similarly to the approach highlighted in this chapter. For the *BIRTH* step they propose new basis parameters from the prior and they also marginalise over $\boldsymbol{\beta}$ to find the Bayes factor in favour of the proposed model. They also suggest including fixed linear terms in the regression by specifying

$$g(\boldsymbol{x}) = \gamma_0 + \sum_{\ell=1}^{p} \gamma_\ell x_\ell + \sum_{i=1}^{k} \beta_i B_i(\boldsymbol{x}),$$

so that the linear combination of basis functions are only used to capture departures from linearity of the underlying regression function.

In the BMARS and piecewise linear models the use of a variable selection procedure was implicitly encoded within the model by using a prior on the number of terms included in any one basis function. Most Bayesian implementations of neural network models instead rely on the use of *automatic relevance determination* (ARD), a phrased coined by Neal (1996). In the ARD framework the basis parameters, $\boldsymbol{t}$, are grouped according to which predictor variable they relate to, so t_{iw} relates to predictor X_w in each of the $i = 1, \ldots, k$ basis functions. Each group of basis parameters are assumed exchangeable from a common prior, typically taken to be normal so that $p(t_{iw} \mid v_w) = N(0, v_w)$, where v_w refers to the prior variance of basis parameters related to the wth predictor. Each of the $v_1, \ldots, v_p$ variances are given independent inverse-gamma hyperpriors. Neal (1996) shows that for irrelevant predictors the corresponding posterior distribution of v_w tends to be tightly distributed around zero, while for important predictors the variance is found to be much higher. In this way, less important predictors are 'shrunk' out of the model by having priors on the corresponding t_{iw}s that are tight around zero. This has the tendency to align the sigmoids to be perpendicular to the axis of irrelevant predictors, e.g. Figure 4.9(b). The ARD framework can be considered as a 'soft' approach as no variable is ever entirely excluded, however, ARD can also be made to completely exclude variables by making inference with the posterior mode rather than mean.

4.5 Time Series Analysis

So far in this book we have assumed that each response, y_i, is independent of all the other responses given its vector of associated predictor variables, $\boldsymbol{x}_i$, and the model. However, this is not the case in some important applications, most notably time series analysis (see, for example, Box and Jenkins 1970; Brockwell and Davis 1991; Chatfield 1996; Hamilton 1994; Priestley 1988). When we have observations over time we expect some dependence between points measured close in time. For instance, if we measured the wind speed at 12 p.m. on a certain day in a specific location, and then measured it again one minute later at 12.01 p.m., we would expect the two results, or responses, to be similar, i.e. we would expect the responses to be highly correlated. If we continued with this process for an hour, we would have a time series of 61 responses (wind speeds) which are mutually *dependent*. However, we should point out that some time series can indeed be modelled as independent data. If, instead of measuring the wind speed once per minute, we obtained a time series of 61 responses by measuring the wind speed every year, we would expect no significant correlation between the responses. In fact, the first dataset we looked at, the Nile discharge data (Section 3.2.1), is a time series which we modelled as independent data.

We shall be concerned with time series that are sequences of responses collected over time with no associated predictors. Further, we shall assume, as is nearly always the case, that the sampling frequency is constant (e.g. one observation per day/hour/minute/second) so that we can write down the complete dataset simply as just

$$\mathcal{D} = (y_1, \dots, y_n).$$

This section is only concerned with the times when the responses are correlated, as if this is not the case we can use the models outlined previously. We must introduce some way of modelling the dependence between these responses. The simplest model is the autoregressive one which assumes that

$$y_t = \beta_0 + \sum_{i=1}^{P} \beta_i y_{t-i} + \epsilon_t, \tag{4.9}$$

for $t = P + 1, \dots, n$ and where the ϵ_t are independent and identically distributed noise terms, commonly taken to be $N(0, \sigma^2)$. This model is often written as AR(P), where $P (\geqslant 0)$ is known as the order of the autoregression. We see straight away that this has the same form as the linear models seen previously and introduced in (2.9). However, now the lagged responses $y_{t-1}, y_{t-2}, \dots$ are used as the predictor variables to model the response y_t.

Even though (4.9) looks like it has the usual linear model formulation, inference cannot take place immediately because of the correlation between the responses, which now also become predictors. However, we see that the likelihood of the data can be written as

$$\begin{aligned} p(\mathcal{D} \mid \boldsymbol{\phi}) &= p(y_{P+1}, \ldots, y_n \mid \boldsymbol{\phi}) \\ &= p(y_n \mid Y_{n-1}, \boldsymbol{\phi}) p(y_{n-1} \mid Y_{n-2}, \boldsymbol{\phi}) \cdots p(y_{P+1} \mid Y_P, \boldsymbol{\phi}), \end{aligned} \tag{4.10}$$

where $Y_t = (y_t, \ldots, y_1)$ represents all the information up to and including the tth time point and $\boldsymbol{\phi} = (\boldsymbol{\beta}, \sigma^2)$. Further, we know that

$$p(y_t \mid Y_{t-1}, \boldsymbol{\phi}) \sim N\left(\beta_0 + \sum_{i=1}^{P} \beta_i y_{t-i}, \sigma^2\right),$$

for $t = P + 1, \ldots, n$. This allows us to write down the likelihood as simply a product of $n - p$ dependent normal distributions, so we find that

$$p(\mathcal{D} \mid \boldsymbol{\phi}) = (2\pi\sigma^2)^{-(n-P)/2} \exp\left\{-\frac{(Y - \boldsymbol{B\beta})'(Y - \boldsymbol{B\beta})}{2\sigma^2}\right\},$$

where $Y = (y_{P+1}, \ldots, y_n)'$ and

$$\boldsymbol{B} = \begin{pmatrix} 1 & y_P & \cdots & y_1 \\ 1 & y_{P+1} & \cdots & y_2 \\ \vdots & \vdots & \ddots & \vdots \\ 1 & y_{n-1} & \cdots & y_{n-P} \end{pmatrix}.$$

This is essentially the same likelihood as in (2.14), where the predictor variables have been replaced by the responses. The similarities also mean that the analytic results in Chapter 2 can be used to integrate out $\boldsymbol{\phi} = (\boldsymbol{\beta}, \sigma^2)$ when a conjugate prior for $\boldsymbol{\phi}$ is adopted.

Inference for the parameters in the autoregressive model of (4.9) is now straightforward as we can think of it as a standard linear model and use the appropriate results in Chapter 2. However, this is all dependent on choosing a suitable value for P. This is not easy to do, *a priori*, so a straightforward extension of the model is to allow for an unknown order P. We can do this by assigning a prior to the model order (Godsill and Rayner 1998). Suppose that we allow positive prior weight on values of P between 0 and $P_{\max}(< n)$, inclusive. We note from (4.10) that the joint density for the responses is of dimension $n - P$. This means that the likelihoods for models with different orders use different sized data samples. To determine Bayes' factors and other potentially interesting quantities, we need to determine the posterior probability of the model order based on a single fixed dataset. One very easy way to ensure this in the time series context is to only consider the likelihood of the responses at times $P_{\max} + 1$ to n: that is, we discard the first $P_{\max}$ points. However, a fully Bayesian alternative to this approach would be to use the complete dataset and sample the unobserved responses $y_0, y_{-1}, \ldots, y_{P_{\max}-1}$ as if they were missing data. Whatever method the choice of maximum model order, $P_{\max}$, is important.

Here we shall concentrate on the easier method which uses the first $P_{\max}$ points in the time series only as predictors. Adopting this method we can easily evaluate the posterior distribution of the model order given by

$$p(P = i \mid \mathcal{D}) = \frac{p(\mathcal{D} \mid P = i)p(P = i)}{\sum_{j=0}^{P_{\max}} p(\mathcal{D} \mid P = j)p(P = j)},$$

for $i = 0, 1, \ldots, P_{\max}$, where the marginal likelihoods here can be determined with reference to (2.23).

The autoregressive model is well used in time series analysis and fitting such models is often included in standard statistical packages. However, it only allows the lagged responses in the past information, Y_{t-1}, to affect the response under consideration, y_t, in a linear fashion. We can generalise this simply by letting the lagged responses enter the model as predictors did in the surface fitting approaches seen earlier in this chapter, i.e. we take

$$y_t = g(y_{t-P_{\max}}, \ldots, y_{t-1}) + \epsilon_t, \tag{4.11}$$

for $t = P_{\max} + 1, \ldots, n$ and where g is some linear combination of coefficients and basis function. We can replace g with any of the models already highlighted in this chapter to estimate the true regression function f. Which one to choose will often depend on the application but, for now, we shall just focus on using the Bayesian MARS model for analysing nonlinear time series.

4.5.1 The BAYSTAR model

The adaptive spline threshold autoregressive (ASTAR) method proposed by Lewis *et al.* (1991) (also called the TSMARS model in Lewis and Ray (1997)) estimates the unknown regression function, f, in (4.11) with a standard MARS model (Friedman 1991). Instead, the Bayesian equivalent of ASTAR, which we shall call the BAYSTAR model, assumes that f is a Bayesian MARS model (Denison and Mallick 1998).

When analysing time series, threshold models like the ASTAR/BAYSTAR ones, are desirable as can naturally accommodate the results of changing physical behaviour (e.g. once x_1 is over c its effect remains constant). Other threshold models are well developed in the literature (see, for example, the threshold autoregressive models of Tong (1983)). However, these models allow discontinuities in the regression function at the threshold values. In some analyses, where switching between states is likely, this may be a reasonable assumption but, in general, the continuous models based on multivariate splines appear more appropriate.

Lewis *et al.* (1991) provide more motivation for adaptive spline models for time series analysis, as well as demonstrating the capabilities of their ASTAR method in identifying simple linear and nonlinear time series models. We take their work as our starting point and here we shall just demonstrate some of the features of the BAYSTAR method.

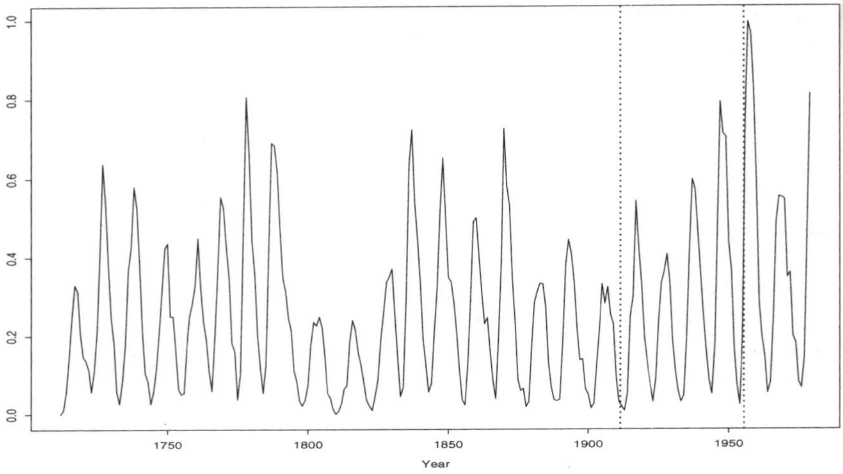

Figure 4.11 Wolf's yearly sunspots dataset. The dotted lines separate the training set and the two test sets.

4.5.2 Example: Wolf's sunspots data

For illustration we present an analysis of Wolf's sunspots dataset using the Bayesian ASTAR model and compare it to the analysis in Lewis *et al.* (1991). The dataset consists of 280 yearly (1700–1979) averages of sunspots (dark patches on the surface of the Sun). Some early analysis of the work was carried out by Yule (1927) and a history of the dataset is given in Izenman (1983). It is generally accepted that the data cannot be well described by a linear time series model.

Following Weigend *et al.* (1992) we split the dataset into three parts: the training set of 221 points (1700–1920); the first test set of the next 35 points (1921–1955) and the second test of the last 24 points (1956–1979). The dataset is shown in Figure 4.11. The periodicity of the data is about 11 years (the solar cycle) but the interesting feature is that the response reaches the peaks quicker than it descends to the troughs. This nonstationary behaviour suggests that a nonlinear model is more appropriate for this particular dataset.

When we ran both the ASTAR and BAYSTAR models with $P_{\text{max}} = 12$ we obtained the results displayed in Table 4.2 given in terms of the mean-squared error in each split of the data. The MSE is given by the average squared error between the mean prediction and the observed response amongst some set of test data so, in general, this is given by

$$\text{MSE} = \frac{1}{n}\sum_{i=1}^{n}\{y_i - E(y_i \mid \boldsymbol{x}_i, \mathcal{D})\}^2.$$

Table 4.2 Comparison of the ASTAR and BAYSTAR models in terms of the mean squared error $\times 10^3$ (MSE).

Model	No. of basis	MSE train	MSE 1st test	MSE 2nd test
ASTAR	18	2.91	3.74	19.00
BAYSTAR-PM	7.6	3.23	4.30	16.75
BAYSTAR-ML	8	3.75	5.40	15.92

Table 4.3 Largest log marginal likelihood estimate to sunspots data.

Basis function	Coefficient
1	0.429
$(0.434 - x_{t-1})+$	−1.192
$(0.183 - x_{t-2})+$	1.174
$(x_{t-1} - 0.000)_+ \times (0.338 - x_{t-3})+$	1.514
$(x_{t-1} - 0.109)_+ \times (0.250 - x_{t-8})+$	−1.314
$(x_{t-1} - 0.225)_+ \times (x_{t-11} - 0.470)_+$	1.085
$(0.230 - x_{t-2})_+ \times (0.366 - x_{t-5})+$	−0.831
$(x_{t-7} - 0.531)_+ \times (x_{t-8} - 0.314)_+$	−0.021

The two BAYSTAR models displayed are the BAYSTAR-PM model which is the posterior mean over all the models in the generated sample, while the BAYSTAR-ML one is the model in the sample with the highest marginal likelihood (we discuss choosing models in this way in greater detail in Section 6.5.2).

From Table 4.2 we see that the classical ASTAR estimate suggests using many basis functions while the Bayesian methods are far more parsimonious. Although the predictions are not overly different this suggests that the one-step ahead search for basis functions adopted by ASTAR can sometimes fail to find parsimonious representations to the true regression function.

In Table 4.3 we give the complete model found with the largest marginal likelihood. If a single model is required, then we believe that this is the one that should be chosen. It is also easier to interpret the properties of this model due to its parsimony.

As we have already mentioned, the number of sunspots increases quicker than it decreases. As we would expect, this is reflected by the model given in Table 4.3. For illustration we shall use the data points at the years 1786 and 1790 which will represent typical data points when the number of sunspots is increasing and decreasing, respectively. Note that these two points have similar responses values (0.4336 in 1786 and 0.4702 in 1790) but very different lagged responses. In Figure 4.12 we display the effect of the first lagged response, y_{t-1}, on the prediction of the number of sunspots

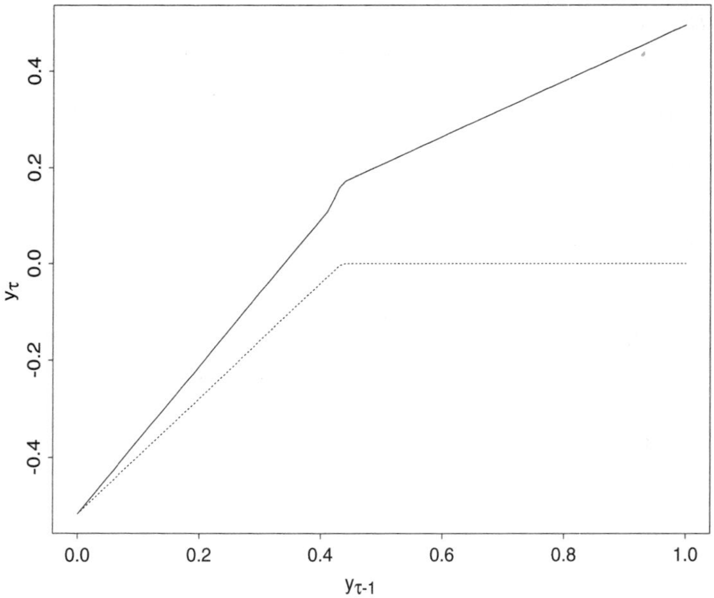

Figure 4.12 Comparison of effect of y_{t-1} on response y_t for a typical point when the number of sunspots is rising (year 1786, solid line) and when it is falling (year 1790, dotted line).

in the next year, y_t, using the predictor values $y_{t-2}, \ldots, y_{t-12}$ from the years 1786 and 1790. The main difference between the predictions at 1786 and 1790 is that if the first lagged response in both years was identical, the response would be increased by more in 1786 than in 1790. This demonstrates the nonstationary nature of the model and how it can capture observed features present in the data.

4.5.3 *Chaotic Time Series*

We can apply the methods adopted in the last subsection to a special class of time series models known as *chaotic time series* (see, for example, Casdagli 1989; Falmer and Sidorowich 1987). These time series, although they can appear under linear assumptions to be random, are in fact completely deterministic. However, the relationship between time points can be difficult to elicit. A famous example of such a series is the Mackey–Glass time series, developed as an equation to model blood cell generation in leukaemia patients (Mackey and Glass 1977). The model suggested gives rise to the responses in the time series being given by

$$y_{t+1} = 0.9y_t + \frac{y_{t-\Delta}}{5(1 + y_{t-\Delta}^{10})},$$

where Δ is a delay parameter. For the starting values of the series $x_1 = \cdots = x_\Delta = 0.9$. Now, to obtain a chaotic time series we can generate values from this deterministic

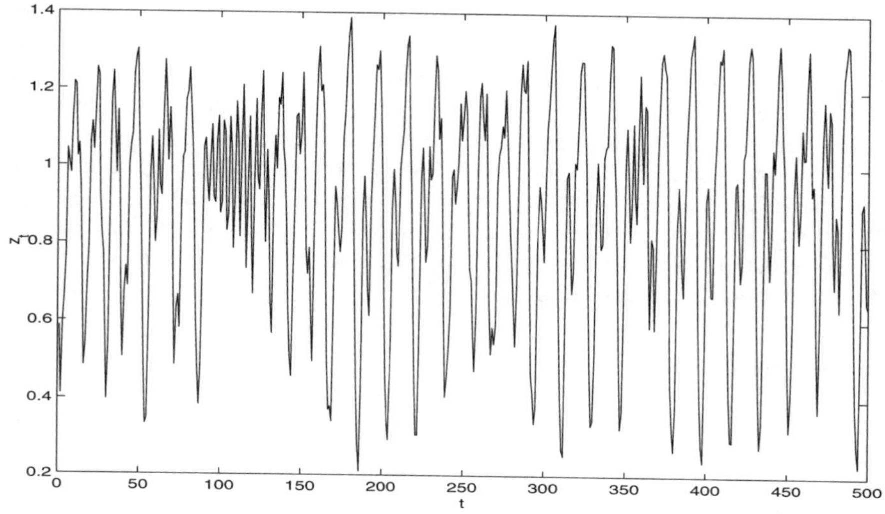

Figure 4.13 An portion of the Mackey–Glass chaotic time series with delay parameter, $\Delta = 30$.

sequence and take every sixth point to be in the final dataset we shall analyse. Hence, these data are given by

$$y_1, y_7, \ldots, y_{6n-5},$$

which, for notational convenience, we shall refer to as the time series with points $z_1, \ldots, z_n$. We plot such a series with the delay parameter taken as $\Delta = 30$ in Figure 4.13.

Although the true sequence is deterministic, we are unlikely to know this if we are just presented with the time series, itself, and need to analyse it. Hence, we model it in the same way as a time series that really is observed with added noise, i.e. using (4.11) but with the y_ts replaced by z_ts. Again we could choose to use any flexible function to estimate f, but Holmes and Mallick (2000a) suggest the class of radial wavelet networks known as the 'Mexican hat' bases (Zhang 1997). These are given by

$$B_j(z) = \left(p - \frac{\|z - \boldsymbol{\mu}_j\|^2}{v_j^2}\right) \exp\left(-\frac{\|z - \boldsymbol{\mu}_j\|^2}{2v_j^2}\right),$$

where p is again the input dimension (i.e. the number of lagged responses used as predictors) and $\boldsymbol{\mu}_j$ and v_j are unknown location and scale parameters associated with B_j. The presence of such location and scale parameters is why such basis functions are known as wavelets, as they are just translations and dilations of some mother wavelet. We also find that, similarly to wavelet bases, $\int B_j(z)\,\mathrm{d}z = 0$. Note that the basis set Holmes and Mallick (2000a) used also included linear terms for each of the lagged response.

The advantage of using continuous wavelets is that they can easily approximate functions when the predictors in the dataset are not given at regular intervals, unlike the discrete wavelet bases we encountered earlier (3.4). However, compared to the more usual wavelets seen earlier, they do suffer from computational disadvantages as fast algorithms do not exist to determine the wavelet coefficients and $\boldsymbol{B}'\boldsymbol{B} \neq \boldsymbol{I}$.

A sampler can be devised (Holmes and Mallick 2000a) to find the number and form of the basis functions. In this way we approximate the posterior $p(\boldsymbol{\mu}_1, v_1, \ldots, \boldsymbol{\mu}_k, v_k \mid \mathcal{D})$ as the coefficients and error variance can again be integrated out. Holmes and Mallick (2000a) found that this method provides good predictive results when compared to a variety of natural competitors.

4.6 Further Reading

Surprisingly, it appears that most of the work on Bayesian nonlinear regression methods has concentrated on the scatterplot smoothing problem. This is a more straightforward problem than that of determining a regression surface, and is much less used in practice. Further, most of the work that does involve nonlinear methods for surface fitting has concentrated on low-dimensional examples, generalising one-dimensional basis functions to work for two or three-dimensional datasets. The earliest examples of such work are the papers by Barry (1983, 1986), who decomposed the regression function as in two-way analysis of variance. Other examples given in this chapter are the work of Smith and Kohn (1997), Holmes and Mallick (1998) and Kohn *et al.* (2001). A good review paper on fitting Bayesian regression models for radial basis functions, but equally applicable to other bases, is given by Andrieu *et al.* (2001a).

Bayesian attempts to tackle high-dimensional regression problems appear to have been mainly tackled by the neural network community. It should be noted that the logistic basis functions used in neural networks lack the interpretability of the BMARS model of Section 4.4.3 and the piecewise linear models in Section 4.4.4. However, the neural network basis functions are smooth and furthermore they asymptote rapidly to 0 or 1 outside of the range of the data. This last point may offer some explanation for the reported excellent performance of neural networks in very high-dimensional smoothing problems as the models are very well behaved when extrapolating outside the convex hull of the data, a situation that we are often in when making predictions with p large. Although we choose to discuss neural networks within the framework of this regression chapter, it should also be noted that many applications of the model are instead concerned with classification of data where y is a class index, for example, see Vehtari and Lampinen (2000) and Vivarelli and Williams (2001).

4.7 Problems

4.1 Consider the switching straight-line regression model that can be written as

$$g(t) = \begin{cases} \beta_{01} + \beta_{11}t, & t \leqslant \theta, \\ \beta_{02} + \beta_{12}t, & t > \theta. \end{cases}$$

Suppose that we make observations, Y_t, which can be assumed to come from the model $Y_t = g(t) + \epsilon_t$, where the ϵ_t are independent and identically distributed normal error variates with unknown variance σ^2.

(i) What constraint do you need to ensure continuity of the estimated regression function, $g(t)$?

(ii) Using this model and the constraint found in (i), write down the likelihood function $p(\mathcal{D} \mid \theta)$, where $\mathcal{D} = (Y_1, \ldots, Y_T)$ and $E(Y_t) = g(t)$. Using uniform prior specifications for the unknown coefficients and $p(\sigma^2) \propto \sigma^{-2}$, find the full conditional distributions of β_{01}, β_{02} and β_{12}.

(iii) Suggest a way to obtain the marginal posterior distribution of θ for a general prior specification for it, $p(\theta)$.

4.2 The fitted values to the responses $Y_1, \ldots, Y_n$ at the predictor locations $\boldsymbol{x}_1, \ldots, \boldsymbol{x}_n$ are given by $\hat{Y} = (g(\boldsymbol{x}_1), \ldots, g(\boldsymbol{x}_n))'$. If $\hat{Y} = \boldsymbol{S}Y$ for some matrix $\boldsymbol{S}$, where $\boldsymbol{S}$ does not depend on the responses Y, then $\boldsymbol{S}$ defines a linear smoother. Any linear smoother can be equivalently presented by $\boldsymbol{S}$ rather than the design matrix $\boldsymbol{B}$.

Consider the linear model,

$$Y = \boldsymbol{B}\boldsymbol{\beta} + \boldsymbol{\epsilon},$$

where $\boldsymbol{\epsilon} \sim N(\boldsymbol{0}, \sigma^2 \boldsymbol{I})$, with σ^2 assumed known. If we take $p(\boldsymbol{\beta}) = N(\boldsymbol{0}, \boldsymbol{V})$, where $\boldsymbol{V} = \text{diag}(v_1, \ldots, v_k)$, and $\boldsymbol{B}$ is the basis function matrix for a fixed set of radial basis functions, $r(\boldsymbol{x})$, at the know points $\boldsymbol{t}_1, \ldots, \boldsymbol{t}_K$. Determine the smoothing matrix, $\boldsymbol{S}$.

The canonical linear model is given when $\boldsymbol{Z} = \boldsymbol{B}\boldsymbol{U}$ and $\boldsymbol{\alpha} = \boldsymbol{U}'\boldsymbol{\beta}$, where $\boldsymbol{U}$ is the matrix of eigenvectors of $\boldsymbol{B}'\boldsymbol{B}$. Recall that

$$\boldsymbol{U}'\boldsymbol{B}'\boldsymbol{B}\boldsymbol{U} = \text{diag}(\lambda_1, \ldots, \lambda_k),$$

where λ_i are the eigenvalues of $\boldsymbol{B}'\boldsymbol{B}$, and also that the degrees of freedom of a linear smoother is given by the trace of the smoothing matrix. By first determining the degrees of freedom of the smoother for the canonical linear model, determine the degrees of freedom of the original regression.

4.3 For the Bayesian linear model with known variance σ^2, we know the distribution of the response vector Y given the coefficients $\boldsymbol{\beta}$ is given by

$$p(Y \mid \boldsymbol{\beta}) = N(\boldsymbol{B}\boldsymbol{\beta}, \sigma^2 \boldsymbol{I}).$$

Taking $p(\boldsymbol{\beta}) = N(\boldsymbol{0}, \sigma^2 \boldsymbol{V})$, show that $E(Y \mid \mathcal{D}) = \boldsymbol{S}Y$ and $\text{var}(Y \mid \mathcal{D}) = \sigma^2(\boldsymbol{S} + \boldsymbol{I})$, for some matrix $\boldsymbol{S}$ which you should determine.

Now show that adopting a prior directly on the response vector Y such that $p(Y) = N(\boldsymbol{0}, \sigma^2 \boldsymbol{B}'\boldsymbol{V}^{-1}\boldsymbol{B})$ is equivalent to the prior used in the above model.

4.4 Show that, for the Bayesian linear model with prior for the coefficients and regression variance $p(\boldsymbol{\beta}, \sigma^2) = \text{NIG}(\boldsymbol{m}, \sigma^2 \boldsymbol{V}, 0, 0)$, the posterior expectation of the coefficients is given by

$$\begin{aligned} E(\boldsymbol{\beta} \mid \mathcal{D}) &= \boldsymbol{m} + (\boldsymbol{B}'\boldsymbol{B} + \boldsymbol{V}^{-1})^{-1}\boldsymbol{B}'\boldsymbol{B}(\hat{\boldsymbol{\beta}} - \boldsymbol{m}) \\ &= \hat{\boldsymbol{\beta}} - (\boldsymbol{B}'\boldsymbol{B})^{-1}\{\boldsymbol{V} + (\boldsymbol{B}'\boldsymbol{B})^{-1}\}^{-1}(\hat{\boldsymbol{\beta}} - \boldsymbol{m}), \end{aligned}$$

where $\hat{\boldsymbol{\beta}}$ is the maximum likelihood estimate to $\boldsymbol{\beta}$. You may find it helpful to use the following identity for suitable matrices $\boldsymbol{E}$, $\boldsymbol{F}$, $\boldsymbol{G}$, $\boldsymbol{H}$,

$$(\boldsymbol{E} + \boldsymbol{F}\boldsymbol{G}\boldsymbol{H})^{-1} \equiv \boldsymbol{E}^{-1} - \boldsymbol{E}^{-1}\boldsymbol{F}(\boldsymbol{G}^{-1} + \boldsymbol{H}\boldsymbol{E}^{-1}\boldsymbol{F})^{-1}\boldsymbol{H}\boldsymbol{E}^{-1}.$$

4.5 By considering the Taylor expansion of $\log g(\boldsymbol{\beta})$ around $\hat{\boldsymbol{\beta}}$, the value of $\boldsymbol{\beta}$ for which $g(\boldsymbol{\beta})$ is a maximum, show that

$$g(\boldsymbol{\beta}) \approx g(\hat{\boldsymbol{\beta}}) \exp\{-\tfrac{1}{2}(\boldsymbol{\beta} - \hat{\boldsymbol{\beta}})'\boldsymbol{V}^{-1}(\boldsymbol{\beta} - \hat{\boldsymbol{\beta}})\},$$

where $-\boldsymbol{V}^{-1}$ is the matrix of second derivatives of $\log g(\boldsymbol{\beta})$.

Hence show that

$$\int g(\boldsymbol{\beta})\, \mathrm{d}\boldsymbol{\beta} \approx g(\hat{\boldsymbol{\beta}})(2\pi)^{p/2}|\boldsymbol{V}|^{1/2},$$

where p is the length of $\boldsymbol{\beta}$.

5

Classification Using Generalised Nonlinear Models

5.1 Introduction

Up to now we have concentrated almost exclusively on models for which we can assume that the relationship between the response y and a vector of measurements on the predictors $\boldsymbol{x}$ follows,

$$p(y \mid \boldsymbol{x}, \boldsymbol{\theta}, \boldsymbol{\beta}, \sigma^2) = N\{g(\boldsymbol{x}), \sigma^2\}, \tag{5.1}$$

where $\boldsymbol{\theta}$ is a set of model parameters and $g(\boldsymbol{x})$ is a linear combination of coefficients and basis functions, i.e. $\sum_1^k \beta_i B_i(\boldsymbol{x}; \boldsymbol{\theta})$. We use $g(\boldsymbol{x})$ as a model of the true relationship between the response and predictors given by $f(\boldsymbol{x}) = E(y \mid \boldsymbol{x})$.

In Chapters 3 and 4 we saw how basis function models provide a flexible approach for nonlinear modelling of the mean of some process. Moreover, it was highlighted that the basis function models can be considered as a linear regression with a nonlinear basis function matrix, $\boldsymbol{B}$. However, for (5.1) to be a reasonable model we need the response to be, at the very least, a real-valued quantity. However, this is not always the case and the focus of this chapter is on such problems.

In many situations the model in (5.1) is inappropriate, most notably in the classification problem where the response is a class label from some known number of categories (as in this book we shall only focus on such *supervised* classification problems). Nevertheless, we find that we can extend the normal model to deal with such a situation through the use of generalised linear model (GLM) methodology (McCullagh and Nelder 1989; Nelder and Wedderburn 1972). In standard GLMs (5.1) is generalised so that now

$$p(y \mid \boldsymbol{x}, \boldsymbol{\theta}, \boldsymbol{\beta}) = F\{g(\boldsymbol{x})\},$$

where F is one of a particular class of distributions known as the exponential family and $g(\boldsymbol{x})$ is taken to be a linear function of the predictors. For the reasons outlined in Chapter 1, we now present methods to extend this model to be handle situations where a linear model is not a good approximation of the truth.

First of all in Section 5.2 we review classification, as it is perhaps the most widespread problem in applied statistics, and describe how Bayesian linear models can be applied to it. The main difficulty we find is that we can no longer assign conjugate priors to the model parameters. This has the potential to make the computation of the models extremely difficult, however, through the judicious use of auxiliary variables we find that efficient sampling methods for these parameters exist. We go on to describe the nonlinear extensions to this model and in Section 5.3 we demonstrate nonlinear classification using the Bayesian MARS model.

Section 5.4 describes how a broadly similar auxiliary variable strategy can be used to aid posterior inference in situations where the data arise as counts, rather than as class labels in the classification context.

We bring together the work of the chapter in Section 5.5, which describes the general framework for statistical modelling when the responses follow a distribution in the exponential family. Finally, Section 5.6 contains helpful references for further reading on Bayesian generalised linear modelling.

5.2 Nonlinear Models for Classification

5.2.1 Classification

As we have already mentioned, classification is a hugely important problem that is used in a wide variety of fields: sometimes it is also known as statistical pattern recognition or discrimination. It involves situations where the response variable labels the observation as belonging to one of C classes, or categories. Typically, there is no natural ordering to these classes yet they are still labelled $1, 2, \ldots, C$. For now we shall concentrate solely on binary, or two-class, classification, so $C = 2$. In this special case the class labels are more commonly taken to be 0 and 1 (rather than 1 and 2) and we shall stick to this convention. Later on we shall show how to cope with multiclass problems for which $C > 2$.

The response of a general data point, $y \in \{0, 1\}$, is influenced by its associated predictor values, $\boldsymbol{x} = (x_1, \ldots, x_p)$. However, we are often unclear about the form of the relationship between predictors and response so this must be determined from the data. Probably the most common approach to classification assumes that, for a general data point $(y, \boldsymbol{x})$, the sampling distribution that generated the response is given by

$$\begin{aligned} p(y \mid \boldsymbol{\beta}) &= \mathrm{Br}\{\mu(\boldsymbol{\beta})\} \\ &= \{\mu(\boldsymbol{\beta})\}^y\{1 - \mu(\boldsymbol{\beta})\}^{1-y}, \qquad y = 0, 1, \end{aligned} \tag{5.2}$$

where

$$\mu(\boldsymbol{\beta}) = h\left(\beta_0 + \sum_{j=1}^{p} \beta_j x_j\right). \tag{5.3}$$

Here h is an invertible function that maps the real line onto the unit interval, $[0, 1]$. This must be the case to ensure that the parameter of the Bernoulli distribution, $\mu(\boldsymbol{\beta})$, lies in the required range. Common choices for h which satisfy these conditions include

$$h(\eta) = \frac{\exp(\eta)}{1 + \exp(\eta)}, \tag{5.4}$$

$$h(\eta) = \Phi(\eta), \tag{5.5}$$
$$h(\eta) = \exp\{-\exp(-\eta)\}, \tag{5.6}$$
$$h(\eta) = 1 - \exp\{-\exp(\eta)\}, \tag{5.7}$$

where Φ is the cumulative distribution function for a standard normal random variable given by

$$\Phi(\eta) = \int_{-\infty}^{\eta} \frac{1}{\sqrt{2\pi}} \exp(-x^2/2)\, dx.$$

The choices of h given in (5.4)–(5.7) give rise to the logistic, probit, log–log and complementary log–log models, respectively.

We have written h as a function of the *linear predictor*, η, given by

$$\eta = \beta_0 + \sum_{j=1}^{p} \beta_j x_j. \tag{5.8}$$

This follows standard notational practice (see, for example, McCullagh and Nelder 1989). The choice of h is a matter of personal taste, though (5.4) and (5.5) are perhaps the most common. We saw in Section 2.4 how normal priors on the coefficients $\boldsymbol{\beta}$ can lead to attractive forms for the posterior distribution of them. This is because these normal priors are conjugate to the likelihood. However, for classification, we find from (5.2) and (5.3) that

$$\begin{aligned} p(\boldsymbol{\beta} \mid \mathcal{D}) &\propto p(\mathcal{D} \mid \boldsymbol{\beta}) p(\boldsymbol{\beta}) \\ &= \left[\prod_{i=1}^{n} \{h(\eta_i)\}^{\sum y_i} \{1 - h(\eta_i)\}^{n - \sum y_i} \right] p(\boldsymbol{\beta}), \end{aligned} \tag{5.9}$$

where η_i is the value of the linear predictor for the ith data point.

Unfortunately, no matter which form for h we take, there is no way for us to assign a prior $p(\boldsymbol{\beta})$ such that the posterior is of the same form, i.e. there is no conjugate prior for this model. This can lead to computational problems, especially when we wish to consider the dimension of the model, in terms of the number of relevant predictors, as unknown. Previously, we have relied heavily on the use of conjugate priors that lead to the posterior distribution, $p(\boldsymbol{\beta} \mid \mathcal{D})$ being of known analytic form. This allowed for a closed form expression for the marginal likelihood which, in turn, made for efficient sampling algorithms.

The lack of a conjugate prior is not computationally significant if the predictor space is considered fixed. In this situation we can rely on standard sampling approaches to draw $p(\boldsymbol{\beta} \mid \mathcal{D})$, such as those described in Dellaportas and Smith (1993). The problem arises when we wish to treat the number of predictors as unknown. Adding or removing a predictor typically causes a large negative change to the log-likelihood of the model, unless accompanied by careful updates to the $\boldsymbol{\beta}$ vector.

5.2.2 *Auxiliary variables method for classification*

In this section we describe a method that allows us to propose 'good' update to the coefficients $\boldsymbol{\beta}$, conditional on a change to the predictor set. Although this is not the only method of sampling the coefficients we find it a particularly attractive one so describe it in some detail.

Let the random variable z be given by

$$z = \eta + \epsilon, \tag{5.10}$$

where η is the linear predictor in (5.8) and ϵ is a standard (i.e. mean zero, unit variance) normal random variable. If we assume that

$$p(y = 1 \mid z, \boldsymbol{\beta}) = \begin{cases} 1, & \text{if } z > 0, \\ 0, & \text{otherwise}, \end{cases} \tag{5.11}$$

and $p(y = 0 \mid z, \boldsymbol{\beta}) = 1 - p(y = 1 \mid z, \boldsymbol{\beta})$. With this set-up we find that the marginal distribution of the response is given by

$$\begin{aligned} p(y = 1 \mid \boldsymbol{\beta}) & \\ &= p(y = 1 \mid z > 0, \boldsymbol{\beta}) p(z > 0 \mid \boldsymbol{\beta}) + p(y = 1 \mid z < 0, \boldsymbol{\beta}) p(z < 0 \mid \boldsymbol{\beta}) \\ &= p(z > 0 \mid \boldsymbol{\beta}) = p(\epsilon > \eta \mid \boldsymbol{\beta}) = \Phi(\eta), \end{aligned}$$

and $p(y = 0 \mid \boldsymbol{\beta}) = 1 - \Phi(\eta)$. Hence the distribution of the response, conditional on the coefficients, is Bernoulli with probability of success $\Phi(\eta)$. We find that this is the same distribution of the response given the coefficients that we would find if we were to take h as in (5.5), which is known as the probit link function. So, including the auxiliary variable z, makes no difference to the original model when we marginalise over this parameter.

If this was all that could be said for this procedure, then we have gained little as we have introduced n additional variables into our model, one for each observation. The importance of the auxiliary variable is that, given z, it is easy to update the coefficient $\boldsymbol{\beta}$. Look again at the form of (5.10) with η given by (5.8). This is just a linear regression in $\boldsymbol{x}$ with known error variance. In a sense, by including the auxiliary variables, we have turned the classification problem into a regression one for which we can use the methods developed earlier. The additional step for classification is that z is itself random and must be updated as part of the overall modelling process.

In detail, consider the full dataset $\mathcal{D} = \{y_i, \boldsymbol{x}_i\}_1^n$. Suppose we place an $N(\boldsymbol{m}, \boldsymbol{V})$ prior on $\boldsymbol{\beta}$ and have a set of auxiliary variables $(z_1, \ldots, z_n) = \boldsymbol{z}$, we now describe how to draw $\boldsymbol{z}$. To update the coefficients conditional on $\boldsymbol{z}$ we note

$$p(\boldsymbol{\beta} \mid z) \propto p(z \mid \boldsymbol{\beta}) p(\boldsymbol{\beta}).$$

However, we see that this is the same as the posterior distribution of the coefficients in a normal linear model where the regression variance is known and the responses are given by the z_i as

$$z_i = \eta_i + \epsilon_i = \beta_0 + \sum_{j=1}^{p} \beta_j x_{ij} + \epsilon_i.$$

Hence, $p(\boldsymbol{\beta} \mid \boldsymbol{z}) = N(\boldsymbol{m}^*, \boldsymbol{V}^*)$, where

$$\boldsymbol{m}^* = (\boldsymbol{B}'\boldsymbol{B} + \boldsymbol{V}^{-1})^{-1}(\boldsymbol{V}^{-1}\boldsymbol{m} + \boldsymbol{B}'\boldsymbol{z}), \tag{5.12}$$

$$\boldsymbol{V}^* = (\boldsymbol{B}'\boldsymbol{B} + \boldsymbol{V}^{-1})^{-1}. \tag{5.13}$$

Also, we find that the marginal likelihood of the data for known error variance is given by

$$p(\mathcal{D} \mid \boldsymbol{z}) \propto \frac{|\boldsymbol{V}^*|^{1/2} \exp(-b^*)}{|\boldsymbol{V}|^{1/2}(2\pi)^{n/2}},$$

where

$$b^* = (\boldsymbol{z}'\boldsymbol{z} + \boldsymbol{m}'\boldsymbol{V}^{-1}\boldsymbol{m} - (\boldsymbol{m}^*)'(\boldsymbol{V}^*)^{-1}\boldsymbol{m}^*)/2.$$

As in the earlier chapters we assume that $\boldsymbol{V} = v\boldsymbol{I}$ and assign a hierarchical prior to v^{-1}, taken to be a gamma distribution with parameters δ_1 and δ_2. Thus, in reality, only $p(\boldsymbol{\beta} \mid \boldsymbol{z}, v)$ is $N(\boldsymbol{m}^*, \boldsymbol{V}^*)$ and we can update v similarly to (3.8) using the fact that

$$p(v^{-1} \mid \boldsymbol{\beta}) = \text{Ga}\left(\delta_1 + \tfrac{1}{2}k, \delta_2 + \frac{1}{2}\sum_{1}^{k} \beta_i^2\right). \tag{5.14}$$

We now know how to sample $\boldsymbol{\beta}$ given $\boldsymbol{z}$, as well as v given $\boldsymbol{\beta}$, for this classification model. However, we now need to marginalise over $\boldsymbol{z}$, which can be accomplished by sampling $\boldsymbol{z}|\boldsymbol{\beta}$. The distribution of $\boldsymbol{z}$ is found by first noting that Bayes' Theorem gives us

$$p(z_i \mid y_i = 1, \boldsymbol{\beta}) \propto p(y_i = 1 \mid z_i, \boldsymbol{\beta})p(z_i \mid \boldsymbol{\beta}),$$

but as we know that for $y_i = 1$, $z_i > 0$ (as the likelihood is zero otherwise) and $p(z_i \mid \boldsymbol{\beta}) = N(\eta_i, 1)$, we see that

$$p(z_i \mid y_i = 1, \boldsymbol{\beta}) \propto \begin{cases} N(\eta_i, 1), & z_i > 0, \\ 0, & \text{otherwise.} \end{cases}$$

Hence, z_i is from a truncated normal distribution. Similarly, we find that

$$p(z_i \mid y_i = 0, \boldsymbol{\beta}) \propto \begin{cases} N(\eta_i, 1), & z_i < 0, \\ 0, & \text{otherwise.} \end{cases}$$

So, generating the z_is given the data involves simulating from truncated normal distributions with mean η_i. Although this can simply be done by generating from the underlying normal and then rejecting the simulated values in the wrong half of the real line this is not particularly efficient. Better algorithms exist, such as the one given in Robert (1995c), and this approach is usually preferable.

So far we have assumed that the terms in η are linear but we can extend this to nonlinear modelling by taking

$$\eta(\boldsymbol{x}) = \beta_0 + \sum_{i=1}^{k} \beta_i B_i(\boldsymbol{x}) = \boldsymbol{B}\boldsymbol{\beta},$$

where the basis functions can be determined from the data and are parametrised by some vector $\boldsymbol{\theta}$, which is possibly of varying dimension. For example, $\boldsymbol{\theta}$ could represent any of the models introduced in the previous chapter, such as the MARS or piecewise linear ones.

We now give the sampling algorithm for this classification model. Recall that the particular form of the proposal subroutines depends on which type of basis function we intend to use.

RJ-sampler:
```
Set θ^(0) = ();                              /*no basis functions*/
Set v = δ_2/δ_1;
Set t = 0;
REPEAT
  Draw z = (z_1, ..., z_n) from appropriate
    truncated normal;
  Draw β^(t) from N(m*, V*);
  Draw v^-1 using (5.14);
  Set u_1 to a draw from a U(0, 1) distribution;
  IF u_1 ≤ b_k
    θ' = Birth-proposal(θ^(t));
  ELSE IF b_k < u_1 ≤ b_k + d_k
    θ' = Death-proposal(θ^(t));
  ELSE
    θ' = Move-proposal(θ^(t));
  ENDIF;
  Set u_2 to a draw from a U(0, 1) distribution;
  IF u_2 < min{1, BF(θ', θ^(t)) × R}          /*acceptance?*/
    θ^(t+1) = θ';
  ELSE
    θ^(t+1) = θ^(t);
  ENDIF;
  t = t + 1;
  Store every mth value of θ^(t) after
    initial burn-in;
END REPEAT;
```

Note that for this model the regression variance is effectively known to be one, given the auxiliary variables $\boldsymbol{z}$. Thus the Bayes factor in favour of model one over model two, with priors $p(\boldsymbol{\beta}_i) = p(\boldsymbol{m}_i, \boldsymbol{V}_i)$ and vectors of parameters $\boldsymbol{\theta}_i$ $(i = 0, 1)$,

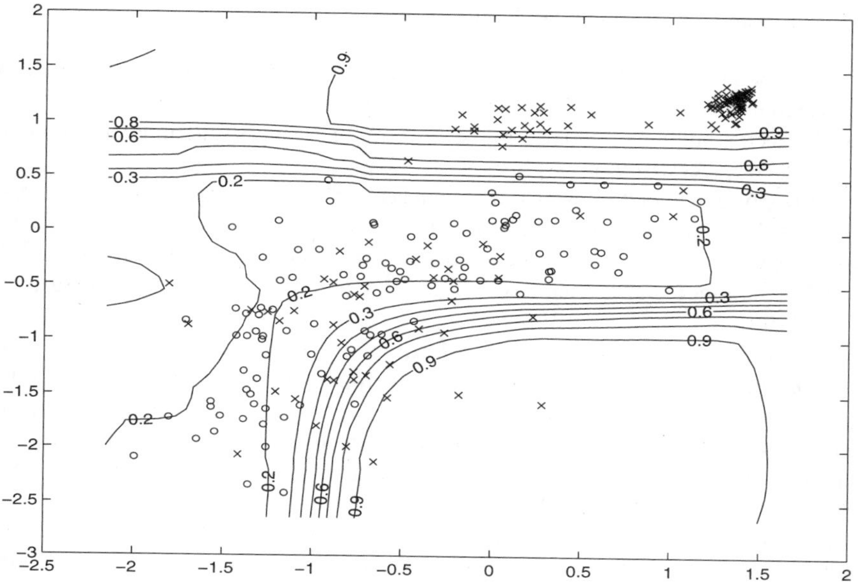

Figure 5.1 Arm tremor data set with class probability contours showing $p(y_i = 1)$ under the Bayesian MARS model. The crosses represent those patients which had Parkinson's disease and the dots are the others.

is given by

$$\mathrm{BF}(\boldsymbol{\theta}_1, \boldsymbol{\theta}_2) = \frac{|\boldsymbol{V}_2|^{1/2}|\boldsymbol{V}_1^*|^{1/2}}{|\boldsymbol{V}_1|^{1/2}|\boldsymbol{V}_2^*|^{1/2}} \exp(b_2^* - b_1^*),$$

where

$$b_i^* = (z'z + \boldsymbol{m}_i'\boldsymbol{V}^{-1}\boldsymbol{m}_i - (\boldsymbol{m}_i^*)'(\boldsymbol{V}_i^*)^{-1}\boldsymbol{m}_i^*)/2,$$

for $i = 1, 2$ and where the $\boldsymbol{m}_i^*$ and $\boldsymbol{V}_i^*$ are calculated with reference to (5.12) and (5.13).

To make predictions with this method we take the generated sample from $p(\boldsymbol{\theta} \mid \mathcal{D})$, which we now relabel as $\boldsymbol{\theta}^{(1)}, \ldots, \boldsymbol{\theta}^{(N)}$, and use

$$\begin{aligned} p(y = 1 \mid \boldsymbol{x}, \mathcal{D}) &\approx \frac{1}{N}\sum_{t=1}^{N} p(y = 1 \mid \boldsymbol{x}, \boldsymbol{\theta}^{(t)}) \\ &= \frac{1}{N}\sum_{t=1}^{N} \Phi(\boldsymbol{B}^{(t)}\boldsymbol{\beta}^{(t)}), \end{aligned}$$

where N is the number of samples and $\boldsymbol{B}^{(t)}$ and $\boldsymbol{\beta}^{(t)}$ relate to the basis design matrix and regression coefficients for the jth element of the generated sample.

5.3 Bayesian MARS for Classification

We now demonstrate the use of a flexible Bayesian model for classification. We choose to use a Bayesian MARS model for the predictors as in Holmes and Denison (2002), even though we could use any other form for the basis functions such as neural networks, or piecewise linear splines.

We firstly consider the arm tremor dataset seen before. Recall that the dataset concerns the presence or absence of Parkinson's disease in 357 individuals. The two predictor variables are measurements of two particular types of arm tremor and the complete dataset is partitioned into 178 training and 179 test points as in Husmeier *et al.* (1999).

We ran the BMARS model for classification discarding the first 50 000 iterations as a burn in, after which time the marginal likelihood of the models in the chain had settled down. The resulting predictive distribution for class membership for the arm tremor dataset is shown in Figure 5.1. Notice how the model has essentially selected a hard linear decision boundary at the top of the figure with another boundary separating off the bottom right of the predictor space.

The benefits of using a nonlinear model in this situation are obvious. Any linear approach separates the predictor space into two region divided by a straight line. However, the MARS model used here splits the predictor space into essentially three regions, two with $p(y = 1)$ greater than 0.5 and one with it less than 0.5. Such a complex decision boundary could only be uncovered by a nonlinear model.

Note how the predictive contours in the bottom of Figure 5.1 appear smooth even though the individual MARS models have axis parallel, non-smooth contours. This is due to the marginalisation over the model space by the simulation algorithm which leads to the averaging over thousands of MARS models.

We noticed when MARS basis functions were used for regression that they can show high variability near the boundaries of the convex hull of the data (Section 4.4.3). This can lead to poor predictions near the edges of the data, although predictions inside the convex hull of the dataset are usually good. However, for classification this tends not to be a problem. Here we are interested in the misclassification rate of the model, defined as being the rate at which false predictions are made. For data where the cost of misclassification is equal this leads to the decision rule, predict class 1 if $p(y_i = 1) > 0.5$, else predict class 0. Hence, in terms of performance, the model needs to fit the data well only in the region around the decision boundary for which $p(y_i = 1) = 0.5$. This region is within the space spanned by the data so the problem of overfitting near the boundaries is of much less significance. However, poor fitting near the boundaries would affect the actual probability assessments at different locations so the MARS model may not perform so well when these are required, e.g. when misclassification costs are very different.

Another reason for the apparent stability of the MARS bases for classification problems relates to the effect of the link function Φ. The probit link function Φ is the cumulative distribution function (CDF) of a standard normal which acts as a squashing function that maps $(-\infty, \infty)$ onto $[0, 1]$. Hence the rate of change in the forecast

$p(y_i = 1)$ for unit change in the linear predictor η is proportional to $\exp(-\eta^2/2)$. Hence large changes in η when its magnitude is large actually cause very little change to $p(y_i = 1)$.

Holmes and Denison (2002) give results that demonstrate that the BMARS model is at least as effective at prediction as some highly trained neural network models (Husmeier *et al.* 1999) as well as the boosting classification algorithm (Schapire *et al.* 1998).

One last word should be made about the auxiliary variable approach we have adopted. We have highlighted this method as we have found that it leads to a particularly tractable MCMC algorithm. In the past we have considered alternative approaches, for example, standard Metropolis–Hastings updates centred around the current values of the parameters, but these led to very poor acceptance rates of around 2%, rather than the 30% achieved with auxiliary variables. Also, when we are not considering models that change dimension other simulation techniques have been suggested (see, for example, Dellaportas and Smith 1993) and we shall discuss these later on.

5.3.1 *Multiclass classification*

When the response is not binary, so that $C > 2$, we are faced with a multiclass classification problem. These can be handled in a similar way to binary classification through the polychotomous classification approach suggested by Albert and Chib (1993). In the spline literature this has been discussed in the paper of Yau *et al.* (2002) and we adopt their approach here.

Let $\boldsymbol{y}_i = (y_{i1}, y_{i2}, \ldots, y_{iC})$ denote the multinomial indicator vector with elements $y_{ij} = 1$ if the ith data point belongs to the jth class and $y_{ij} = 0$ otherwise. Let $\boldsymbol{Y}$ denotes the $n \times C$ matrix of these indicators. The likelihood of the data given a single spline basis set $\boldsymbol{B}$ and a set of coefficients $\boldsymbol{\beta}_1, \ldots, \boldsymbol{\beta}_C$, one for each class, is given by

$$p(\boldsymbol{y}_i \mid \boldsymbol{\beta}) = \mu(\boldsymbol{\beta}_1)^{y_{i1}} \mu(\boldsymbol{\beta}_2)^{y_{i2}} \cdots \mu(\boldsymbol{\beta}_C)^{y_{iC}},$$

where $\mu(\boldsymbol{\beta}_j)$ is the probability that the point came from class j. To model this probability we refer back to the previous section and define an auxiliary variable, z_{ij}, for each y_{ij} such that

$$z_{ij} = \beta_{0j} + \sum_{l=1}^{k} \beta_{lj} B_l(\boldsymbol{x}_i) + \epsilon_{ij},$$

where we assume $\epsilon_{ij} \sim N(0, 1)$ for all i, j. Then we define the response as

$$p(y_{ij} = 1 \mid \boldsymbol{z}, \boldsymbol{\beta}) = \begin{cases} 1, & \text{if } z_{ij} > z_{il} \text{ for all } l \neq j, \\ 0, & \text{otherwise.} \end{cases} \tag{5.15}$$

Conditional on the current model the auxiliary variables z_{ij} are normally distributed, $z_{ij} \sim N(\eta_{ij}, 1)$ subject to $z_{ij} > z_{il}$ for all $l \neq j$ if the ith data point is from the jth category, where η_{ij} is the linear predictor, $\eta_{ij} = \beta_{0j} + \sum_{l=1}^{k} B_l(\boldsymbol{x}_i)\beta_{lj}$.

Conditional on $z = (z_{ij})_{ij}$ the basis functions can be sampled by a multivariate extension of the algorithm of Section 5.2.2. In particular, suppose we place an $N(\boldsymbol{m}, \boldsymbol{V})$ prior on each set of $\boldsymbol{\beta}^{(j)}$, for $j = 1, \ldots, C$, then we find that the posterior distribution of the coefficients given $\boldsymbol{z}_j = (z_{1j}, \ldots, z_{nj})$ is $p(\boldsymbol{\beta}_j \mid \boldsymbol{z}_j) = N(\boldsymbol{m}_j^*, \boldsymbol{V}^*)$, where

$$\boldsymbol{m}_j^* = (\boldsymbol{B}'\boldsymbol{B} + \boldsymbol{V}^{-1})^{-1}(\boldsymbol{V}^{-1}\boldsymbol{m} + \boldsymbol{B}'\boldsymbol{z}_j), \tag{5.16}$$

$$\boldsymbol{V}^* = (\boldsymbol{B}'\boldsymbol{B} + \boldsymbol{V}^{-1})^{-1}. \tag{5.17}$$

Also, we find that the marginal likelihood of the data for known error variance is given by

$$p(\mathcal{D} \mid z) \propto \prod_{j=1}^{C} \frac{|\boldsymbol{V}^*|^{1/2} \exp(-b_j^*)}{|\boldsymbol{V}|^{1/2}(2\pi)^{n/2}},$$

where

$$b_j^* = (z_j'z_j + \boldsymbol{m}'\boldsymbol{V}^{-1}\boldsymbol{m} - (\boldsymbol{m}_j^*)'(\boldsymbol{V}^*)^{-1}\boldsymbol{m}_j^*)/2.$$

As in the binary classification case we assume that $\boldsymbol{V} = v\boldsymbol{I}$ and assign a hierarchical prior to v^{-1}, taken to be a gamma distribution with parameters δ_1 and δ_2. Thus, we can update v similarly to (3.8) using the fact that

$$p(v^{-1} \mid \boldsymbol{\beta}) = \mathrm{Ga}\left(\delta_1 + \tfrac{1}{2}kC, \delta_2 + \frac{1}{2}\sum_{1}^{k}\sum_{j=1}^{C}\beta_{ij}^2\right). \tag{5.18}$$

Hence, the algorithm follows the same as that in Section 5.2.2 but now with multivariate updates to the z with the marginal likelihood specified above.

5.4 Count Data

Suppose that instead of needing to classify data into one of a number of categories the responses are counts of a particular event. This means that all the responses are greater than or equal to zero and that they are integers. For example, count data are commonly found in epidemiological studies where the counts relate to the number of individuals in an area exhibiting a particular disease.

For classification data the appropriate sampling distribution was Bernoulli, or multinomial, while for count data we often assume that the observations are drawn from a Poisson distribution with unknown mean. That is, if $X \sim \mathrm{Poi}(\lambda)$, then its probability mass function is given by

$$p(X = x \mid \lambda) = \frac{\lambda^x \exp(-\lambda)}{x!}, \qquad x = 0, 1, 2, \ldots.$$

Also, when the response represents a count with some prespecified maximum, so that the responses follow a binomial distribution, the Poisson distribution can be used as a good approximation as

$$\mathrm{Bi}(N, \xi) \overset{\mathrm{d}}{\approx} \mathrm{Poi}(N\xi),$$

for large N and small $\xi(> 0)$. A good example of such a situation is given by disease mapping examples. Here there is some maximum number, N, of susceptibles in a population and some probability of contracting a disease, ξ. For rare diseases the Poisson approximation to the binomial is nearly always adopted.

It follows that it would be useful to be able to model Poisson data, if possible by adopting a similar approach to the one used previously for classification. Fortunately, even though an identical method cannot be employed, a refinement of the auxiliary variable method is applicable. In fact we find that this model more closely resembles the Bayesian linear model with unknown regression variance.

We can assume that each response, given the coefficients $\boldsymbol{\beta}$ and basis function matrix $\boldsymbol{B}$, is distributed according to a Poisson distribution with mean parameter $h(\eta_i)$, where function h is an invertible link function that maps the real line to the positive half-line, e.g. $h(\eta_i) = \exp(\eta_i)$. For our chosen basis design matrix we have

$$p(y_i = c) = \text{Poi}\{\exp(\eta_i)\}, \qquad c = 0, 1, \ldots,$$

$$\eta_i = \beta_0 + \sum_{j=1}^{k} \beta_j B_j(\boldsymbol{x}_i) + \epsilon_i, \tag{5.19}$$

where we have introduced a residual variance ϵ_i, which we assume to be normal with mean 0 and unknown variance σ^2. The use of a residual component is consistent with the belief that there may be other unknown sources of variation that are not accounted for by the basis function model perhaps due to predictors that were not recorded. The model in (5.19) is an example of a generalised nonlinear mixed model (Sun *et al.* 2000) and is a special case of the class of generalized linear mixed models (Breslow and Clayton 1993; Clayton 1996; Zeger and Karim 1991).

By adopting a normal residual effect we again find that the model parameters are now of standard form, which greatly aids in the computations. To be specific, conditional on η_i the model defined by $\boldsymbol{B}$ and $\boldsymbol{\beta}$ (5.19) is independent of y_i and can be written as a standard Bayes linear regression of z on the basis space defined by $\boldsymbol{B}$. Hence, we can follow Holmes and Mallick (2000b) and take the prior on the parameters that define $\boldsymbol{B}$ to be model specific, with the other priors given by

$$p(\boldsymbol{\beta}, \sigma^2 \mid v, \boldsymbol{B}) = \text{NIG}(\boldsymbol{0}, v\boldsymbol{I}, \gamma_1, \gamma_2), \tag{5.20}$$

$$p(v) = \text{Ga}(\delta_1, \delta_2), \tag{5.21}$$

$$p(z \mid \boldsymbol{\beta}, \sigma^2, \boldsymbol{B}) = N(\boldsymbol{B}\boldsymbol{\beta}, \sigma^2\boldsymbol{I}), \tag{5.22}$$

so that the model is

$$z_i = \boldsymbol{B}\boldsymbol{\beta} + \epsilon_i,$$

where $\epsilon_i \sim N(0, \sigma^2)$. The simulation algorithm is then identical to the one for classification except that the distributions from which we sample z is different and that we need to sample both $\boldsymbol{\beta}$ and σ^2 jointly, from a normal-inverse gamma distribution this time. As we are assuming an unknown variance linear model for z_i the acceptance

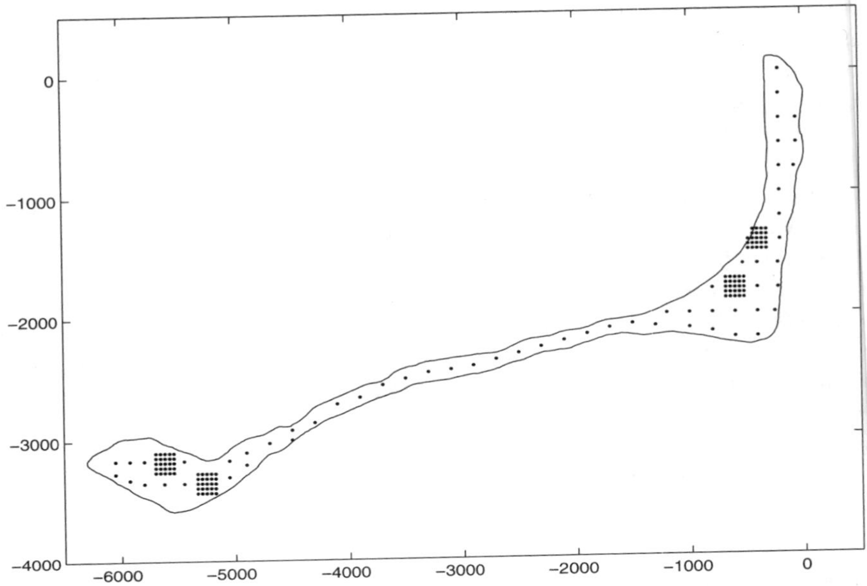

Figure 5.2 Rongelap Island showing the locations of 157 sampling points.

probabilities for changing the model are identical to (2.43), except that now the responses are given by the vector z.

We saw for the classification example that we had to sample z from a nonstandard distribution (i.e. a truncated normal one). Here we find the same problem but this time the actual distribution, $p(z_i \mid z_{-i}, \boldsymbol{\beta}, \sigma^2, \boldsymbol{B}, \mathcal{D})$, is not even of known form. This problem is easily overcome by using a Metropolis–Hastings sampler for z and noting as before that

$$p(z_i \mid y_i = c, \boldsymbol{\beta}, \sigma^2) \propto p(y_i = c \mid z_i) p(z_i \mid \boldsymbol{\beta}, \sigma^2),$$

where $p(y_i = c \mid z_i)$ is the Poisson likelihood and the prior $p(z_i \mid \boldsymbol{\beta}, \sigma^2)$ is $N(\boldsymbol{B}\boldsymbol{\beta}, \sigma^2)$.

5.4.1 *Example: Rongelap Island dataset*

We illustrate the method of Poisson regression using an example from Diggle *et al.* (1998). Diggle *et al.* (1998) analyse Geiger counter readings taken from Rongelap Island. Rongelap Island lies in the Pacific Ocean around 2500 miles southwest of Hawaii. The island was exposed to radionuclide contamination due to fall-out from the Bikini Atoll nuclear weapons testing programme in the 1950s.

A survey was undertaken on the island at 157 sampling locations shown in Figure 5.2. Measurements of ^{137}Cs radioactivity were recorded at each location. The task is to provide a spatial map of contamination levels across the island. Diggle *et al.*

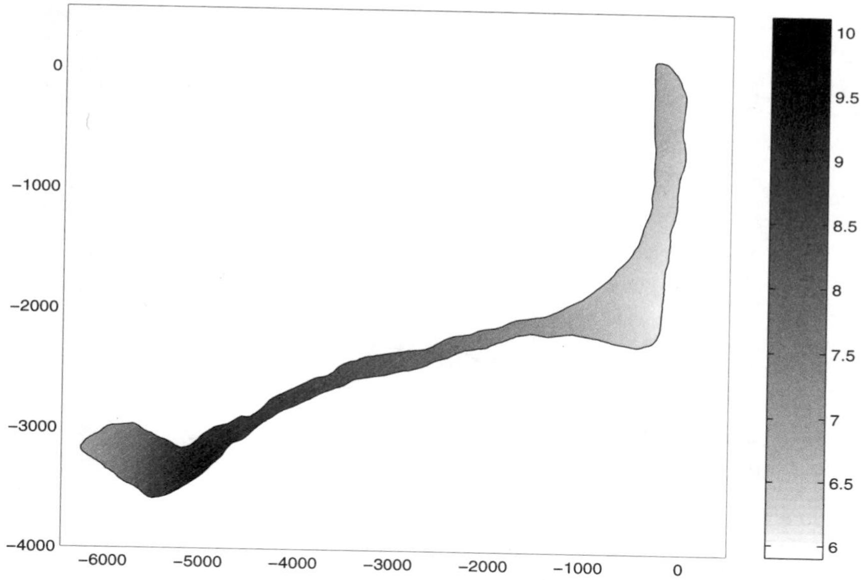

Figure 5.3 Posterior expectation of the number of counts, $E(y \mid \boldsymbol{x}, \mathcal{D})$, over Rongelap Island.

(1998) examine the use of kriging methods (Cressie 1993) to model this data. Due to the spatial nature of this problem, we adopt thin-plate regression splines. Recall that the basis function is defined as

$$B_j(\boldsymbol{x}) = \|\boldsymbol{x} - \boldsymbol{t}_j\|^2 \log(\|\boldsymbol{x} - \boldsymbol{t}_j\|)$$

for knot location $\boldsymbol{t}_j$.

The sampling algorithm for this model is the same as that for two-class classification except that now we also sample the residual variance from the conjugate inverse-gamma distribution.

The expected count level $E(y \mid \boldsymbol{x}, \mathcal{D})$ is shown in Figure 5.3, where we see a high level of contamination in the southwest region of the island. The posterior variance of the mean level is shown in Figure 5.4. Note that the variance increases within the middle portion of the island where there are less sampling locations. We also observe greater variations at the outer tips which is to be expected as there is less information about the contamination here.

5.5 The Generalised Linear Model Framework

Both the count data and classification problem using nonlinear bases have much in common with the family of generalised linear models (GLM) originally introduced

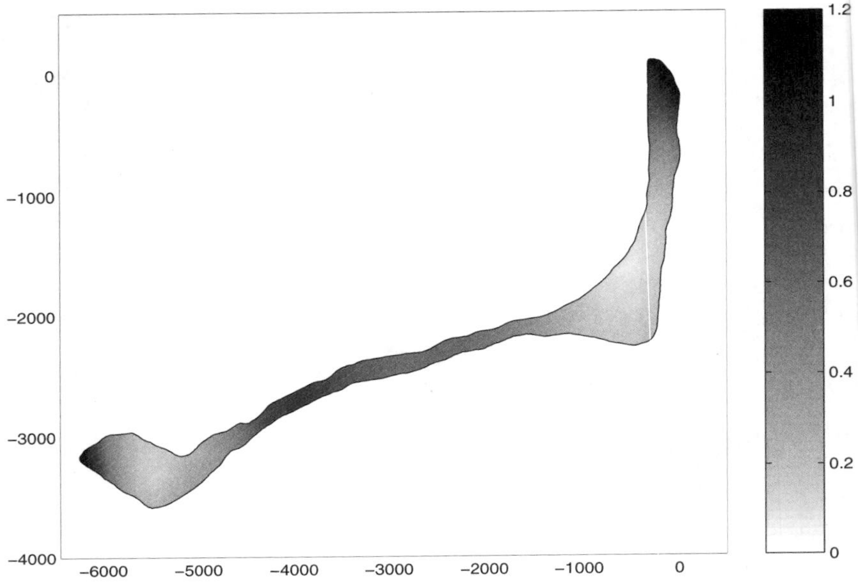

Figure 5.4 Variance of the posterior mean count level, $\text{var}\{E(y \mid \boldsymbol{x}, \mathcal{D})\}$, over Rongelap Island.

by Nelder and Wedderburn (1972) and covered in detail in McCullagh and Nelder (1989).

The GLM assumes that the data are generated independently and a general response, y, has a distribution that depends on the measurements of its associated predictor variables, $\boldsymbol{x} = (x_1, \ldots, x_p)$. The relationship between the predictors and the distribution of the response is assumed to depend explicitly on a linear predictor of the form in (5.8). However, we suggest the qualitatively similar model that uses the nonlinear predictor, η, which can be written

$$\eta = \beta_0 + \sum_{j=1}^{p} \beta_j B_j(\boldsymbol{x}).$$

Further, the distribution of y must be able to be written in the form,

$$p(y \mid \boldsymbol{x}) = \exp[(A/\phi)\{y\eta - \psi(\eta)\} + \tau(y, \phi/A)],$$

where ϕ is a scale parameter which is often known, A is a known constant, η depends on the linear predictor in some way and τ is an arbitrary function. Note that distributions of this form are known as exponential family distributions. The last requirement of generalised linear models is that the mean of the distribution of the response is related to the linear predictor according to

$$\mu = h(\eta) \quad \text{or} \quad \eta = h^{-1}(\mu),$$

where h is invertible. In the terminology of GLMs the function h^{-1} is known as the *link* function.

We now give a few examples of GLMs and see that they form the basis of all the models we have used so far in the book, not just in this chapter.

Normal. For a response, y, from a normal distribution with mean μ and variance σ^2 we can write

$$p(y) = \frac{1}{\sqrt{2\pi\sigma^2}} \exp\left\{-\frac{1}{2\sigma^2}(y-\mu)^2\right\},$$

for all real y. Now, if we set $\phi = \sigma^2$ and take logarithms of both sides we see that

$$\log p(y) = \frac{1}{\phi}(y\mu - \tfrac{1}{2}\mu^2) - \tfrac{1}{2}\{y^2 + \log(2\pi\phi)\}.$$

Hence the normal distribution is a member of the exponential family and is given when we take $A = 1$, $\eta = \mu$, $\psi(\eta) = \eta^2/2$ and the other terms give the form of τ. Comparing this to the standard regression model of Section 2.3 we see that this particular GLM is identical when we take h to be the identity function, so that $h(\eta) = \eta$.

Bernoulli. In the case when $y \sim Br(\mu)$ with $y \in \{0, 1\}$ then

$$\log p(y) = y \log\left(\frac{\mu}{1-\mu}\right) + \log(1-\mu),$$

for $y = 0, 1$. Hence $A = \phi = 1$, $\eta = \log\{\mu/(1-\mu)\}$ and

$$\psi(\eta) = -\log(1-\mu) = \log(1+\mathrm{e}^{\eta}),$$

as $\mu = \mathrm{e}^{\eta}/(1+\mathrm{e}^{\eta})$.

Again we find, by comparing this model with the one for classification introduced earlier, that the classification problem fits into the GLM framework.

Poisson. Similarly to before we find that the Poisson model for count data is in the exponential family when we take $\eta = \log\mu$, $\phi = 1$ and $\psi(\eta) = \mathrm{e}^{\eta}$.

We have now seen how all the models we have come across so far can be fitted into the generalised linear model framework. In all these cases we modelled the mean of the distribution of the response with some function of the nonlinear predictor so that $E(y) = h(\eta)$. However, we have not yet discussed ways to pick h. Essentially, we just need to pick some invertible function that maps the real line to the range of values that $E(y)$ can possibly take. However, one common way to pick h is suggested by standard properties of exponential family distribution (McCullagh and Nelder 1989). We find that

$$E(y) = \frac{\mathrm{d}\psi}{\mathrm{d}\eta} = \psi'(\eta),$$

so a good choice is to take $h(\eta) = \psi'(\eta)$. This is known as the canonical choice with the canonical link given by $(\psi')^{-1}$.

We see that for the normal and Poisson models already encountered we have used the canonical link. However, we suggested taking $h(\eta) = \Phi(\eta)$ when performing classification, even though the canonical link suggests $h(\eta) = e^{\eta}/(1 + e^{\eta})$. We did not adopt the canonical link here because this does not allow us to use auxiliary variables to aid our sampling.

5.5.1 *Bayesian generalised linear models*

Similarly to the difference between linear and Bayesian linear models, Bayesian GLMs allow for uncertainty in the regression parameters. Further, when using a non-linear predictor to model the mean of the response, we can also allow for uncertainty in the set of basis functions to use. These are the sort of models we presented earlier for classification and analysing count data.

However, there has been a lot of work on Bayesian GLMs with the predictor fixed to be linear. Effectively, the only difference between these models and the GLM method of McCullagh and Nelder (1989) is that prior distributions are assigned to the coefficient vector, $\boldsymbol{\beta}$. In common with the classification example, no conjugate prior for $\boldsymbol{\beta}$ exists for all Bayesian GLMs, so a method needs to be found to sample them efficiently to obtain $p(\boldsymbol{\beta} \mid \mathcal{D})$.

An early attempt at making inference from a GLM using Bayesian methods was given by Knuiman and Speed (1988). Here they noted that the posterior distribution of the coefficients was not of standard form and would require a three-dimensional numerical integration to obtain the normalising constant of the posterior. Instead they suggested approximating the posterior by a normal distribution with mean the posterior mode and variance given by the negative Hessian, where the Hessian is given by

$$\boldsymbol{H}(\boldsymbol{\beta}) = \left[\frac{\mathrm{d}^2\{\log p(\boldsymbol{\beta} \mid \mathcal{D})\}}{\mathrm{d}\boldsymbol{\beta}\, \mathrm{d}\boldsymbol{\beta}'} \right]^{-1},$$

evaluated at the posterior mode. We find that this posterior mode is unique and the Hessian has some attractive properties which help make GLMs such useful models.

5.5.2 *Log-concavity*

A positive function f on a (open convex) set $\mathcal{X} \in \mathbb{R}^p$ is called log-concave if its logarithm is twice differentiable on $\mathcal{X}$ and its Hessian is negative semi-definite, so that all of its eigenvalues are less than or equal to zero. Strict log-concavity of f occurs when none of these eigenvalues are zero. The importance of log-concavity is that a strictly log-concave function has a unique maximum (Wedderburn 1976).

We find that the likelihood functions, $p(\mathcal{D} \mid \boldsymbol{\beta})$, of generalised linear models are log-concave and, when combined with a log-concave prior distributions $p(\boldsymbol{\beta})$, we find that the posterior is itself log-concave. The fact that $p(\boldsymbol{\beta} \mid \mathcal{D})$ has a unique maximum and a Hessian which can be determined allows us to use the adaptive rejection sampling method of Gilks and Wild (1992) to sample the coefficients, as suggested by Dellaportas and Smith (1993). The attractiveness of adaptive rejection sampling is that it samples from the required posterior without the need to determine the mode of the density, which can be a computationally expensive step. This is in contrast to other methods of sampling from log-concave densities (see, for example, Devroye 1986). Also, by simulating from the posterior of interest via Gibbs sampling Dellaportas and Smith (1993) do not need to make approximations to the functional form of the posterior. However, adaptive rejection sampling is much better suited to making inference in GLMs when the basis set is fixed. When the basis set varies the auxiliary sampling method is perhaps more attractive and easier to implement.

5.6 Further Reading

The Bayesian generalised linear model has received considerable attention. Gelman *et al.* (1995) is an excellent reference to such models, as is the collection of papers in Dey *et al.* (2000). Apart from the papers on Bayesian GLMs we have already referenced other work includes Ibrahim and Laud (1991), Zeger and Karim (1991), Bedrick *et al.* (1996), Oh (1997), Ntzoufras *et al.* (2001) and Dellaportas *et al.* (2002). However, we now restrict ourselves to giving references relating to Bayesian GLMs that allow for nonlinearities in the model.

We have suggested adopting an approach based on standard Bayesian GLMs but allowing for nonlinearity by taking

$$\eta(\boldsymbol{x}) = \beta_0 + \sum_{j=1}^{p} \beta_j B_j(\boldsymbol{x}),$$

with the basis functions determined by the data. In this chapter we discuss taking the basis functions as MARS-type ones or thin-plate splines as in Holmes and Denison (2002) and Holmes and Mallick (2000b).

Other work that uses nonlinear functions of $\eta(\boldsymbol{x})$ to model the mean includes the work on generalised additive models (Hastie and Tibshirani 1990) for which we can write

$$\eta(\boldsymbol{x}) = \beta_0 + g_1(x_1) + \cdots + g_p(x_p).$$

Bayesian work on such models is given in Wood and Kohn (1998), Hastie and Tibshirani (2000) and Fahrmeir and Lang (2001). Using neural networks basis functions for the mean function has been suggested by Ishwaran (1999) and Biller (2000) instead uses B-spline bases and describes a reversible jump MCMC sampling strategy that does not use auxiliary variables.

Generalised linear models have also been tackled by Bayesian nonparametric methods. For binary classification Newton *et al.* (1996) suggested taking

$$p(y = 1 \mid \boldsymbol{x}, \boldsymbol{\beta}) = F\left(\beta_0 + \sum_{1}^{p} \beta_j x_j\right) = F(\eta),$$

where F is a random draw from a Dirichlet process prior. This is in contrast to the standard GLM which for which the link function h^{-1} is constant. Other work that has advocated modelling the link function together with the coefficients includes Mallick and Gelfand (1994) and Basu and Mukhopadhyay (2000).

Adding terms to the linear predictor to take into account the effects of unobserved predictors has also proved popular. Here the linear predictor is given by

$$\eta(\boldsymbol{x}) = \gamma(\boldsymbol{x}) + \sum_{j=1}^{p} \beta_j x_j,$$

where $\gamma(\boldsymbol{x})$ is a random intercept term that may only depend on the index of the data point, rather than the actual covariate values as displayed here. This extra term is known as a *random effect* and it is intended to account for random variation that cannot be explained by the predictors. Modelling the random effects by nonparametric mixtures goes back to the work of Laird (1978) and Follman and Lambert (1989). In a Bayesian setting the random effect term has been modelled by a Dirichlet process (Mukhopadhyay and Gelfand 1997), by a Polya tree (Walker and Mallick 1997) and also by a Gaussian process (Gutiérrez-Peña and Smith 1998).

5.7 Problems

5.1 Suppose that we wish to generate samples from a density of the form $p(\theta) = \ell(\theta)\pi(\theta)$ for which it is not possible to sample directly from $p(\theta)$. Damien *et al.* (1999) suggest introducing an auxiliary variable ϕ, which follows a uniform distribution, such that $p(\theta, \phi) \propto I\{\phi < \ell(\theta)\}\pi(\theta)$.

Find the marginal density, $p(\theta)$, by integrating out ϕ. Now derive the conditional distributions $p(\theta \mid \phi)$ and $p(\phi \mid \theta)$ and propose a Gibbs sampling algorithm to sample from $p(\theta, \phi)$.

This method is known as slice sampling when we think of ℓ as defining a likelihood function and π a prior. Now if $\ell(\theta) = \exp\{-\exp(\theta)\}$, find the joint distribution of θ and ϕ as well as the conditional distributions for any prior $\pi(\theta)$.

5.2 Consider a classification model for binary response data $y_1, \ldots, y_n$ with fixed basis function matrix $\boldsymbol{B}$ and random coefficients $\boldsymbol{\beta}$. Describe how it is possible to introduce independent latent variables $z = (z_1, \ldots, z_n)$ such that the marginal model of the responses is unchanged. Determine the complete conditional distributions of the z_i and then show that if we take the prior on the coefficients as $p(\boldsymbol{\beta}) = N(\boldsymbol{0}, \boldsymbol{V})$ then

$$p(\boldsymbol{\beta} \mid \mathcal{D}, z) = N\{(\boldsymbol{B}'\boldsymbol{B} + \boldsymbol{V})^{-1}\boldsymbol{B}'z, (\boldsymbol{B}'\boldsymbol{B})^{-1}\}.$$

5.3 Consider a classification model for binary response data $y_1, \ldots, y_n$ with fixed basis function matrix $\boldsymbol{B}$ and random coefficients $\boldsymbol{\beta}$. We can introduce two sets of auxiliary variables, $z = (z_1, \ldots, z_n)$ and $\boldsymbol{\lambda} = (\lambda_1, \ldots, \lambda_n)$, to aid us in simulating from the posterior of interest $p(\boldsymbol{\beta} \mid \mathcal{D})$. If we assign a joint prior to z and $\boldsymbol{\lambda}$ such that

$$\begin{aligned} p(z, \boldsymbol{\lambda}) &= p(z \mid \boldsymbol{\lambda}) p(\boldsymbol{\lambda}) \\ &= N\{z \mid \boldsymbol{B}\boldsymbol{\beta}, \operatorname{diag}(\boldsymbol{\lambda})\} \prod_{i=1}^{n} \operatorname{Ga}(\lambda_i \mid \gamma, \gamma), \end{aligned}$$

show that after integrating out z and $\boldsymbol{\lambda}$ the marginal model is similar to the probit one with the normal CDF Φ replaced by the CDF of a t-distribution.

Derive the complete conditional posterior distributions for $\boldsymbol{\beta}$, z and $\boldsymbol{\lambda}$ assuming a uniform prior for $\boldsymbol{\beta}$.

6

Bayesian Tree Models

6.1 Introduction

This chapter highlights a particular model which can be used to perform both regression as well as classification. Although tree models can be thought of in quite general terms the presentation in this chapter concerns the Bayesian generalisation of the approach to trees given by Breiman *et al.* (1984). Thus we shall discuss tree models formed by a binary partition of the predictor space. Although the tree models of Breiman *et al.* (1984) are well established in the statistical community there has also been much work on trees amongst the machine learning and computer science communities, in particular. In these fields they are more commonly referred to as decision trees and the focus is usually on classification. We shall use the statistical convention of referring specifically to classification and regression trees, depending on the problem at hand.

We devote a full chapter to Bayesian trees as their analysis demonstrates what Bayesian ideas can bring to the well-developed field of tree-based modelling. Further, some special features of tree models mean that they only admit a full Bayesian analysis in the simplest of cases allowing us to demonstrate how we can use a Markov chain to perform a search, not a full exploration, of the posterior distribution of the model.

Although tree-based models are usually thought of in terms of the tree structure, we shall describe how they can be formulated in terms of basis functions, similarly to all the previous Bayesian models we have introduced. We find that these basis functions have some undesirable features, most notably their dependence on each other. This makes sampling from the full posterior distribution of the model parameters infeasible with current techniques in most situations. Instead we concern ourselves with only searching the target posterior space, not complete exploring it. This demonstrates the usefulness of Bayesian methods generally but does also raise other important issues to do with summarising the generated 'sample' of models. This is especially important with tree models as single trees are favoured due to their interpretability and ease of use, whereas averages over trees do not admit representation by a simple structure.

Note that we shall use the abbreviation C&RT for the classification and regression tree method of Breiman *et al.* (1984) as CART itself is a registered trademark of California Statistical Software, Inc.

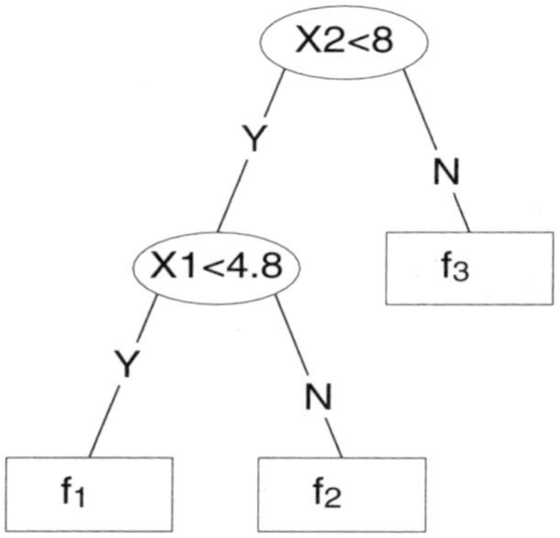

Figure 6.1 A simple example of a tree structure with two splitting and three terminal nodes.

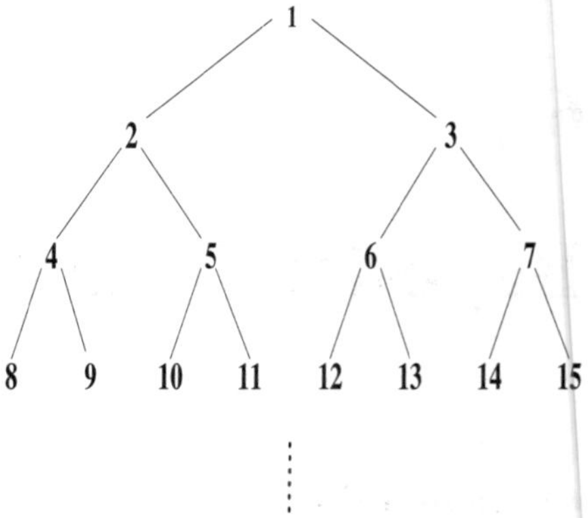

Figure 6.2 The labelling used to determine the s_i^{pos} values.

6.1.1 Motivation for trees

Tree models date back to the recursive partitioning ideas of Morgan and Sonquist (1963) and Friedman (1979), with the main exposition given in Breiman *et al.* (1984). The main idea behind the model is to provide a method that is locally adaptive and admits a simple rule-based representation which is easy to use. This has particular attractions to non-statisticians, who can make decisions using trees without having to fully understand the statistical ideas used to find their structure.

Another advantage of trees is the way they naturally perform variable selection. We shall see how unimportant variables can be excluded from the final classifier, or tree structure, ensuring that their presence in the dataset does not have any effect on the final predictions, except to increase the computation time to some extent.

From the first chapter of Breiman *et al.* (1984) we learn that these two ideas spurred much of the work on tree models. The motivating dataset they present concerns measurements made on patients who have been admitted to hospital after a myocardial infarction (heart attack). When the patients are first admitted many measurements on them are determined (age, sex, systolic blood pressure, etc.) and the aim is to predict which patients are likely to die within 24 hours of admission. A classification tree was ideal in this situation as it provided a simple model with reasonably low misclassification error which doctors could easily follow.

6.1.2 Binary-tree structure

The general binary-tree structure is made up of both *splitting* and *terminal* nodes. Each splitting node asks a yes/no question and, depending on the answer, data points are assigned to the left or right branch from this node. The terminal nodes do not have associated questions but just assign all points in them to a common probability density function. In the terminology of trees the splitting and terminal nodes are also known as branching and leaf nodes, respectively. For similar reasons the node at the top of the tree is called the root node.

In Figure 6.1 we display a simple (binary-)tree. In this example the basis functions that define this tree are

$$\begin{aligned} B_1(\boldsymbol{x}) &= I(x_2 < 8)I(x_1 < 4.8), \\ B_2(\boldsymbol{x}) &= I(x_2 < 8)I(x_1 \geqslant 4.8), \\ B_3(\boldsymbol{x}) &= I(x_2 \geqslant 8), \end{aligned}$$

where $I(\cdot)$ is the usual indicator function. The first thing to notice about these basis functions, and those more generally associated with trees, is that they cover the predictor space and at a given value, $x \in \mathcal{X}$, only one of the basis functions is non-zero. Thus they split up the predictor space into disjoint regions, where within each region the response values of the data are assumed to come from the same distribution. Note that in the C&RT methodology of Breiman *et al.* (1984), and with most non-Bayesian generalisations of this, the distributions in the terminal nodes, $f_1, \ldots, f_k$, are all degenerate and have mass one either on a single category or a single real number. That is, all points in a single terminal node are classified as either the most abundant category in that node, or as the mean value of the data points associated with the node.

One important point to note about the basis functions in a regression context is their interdependence. They need to be defined relative to each other as one of them must be non-zero at every location in predictor space. When performing regression problems more generally, we find orthogonal basis functions particularly appealing, so this high dependence is undesirable and a significant drawback in the model.

In a Bayesian context we wish to assign non-degenerate distributions to each terminal node so we can assess the variability within each node. Usually, the distributions f_i are taken to be from the same class indexed by different vectors of parameters, $\boldsymbol{\phi}_i$ $(i = 1, \ldots, k)$. For example, regression problems assuming normal errors take the f_i as normal distributions with mean μ_i and common variance σ^2.

The description of trees so far has focused on the treatment of data in the terminal nodes of the tree, this assumes that the structure above is known. However, the conceptual simplicity at the lowest level of the structure hides the greatest difficulty in fitting trees, that is determining the basis functions which define the structure. This is in common with all the nonlinear methods we have introduced so far but, in this case, restrictions need to be placed on the set of bases to ensure that they define a tree. This makes the problem of Bayesian analysis of C&RT-like structures non-trivial. The basic simulation algorithms which we shall describe below only perform a stochastic search of the target posterior, not a full exploration of it.

Note that the usual tree algorithm uses a stepwise search strategy to determine the tree structure (Breiman *et al.* 1984). Due to the inadequacies of this approach, as described earlier, there has been much work on other tree-growing algorithms, most involving some form of random component to allow the tree growing algorithm to escape from local maxima. A good example of this is Murthy *et al.* (1994), who not only compare some current tree-growing algorithms but also provide code for their method (OC1) along with the more usual C&RT methodology (see `ftp.cs.jhu/pub/oc1/`).

6.2 Bayesian Trees

The earliest work we have discovered on relating Bayesian ideas to the construction and selection of tree models appears to be the excellent thesis by Buntine (1992b), particularly Chapter 6, with many of the key ideas of this chapter summarised in Buntine (1992a). This work introduces Bayesian-motivated ideas for trees as well as providing extensive numerical comparisons between the Bayesian learning techniques Buntine suggested and the others commonly used in practice.

At the time of writing there have been two main papers on Bayesian tree modelling, Chipman *et al.* (1998a) and Denison *et al.* (1998b) but the description of Bayesian trees given here will more closely follow that outlined by Chipman *et al.* (1998a). Denison *et al.* (1998b) use maximum likelihood estimates in their model, which correspond to reference prior choices, but these are not well motivated in a fully Bayesian setting, so we avoid them. However, we shall describe the simulation algorithms given by both authors and compare their relative merits. These are also discussed in the review paper of Denison and Mallick (2000).

Other related work on trees involves using a selection of trees and then making predictions by 'averaging' over them (see, for example, Breiman 1996; Kwok and Carter 1990; Oliver and Hand 1994, 1995, 1996; Shannon and Banks 1999). Posterior averaging is a sensible way of combining a collection of trees, as noted in Buntine (1992a), but this is possible only when considering simple tree structures (see Section 6.3) so none of these papers take a completely Bayesian viewpoint (i.e. assigning a sampling distribution to the data and then combining this likelihood with prior distributions), in contrast to the work of Buntine.

Once we have set up the Bayesian tree models, this chapter will review the work of Buntine (1992a,b) and then relate this to the more recent work on Bayesian trees by Chipman *et al.* (1998a) and Denison *et al.* (1998b). The latter papers attempt to put the complete tree finding method into a fully Bayesian framework and the majority of this chapter will focus on this work.

6.2.1 The random tree structure

First we propose a model that can be used to set up a probability distribution over the space of possible tree structures. We refrain from defining the model using basis functions, as given above, to ease the interpretability of the random variables describing the tree.

Any binary-tree model can be uniquely defined by the positions of the splitting nodes present, together with the variables and points where these variables are split. We define these variables respectively as s_i^{pos}, s_i^{var} and s_i^{rule} for $i = 1, \ldots, k-1$. Hence, $k-1$ here gives us the number of splitting nodes present in a tree structure.

Thus k actually represents the number of terminal nodes in the tree. Defining k in terms of the number of terminal nodes is the convention followed by Chipman *et al.* (1998a) and Denison *et al.* (1998b), which is why we follow it here.

The tree in Figure 6.1 and diagram in Figure 6.2 are useful in helping to understand the role of the variables that make up $\boldsymbol{\theta}$. In this case we know $k = 3$ and we can write

$$\boldsymbol{\theta} = (s_1^{\text{pos}}, s_1^{\text{var}}, s_1^{\text{rule}}, s_2^{\text{pos}}, s_2^{\text{var}}, s_2^{\text{rule}}) = (1, 2, 8, 2, 1, 4.8),$$

using the convention of Figure 6.2 to assign the s_i^{pos} with the other variables easily defined.

Similarly to previous chapters we are interested in making inference about the tree structure given the data, so our target posterior distribution is $p(\boldsymbol{\theta} \mid \mathcal{D})$. Again we find that we can assign conjugate distributions to some further parameters $\boldsymbol{\phi}$ allowing us to analytically evaluate the marginal likelihood, $p(\mathcal{D} \mid \boldsymbol{\theta})$. The next two subsections describe how we set up the model to allow direct calculation of the marginal likelihoods for Bayesian C&RT. After this we shall describe ideas on how to assign the prior to the parameters which define the tree structure.

6.2.2 *Classification trees*

The general classification problem involves data of the form $(y_i, \boldsymbol{x}_i)_1^n$, where the responses are categorical, e.g. $y_i \in \{1, \ldots, C\}$. In this case we must determine the probabilities of assignment to each class in each terminal node. Initially we assign $(C-1)$-dimensional Dirichlet priors to the class assignments in each node so that

$$\begin{aligned} p(\boldsymbol{\phi}_i \mid \boldsymbol{\theta}) &= \text{Di}_{C-1}(\boldsymbol{\phi}_i \mid \boldsymbol{\alpha}) \\ &= \frac{\Gamma(\sum_1^C \alpha_j)}{\prod_1^C \Gamma(\alpha_j)} \phi_{i1}^{\alpha_1 - 1} \cdots \phi_{iC}^{\alpha_C - 1}, \end{aligned} \tag{6.1}$$

for $i = 1, \ldots, k$, and with each $0 < \phi_{ij} < 1$ with $\phi_{iC} = 1 - \sum_{j=1}^{C-1} \phi_{ij} < 1$. Thus the density is of dimension $C - 1$ as ϕ_{iC} is a deterministic function of the other ϕ_{ij}. We think of each of the ϕ_{ij} as the probability of a datum being in class j if it is assigned to the ith terminal node, and write $\boldsymbol{\phi}_i = (\phi_{i1}, \ldots, \phi_{iC})$ with the complete vector of unknown probabilities given by $\boldsymbol{\phi} = (\boldsymbol{\phi}_1, \ldots, \boldsymbol{\phi}_k)$. We also denote by $\boldsymbol{\alpha} = (\alpha_1, \ldots, \alpha_C)$ the vector of prior constants which are needed to parametrise the Dirichlet distribution.

The prior in (6.1) is very general and requires setting one prior constant, α_i, for each category. This may be appropriate when there are different costs for misclassification of each category but, typically, this information is either not available or unknown. In this case choosing $\alpha_1 = \cdots = \alpha_C = \alpha$ is preferable. This specification requires the setting of only one prior constant, α, and assumes that the misclassification costs are equal. Note that by taking $\alpha = 1$ the Dirichlet prior for the class probabilities corresponds to a uniform prior density on the unit simplex of dimension $C - 1$, i.e. $U[0, 1]^{C-1}$. In Figure 6.3 we display some Dirichlet prior densities for the two class

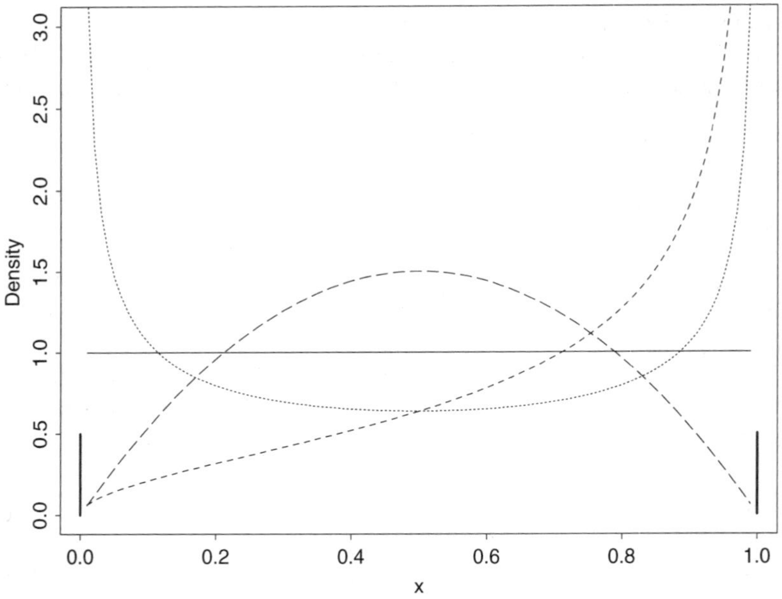

Figure 6.3 Some prior densities for the binary classification problem: $\mathrm{Di}_1(x \mid 1, 1)$ (solid line); $\mathrm{Di}_1(x \mid \frac{1}{2}, \frac{1}{2})$ (dotted); $\mathrm{Di}_1(x \mid 2, 2)$ (long-dashed); $\mathrm{Di}_1(x \mid \frac{3}{2}, \frac{1}{2})$ (short-dashed) and $\mathrm{Di}_1(x \mid 0, 0)$ (thick vertical lines).

problem ($C = 2$) for various values of α. These are identical to the more well-known beta densities.

The likelihood for the data given the model is a multinomial distribution, which we can write down as

$$p(\mathcal{D} \mid \boldsymbol{\theta}, \boldsymbol{\phi}) = \prod_{i=1}^{k} \frac{n_i!}{\prod_j m_{ij}!} \prod_{j=1}^{C} (\phi_{ij})^{m_{ij}},$$

where n_i is the number of data points in the ith terminal node with m_{ij} the number in that node of class j. Again, as we use a conjugate prior specification for $\boldsymbol{\phi}$ we find we can integrate it out giving the marginal likelihood for this model as

$$\begin{aligned} p(\mathcal{D} \mid \boldsymbol{\theta}) &= \int_{\boldsymbol{\phi}} p(\mathcal{D} \mid \boldsymbol{\theta}, \boldsymbol{\phi}) p(\boldsymbol{\phi} \mid \boldsymbol{\theta})\, \mathrm{d}\boldsymbol{\phi} \\ &= \left[\frac{\Gamma\{\alpha C\}}{\{\Gamma(\alpha)\}^C} \right]^k \prod_{i=1}^{k} \frac{\prod_j \Gamma\{m_{ij} + \alpha_j\}}{\Gamma\{n_i + \sum_{j=1}^{C} \alpha_j\}}. \end{aligned} \tag{6.2}$$

This is an example of the multinomial-Dirichlet inferential process which is a generalisation of the more well-known binomial-beta model.

We may not want any tree structure to include empty terminal nodes in which no data points reside. To allow for this we can simply impose a restriction on the above equation such that the marginal likelihood is given by (6.2) if $n_i > 0$ for all i, otherwise we can define it as zero. In this way there will be no posterior weight on such trees. This 'prior' information could be specified through the tree prior but we find that this leads to more complex acceptance probabilities in the sampling algorithm. We can extend this approach to allow no fewer than any number of data points in a terminal node in the obvious way: this value is often taken to be five.

6.2.3 *Regression trees*

As demonstrated before we can think of regression trees in exactly the same way as other basis function regression methods. Similarly, we shall assume that the responses are given by the addition of the true regression function with zero-mean normal errors of unknown variance. Thus the prior on $(\boldsymbol{\phi} \mid k, \boldsymbol{\theta})$ for this model is

$$p(\boldsymbol{\phi} \mid \boldsymbol{\theta}) = p(\beta_1, \ldots, \beta_k, \sigma^2) \tag{6.3}$$

$$= \mathrm{NIG}(\mathbf{0}, \boldsymbol{V}, a, b) \tag{6.4}$$

as before (see (2.12)). Each of the coefficients relates to a constant level in each region so both the ridge and g-priors of Section 3.5.3 give diagonal forms for $\boldsymbol{V}$. However, we prefer to use the ridge prior which relates to $\boldsymbol{V} = v\boldsymbol{I}$ rather than the g-prior which gives $\boldsymbol{V} = \mathrm{diag}(c/n_1, \ldots, c/n_k)$ for some suitably chosen constant c.

Whichever specification is used it is best to routinely demean the response data before the analysis so the posterior means in each node are shrunk towards the overall mean of zero. A similar, but more algebraically complex, prior is given in Chipman *et al.* (1998a), where the coefficients are shrunk to $\overline{y}$ rather than zero.

As before, the normal inverse-gamma prior specification we adopted for $\boldsymbol{\phi} = (\boldsymbol{\beta}, \sigma^2)$ allows us to evaluate the marginal likelihood of the data given the tree structure analytically using (2.23). We find that when we take $\boldsymbol{V} = v\boldsymbol{I}$ the marginal likelihood is given by

$$p(\mathcal{D} \mid \boldsymbol{\theta}) = \frac{(b)^a \Gamma(a^*)}{\pi^{n/2}\Gamma(a)} \frac{(b^*)^{-a^*}}{\prod_1^k (n_i v + 1)^{1/2}}, \tag{6.5}$$

where

$$b^* = b + \frac{1}{2}\sum_{i=1}^{k}\left\{\sum_{j=1}^{n_i} y_{ij}^2 - \frac{v n_i^2}{1 + v n_i}\overline{y}_i^2\right\},$$

where $\overline{y}_i$ is the average of points in node i and y_{ij} is the response of the jth point in region i. Again, we may wish to ensure that there are a certain number of points in each terminal node, assigning a zero likelihood to situations where too few points occur in any of the bottom nodes.

A simple generalisation of this model is to allow for different variances within each node in the same spirit as the mean-variance shift model in Section 3.2.1. In this case $\boldsymbol{\phi} = (\beta_1, \sigma_1^2, \ldots, \beta_k, \sigma_k^2)$ and the priors in each terminal node can be taken to be independent normal-inverse gamma distributions so that

$$p(\beta_i, \sigma_i^2) = \mathrm{NIG}(0, v, a, b),$$

for $i = 1, \ldots, k$. Essentially, this model is only justified in a Bayesian framework if we believe that each data point is generated according to the model

$$y_i = g(\boldsymbol{x}_i) + \epsilon_i(\boldsymbol{x}_i),$$

for $i = 1, \ldots, n$, where g is the tree estimate to the true regression function and we have stressed the dependence of the errors on the predictor location. If we were modelling f with a good approximation to it, then we would be more inclined to believe that the errors were identically distributed and not dependent on location in the predictor space. However, this model tacitly concedes that the true regression function cannot be well estimated by a regression tree, which is why the errors are taken to be dependent on the predictor values. However, when this model is appropriate we find that the marginal likelihood is given by

$$p(\mathcal{D} \mid \boldsymbol{\theta}) = \prod_{i=1}^{k} \left\{ \pi^{-n_i/2} (b)^a \frac{\Gamma(a^*)}{(v n_i + 1)^{1/2} \Gamma(a)} (b_i^*)^{-a^*} \right\},$$

where

$$b_i^* = b + \frac{1}{2} \sum_{j=1}^{n_i} y_{ij}^2 - \frac{v n_i^2}{1 + v n_i} \bar{y}_i^2.$$

6.2.4 *Prior on trees*

So far we have only discussed the priors on the parameters, $\boldsymbol{\phi}$, which we can integrate out given the model parameters. Now, to complete the prior specification, we focus on the prior for the parameters $\boldsymbol{\theta}$ that define the tree structure.

First of all we must explicitly define the possible values that the unknowns in $\boldsymbol{\theta}$, $s_i^{\mathrm{var}}, s_i^{\mathrm{rule}}, s_i^{\mathrm{pos}}$ $(i = 1, \ldots, k-1)$, can take. We take $s_i^{\mathrm{var}} \in \{1, \ldots, p\}$, so it can refer to any of the p predictors, $X_1, \ldots, X_p$. As we want to partition the predictor space in as many ways as possible the natural way to choose the candidate values for s_i^{rule} is from the appropriate set of distinct marginal predictor values when the predictor is ordinal. However, for categorical predictors we take s_i^{rule} to be taken from the set of (non-empty) category subsets of the required predictor. For notational convenience we set $N(j)$ to be the total number of possible splitting rules for predictor X_j, whether it is categorical or ordinal. Note that this specification depends on the dataset, in a similar way to the curve and surface fitting methods of Chapters 3 and 4. This dependence was noted by Buntine (1992a), who also discussed potential advantages of the approach.

Finally, to ensure a finite number of possible trees, we choose some maximum number of splits allowed in one route down the tree, say $S_{\max}$, so

$$s_i^{\mathrm{pos}} \in \{1, \ldots, 2^{S_{\max}+1} - 1\}.$$

This is a technical matter included only to ensure the total number of possible trees is finite. It has no real effect on the algorithm as we can choose $S_{\max} = n - 1$ so that the space of possible trees even includes those that partition each data point into its own terminal node.

Throughout this book we have argued that simple prior distributions are most appropriate when fitting nonlinear models. As such we adopt a similar prior specification for tree structures as we did for the other models introduced earlier. This takes the form of assuming that each model structure with the same number of terminal nodes is equally likely. Hence the prior for a complete tree structure is given by

$$\begin{aligned} p(\boldsymbol{\theta}) &= \left\{ \prod_{i=1}^{k-1} p(s_i^{\mathrm{rule}} \mid s_i^{\mathrm{var}}) p(s_i^{\mathrm{var}}) \right\} p(\{s_i^{\mathrm{pos}}\}_1^{k-1}) \\ &= \left\{ \prod_{i=1}^{k-1} \frac{1}{N(s_i^{\mathrm{var}})} \frac{1}{p} \right\} \frac{k!}{S_k} \frac{1}{K}, \end{aligned} \tag{6.6}$$

where S_k is the number of ways of choosing the $\{s_i^{\mathrm{pos}}\}_1^{k-1}$ which produce a tree with k terminal nodes and K is the maximum number of terminal nodes possible. The factorial term comes into the prior as the index of the basis functions is unimportant.

The main papers on Bayesian C&RT all use different prior specifications than that given above. However, in a similar spirit to that described here Denison *et al.* (1998b) use essentially the same priors expect that they impose a Poisson prior, restricted to be greater than zero, on the number of terminal nodes. This was required to penalise the complexity of the tree as they did not integrate out the parameters in $\boldsymbol{\phi}$.

Chipman *et al.* (1998a) suggest a generalisation of the prior structure described, which is appropriate when there is knowledge of the favoured topology of the tree before the analysis is carried out. The only difference with that given is in the prior on $(\{s_i^{\mathrm{pos}}\}_1^{k-1})$. They define this recursively and suggest that the prior probability that a terminal node in a tree should be split further depends on how many splits have been made above it. Hence, for a specific terminal node η, the probability that it should be further split is given by

$$p_{\mathrm{SPLIT}}(\eta, \boldsymbol{\theta}) = \gamma(1 + d_\eta)^{-\delta}, \tag{6.7}$$

where d_η represents the number of splits made above η (i.e. the depth of the node) and $\delta \geqslant 0$. The larger the value of δ the more the prior favours 'bushy' trees whose terminal nodes have similar depth, whereas choosing $\delta = 0$ corresponds to assuming that each tree with the same number of terminal nodes has the same prior probability. Setting the vector (γ, δ) to reasonable values can be achieved by examining the marginal priors they induce, most notably the prior distribution on the number of terminal nodes (see Figure 6.4).

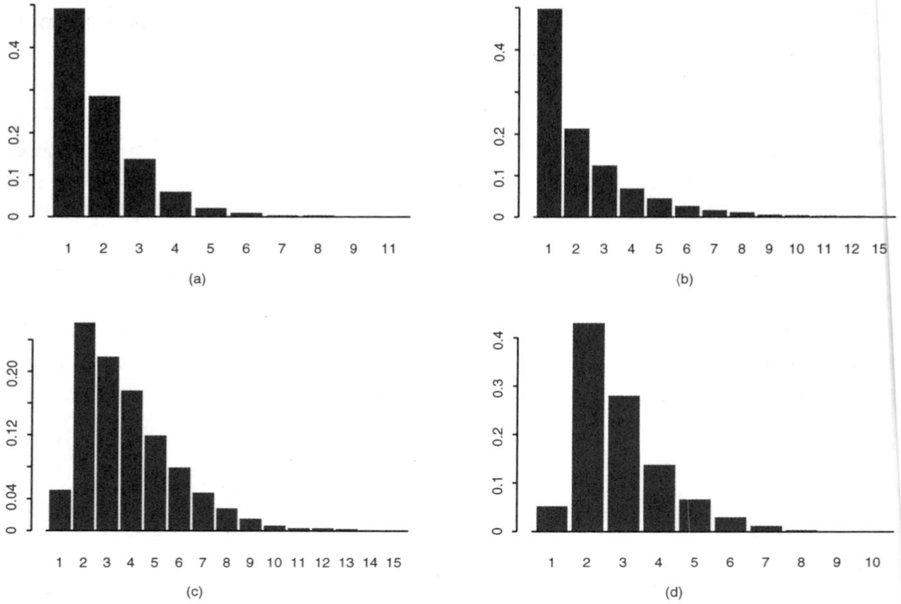

Figure 6.4 Examples of the marginal prior distribution for different setting of γ and δ in (6.7): (a) $\gamma = 0.5, \delta = 1$; (b) $\gamma = 0.5, \delta = 0.5$; (c) $\gamma = 0.95, \delta = 1$; (d) $\gamma = 0.95, \delta = 1.5$.

In a further paper Chipman *et al.* (2000b) discuss setting (γ, δ) in more detail as well as introducing further prior structure to the $\boldsymbol{\phi}$ parameters in the regression tree case. Instead of assuming independent priors on the mean levels within each terminal node, they propose a hierarchical prior structure which allows shrinkage between terminal nodes near each other in the tree structure. They devise these priors to retain conjugacy so the marginal likelihood of this more general model can still be determined explicitly.

6.3 Simple Trees

This section is concerned with simple tree structures for which a full Bayesian analysis can be performed. As in many of the previous chapters when models are relatively straightforward (e.g. the analysis of the Nile river dataset in Section 3.2.1) analytic integration over the complete posterior distribution of the model parameters, $\boldsymbol{\theta}$, can be undertaken. This can be achieved when the set of possible values of $\boldsymbol{\theta} \in \boldsymbol{\Theta}$ is not too large so that we can determine the posterior predictive distribution at any point $\boldsymbol{x} \in \mathcal{X}$, via the summation

$$p(y \mid \boldsymbol{x}, \mathcal{D}) = \sum_{\boldsymbol{\theta} \in \boldsymbol{\Theta}} p(y \mid \boldsymbol{x}, \boldsymbol{\theta}, \mathcal{D}) p(\boldsymbol{\theta} \mid \mathcal{D}). \tag{6.8}$$

The unmarginalised predictive distributions for the classification and regression case are simple to calculate. For regression trees we know from (2.30) that the predictive densities are Student distributions. Substituting these in (6.8) allows us to calculate the predictive marginalised over the model space.

Now let us look specifically at evaluating (6.8) for classification trees. From Bayes' Theorem we know that the posterior predictive distribution, conditioned on $\boldsymbol{\theta}$, is given by

$$p(y = j \mid \boldsymbol{x} \in R_i, \boldsymbol{\theta}, \mathcal{D}) \propto p(y = j, \mathcal{D} \mid \boldsymbol{x} \in R_i, \boldsymbol{\theta}) p(y = j \mid \boldsymbol{x} \in R_i, \boldsymbol{\theta}),$$

where R_i is the region corresponding to the ith terminal node.

With a prior that assumes that each class is equally likely in each terminal node then $p(y = j \mid \boldsymbol{x} \in R_i, \boldsymbol{\theta}) \propto 1$. Hence the distribution we are interested in is proportional to (6.2) with $\mathcal{D}$ replaced by $\mathcal{D}_j$, where $\mathcal{D}_j = \{\mathcal{D}, (j, \boldsymbol{x})\}$, i.e. the original dataset with an extra data point with response j at predictor location $\boldsymbol{x}$. To find the full distribution we need to normalise over the possible values of j so that

$$p(y = j \mid \boldsymbol{x} \in R_i, \boldsymbol{\theta}, \mathcal{D}) = \frac{p(\mathcal{D}_j \mid \boldsymbol{\theta})}{p(\mathcal{D}_1 \mid \boldsymbol{\theta}) + \cdots + p(\mathcal{D}_C \mid \boldsymbol{\theta})}.$$

This turns out to be a particularly simple expression for classification trees as we find that

$$p(y = j \mid \boldsymbol{x} \in R_i, \boldsymbol{\theta}) = \frac{m_{ij} + \alpha_j}{n_i + \sum_{j=1}^{C} \alpha_j}. \tag{6.9}$$

This fraction approaches the empirical distribution of the classes as the dataset becomes large but for low values of m_{ij} the class probabilities are shrunk towards their prior probabilities, $1/C$. Note that the maximum likelihood estimate of this probability is given when $\alpha = 0$, the degenerate prior case where the prior probability of being in each class is either zero or one (see Figure 6.3).

6.3.1 Stumps

A tree with just one splitting node is known as a stump. This is the simplest non-trivial tree structure and has been used in the past as a building block for a classifier. The usual vector of unknowns for a Bayesian tree is given by $\boldsymbol{\theta} = (\{s_i^{\text{pos}}, s_i^{\text{var}}, s_i^{\text{rule}}\}_{i=1}^{k-1})$ but for a stump we know that there are two terminal nodes and that the single splitting node is at position 1 (see Figure 6.2), i.e. $k = 2$ and $s_1^{\text{pos}} = 1$. These quantities are not random once we decide that we are fitting a stump so the vector of unknowns becomes just $\boldsymbol{\theta} = (s^{\text{var}}, s^{\text{rule}})$. With such a simple structure the summation in (6.8) is easily computable as $\boldsymbol{\Theta}$ contains only $\sum_1^p N(i)$ terms as $s^{\text{var}} \in \{1, \ldots, p\}$.

Unless the structure in the data is extremely simple, a single stump will provide very poor predictive performance. However, by sensibly combining many stumps together we can provide a model with acceptable predictive properties. As we have already seen in other contexts, Bayesian model averaging can be used to account for the misspecification of the truth by a simple stump. Hence we can assign tree priors, such as those outlined in Section 6.2.4, and provide predictions via the posterior predictive distribution.

In a similar spirit Oliver and Hand (1994) combined trees but without explicitly defining tree priors beforehand. Instead they noted that a sensible posterior distribution for the tree parameters is

$$p(\boldsymbol{\theta} \mid \mathcal{D}) \propto 2^{-\mathrm{MML}(\boldsymbol{\theta})},$$

where $\mathrm{MML}(\boldsymbol{\theta})$ is the minimum message length (Wallace and Freeman 1987; Wallace and Patrick 1993) of the explanation provided by the stump defined by $\boldsymbol{\theta}$. The minimum message length has been suggested as a model choice criteria in a similar spirit to the Akaike and Schwarz information criteria (Akaike 1974; Schwarz 1978) and in the context of tree models a similar criterion was used by Quinlan and Rivest (1989). Nevertheless, this model can not really be considered Bayesian as no notion of priors is adopted. Instead, we think of this 'posterior' as just a good, empirically driven, method for combining forecasts.

6.3.2 *A Bayesian splitting criterion*

Suppose that we want to describe the data with just the single 'best' stump model we can find. To do this we would choose the split that leads to the stump with the largest posterior weight, and we write the parameter vector that defines this stump as $\tilde{\boldsymbol{\theta}}$. Using the Bayesian machinery, we can determine $\tilde{\boldsymbol{\theta}}$ using

$$\begin{aligned}\tilde{\boldsymbol{\theta}} &= \arg\max_{\boldsymbol{\theta}}\{p(\boldsymbol{\theta} \mid \mathcal{D})\} \\ &= \arg\max_{\boldsymbol{\theta}}\{\log p(\boldsymbol{\theta} \mid \mathcal{D})\} \\ &= \arg\max_{\boldsymbol{\theta}}\{\log p(\mathcal{D} \mid \boldsymbol{\theta}) + \log p(\boldsymbol{\theta})\}. \qquad (6.10)\end{aligned}$$

When we assume that each possible split is equally likely then the $\log p(\boldsymbol{\theta})$ terms disappear and we see we are led to picking the stump with the largest marginal likelihood.

Buntine (1992b) shows that the log marginal likelihood of the stump approximates to Quinlan's information gain (Quinlan 1986) splitting rule when the total number of examples in each node is large so the Bayesian splitting rule has similar characteristics. Buntine assumes that $p(\boldsymbol{\theta})$ is constant for different stumps so, in this case, the Bayesian splitting criterion found by simple probabilistic arguments corresponds to a rule motivated empirically by good results. Two characteristics of this Bayesian rule, and Quinlan's information gain one, are highlighted in Buntine (1992b). Firstly, it favours splits made on predictors with many possible split points rather than those with only a few (e.g. a binary variable). Also, splits of the data that produce one large and one small group are preferred to ones that partition the data into equal sized blocks.

Instead of the prior suggested in Buntine (1992b) we could assign the prior given in (6.6). This compensates for the problem of favouring splits on predictors with many possible split points. The problem of splitting so more points are in one group than the other is not tackled by this prior but we could do this by introducing an extra term in expression (6.6) so that

$$p(\boldsymbol{\theta}) \propto \frac{2}{N(s^{\text{var}})p} \times (n_1 n_2)^c. \tag{6.11}$$

This prior penalises non-equal splits for $c > 0$ and corresponds to the usual prior of (6.6) for $c = 0$.

6.4 Searching for Large Trees

6.4.1 The sampling algorithm

We defined our Bayesian C&RT models in Section 6.2 and saw how to perform a full Bayesian analysis of simple trees in the last section. However, many datasets require quite complex trees so we need to perform a simulation algorithm to draw from the posterior distribution of the tree structures as the summation in (6.8) is computationally infeasible. As before, we must think of a set of move types which will traverse the probability space of interest. The following proposals have been suggested in the literature.

BIRTH. Randomly split one of the terminal nodes into two new ones by assigning it a splitting rule drawn from the prior.

DEATH. Randomly pick a pair of two terminal nodes with a common parent and recombine them.

CHANGE-SPLIT. Randomly pick a splitting node and assign it a new splitting rule drawn from the prior.

CHANGE-RULE. Randomly pick a splitting node and draw a new s_i^{rule}, from the prior, for the splitting question there.

SWAP. Randomly pick a parent–child that are both splitting nodes. Swap their splitting rules unless the other child has the identical rule, in which case swap the splitting rule of the parent with that of both children.

As required, the *BIRTH* and *DEATH* steps are the reverse of each other and change the dimension of $\boldsymbol{\theta}$, while the other moves propose jumps within a dimension. Note that *CHANGE-RULE* is a special case of *CHANGE-SPLIT* but is included to make more local jumps about the posterior probability space. The effectiveness of the *SWAP* step is demonstrated in Chipman *et al.* (1998a) and commented on in the accompanying discussion to their paper.

To simulate from the posterior of the tree structure it is obvious that the *BIRTH* and *DEATH* steps are required. Which of the within-dimension moves to allow is more difficult, but often it is wise to include one move that makes 'large' jumps and one that favours much more local moves. The local moves allow local modes to be found, whereas the larger jumps allow the sampler to escape from local modes so others can be discovered. With this in mind Denison *et al.* (1998b) only employ the *CHANGE-SPLIT* and *CHANGE-RULE* within-dimension moves while Chipman *et al.* (1998a) favour *CHANGE-SPLIT* and *SWAP*. Both approaches would appear satisfactory although incorporating all three move types is also possible. In any case a hybrid sampler is constructed to simulate draws from the posterior distribution of interest.

Having set out the Bayesian C&RT model in this chapter, it would appear straightforward to make posterior inferences about the tree structures using the generated sample of trees found from the simulation algorithm, as demonstrated in previous chapters. However, the nature of the model means that the analysis of the results of running this standard simulation algorithm is not so easy.

Firstly, for the model outlined so far, the acceptance probability for the moves that change dimension is not just given by (2.39), the minimum of one and the Bayes factor. This equation still holds for the moves within dimension but for those that change dimension it is not true. Consider a *BIRTH* step from current state $\boldsymbol{\theta}$, with k terminal nodes, to proposed new state $\boldsymbol{\theta}'$, with $k+1$ terminal nodes. We need to determine the ratio

$$R = \frac{q(\boldsymbol{\theta}' \mid \boldsymbol{\theta}')p(\boldsymbol{\theta}')}{q(\boldsymbol{\theta}' \mid \boldsymbol{\theta})p(\boldsymbol{\theta})}, \tag{6.12}$$

in the usual acceptance probability in the algorithm

$$\alpha = \min\left\{1, \frac{p(\mathcal{D} \mid \boldsymbol{\theta}')}{p(\mathcal{D} \mid \boldsymbol{\theta})} \times R\right\},$$

where the marginal likelihoods are given by either (6.2) or (6.5).

The prior terms in R are given in (6.6) but the proposal ratio in the C&RT case is not as straightforward as those met before. In the other models we have met when removing a basis function we have been able to propose any of those currently in the model. However, with a tree structure this is not possible as we only want to remove one terminal node in the *DEATH* step. To do this we need to find splitting nodes that have two terminal nodes as direct descendants and recombine them (see Figure 6.5), as mentioned in the description of the *DEATH* step. This figure also demonstrates the reversible nature of the dimension-changing steps. Whenever we perform a *BIRTH* step we can always get back to the original tree in one *DEATH* step.

We find that the proposal ratio depends on the number of locations in the tree at which there are pairs of terminal nodes that can be recombined into one terminal node. We write this number as $D_{\boldsymbol{\theta}}$. Now we can write down the probability of moving from the proposed to the current model as

$$q(\boldsymbol{\theta} \mid \boldsymbol{\theta}') = \frac{d_{k+1}}{D_{\boldsymbol{\theta}'}}, \tag{6.13}$$

as this is just the probability of performing a *DEATH* step together with the probability of choosing the pair of terminal nodes to recombine which were added from $\boldsymbol{\theta}$ to

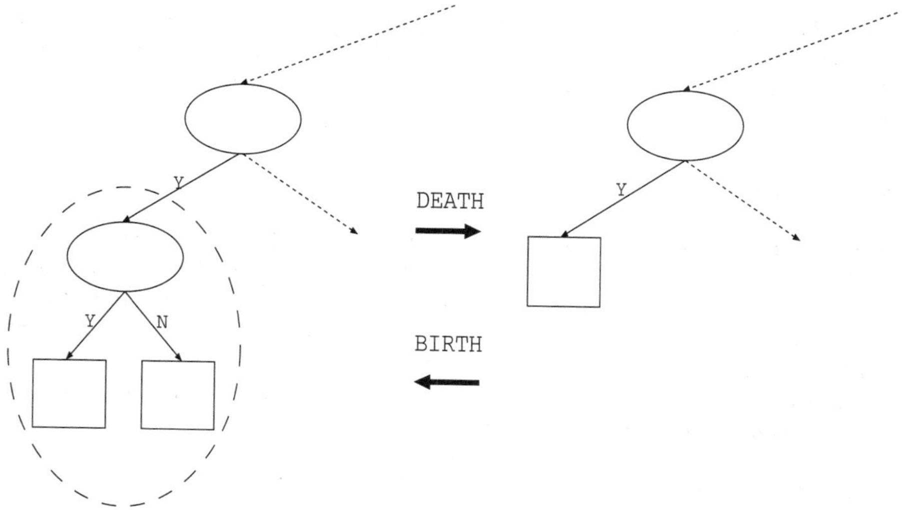

Figure 6.5 The *BIRTH* and *DEATH* steps allowed. Note that *DEATH* steps can only occur at locations where two terminal nodes have a common parent and that this structure is naturally produced by the *BIRTH* step. The node labels refer to the splitting nodes defined by the members of $\boldsymbol{\theta}$ with the appropriate index.

$\boldsymbol{\theta}'$. Similarly, the probability of moving from the current to the proposed model is the product of the probability of proposing a specific terminal node to split and the probability of assigning a specific question to this new split, i.e.

$$q(\boldsymbol{\theta}' \mid \boldsymbol{\theta}) = \frac{b_k}{k} \times \frac{1}{N(s_k^{\text{var}})p}. \tag{6.14}$$

We can now determine R using (6.6), (6.13) and (6.14) which turns out to be

$$R = \frac{k}{D_{\boldsymbol{\theta}'}} \times \frac{S_k}{S_{k+1}} \times \frac{d_{k+1}}{b_k},$$

rather than simply d_{k+1}/b_k as we have found in previous examples. Recall that S_k is the number of ways of constructing a tree with k terminal nodes.

In the simulation algorithms we met earlier we have taken the b_k and d_k to be constant in k for values of k except zero and the maximum number of bases, K. This ensures that R usually equals one and the acceptance probability is the minimum of one and the Bayes factor in favour of the proposed model. To allow us to accept with the same expression when using trees we could think of taking

$$b_k = \min\left(1, \frac{kS_k}{D_{\boldsymbol{\theta}'}S_{k+1}}\right), \qquad d_k = \min\left(1, \frac{D_{\boldsymbol{\theta}}S_k}{(k-1)S_{k-1}}\right),$$

as then R would equal one for nearly all values of k. However, the birth probabilities here depend on $D_{\boldsymbol{\theta}'}$, the proposed model. As we have not proposed a model yet this cannot be the case. Instead, as we find that in most cases $D_{\theta'} = D_{\theta} + 1$ we substitute this into the expression for b_k. In this way $R = 1$ all the times for which $D_{\theta'} = D_{\theta} + 1$. It is this algorithm which we shall use when sampling from the posterior in the rest of this chapter.

6.4.2 *Problems with sampling*

As we saw in the last section, the *BIRTH* and *DEATH* steps for the tree models depend on the current model, in contrast to the samplers we introduced before. Previously, we were used to deleting any of the basis functions present in the current model but for trees only terminal nodes that have a common parent can be combined. The number of such sites depends entirely on the shape of the tree. If we did not have this restriction, the sampler as described above would not be reversible. Also, we have to recombine terminal nodes as otherwise split nodes may not always have two descendants, leading to the tree structure not covering the whole of the predictor space. Thus, the basis functions which describe the tree are strongly interrelated as together they must completely span the predictor space.

We also find that the within-dimension moves in the sampler can change more than one basis function at a time. For instance, if a change in the splitting question at the root node is accepted, every single basis function is altered (think of the example in Figure 6.1). Again, this demonstrates the strong relationship between the parameters that define the tree in $\boldsymbol{\theta}$.

The interdependence between the parameters in $\boldsymbol{\theta}$ has a significant effect on the efficiency of the sampling algorithm. Firstly, within-dimension moves that affect many basis functions (i.e. those that propose changes near the root node) are unlikely to be accepted as they alter the model so much, possibly leading to empty, or nearly empty, terminal nodes. Also, as the current tree was chosen as it is a fairly good representation of the data, we find that it is difficult to construct steps that alter the model significantly yet still remain in an area of high probability. Thus, a tree of over about 10 terminal nodes will hardly accept any changes in splitting questions near the root. Further, if a tree of reasonable complexity is required to approximate the data, once grown, the tree is unlikely to be pruned back far enough so that top node changes become much more likely. Together these effects mean that throughout the algorithm the top few nodes are rarely altered. This has a great affect on the sample of models produced. Instead of the sample being drawn from the full posterior distribution, $p(\boldsymbol{\theta} \mid \mathcal{D})$, it is drawn from an approximate posterior distribution which is effectively conditioned on the top few nodes' splitting questions.

The most usual method of correcting this mixing problem is to devise move strategies that can traverse the posterior space better. The most obvious idea would be to add/delete whole branches of the tree, not just use the simple split/combine moves defined by *BIRTH* and *DEATH*. It would be simple to construct a step that removes a branch of a tree. However, the converse step, which must be included, will have to be able to add a similar branch and it is not obvious how to construct this step to maintain reversibility of the algorithm. Even with good proposals of this sort it is unlikely that the general problem with sampling will be completely overcome so, it appears, we have to accept that approximately independent draws from the full posterior will not be possible in practice. Note that as we have set up the sampler and model in a completely standard and probabilistically sound way, in theory none of this is a problem. However, in practice it is because the waiting times in local modes are so large that full sampling is infeasible.

Having set up a Bayesian model and not being able to fully sample from it is certainly frustrating. However, the Bayesian methods already seen can be utilised to provide a useful contribution to the tree-searching literature. It is to be expected that the stochastic search for trees would improve on the usual deterministic stepwise search, and this is certainly the case in many situations. Further, being able to include prior information to affect the selection of trees can be highly desirable in some circumstances.

The specific problems introduced by the Bayesian C&RT model allow us to explore Bayesian model selection criteria. This is particularly important for tree models as commonly a single tree will be required as a summary of the sample. This is because single trees provide a great deal of interpretability, for both data analysers as well as those who know little of the method. In contrast, averages over trees are difficult to interpret as they cannot be represented by a single tree structure. Also, with the difficulty in sampling from the posterior, it is not clear how to think of the 'posterior mean' found using the output of the stochastic search. Nevertheless it would be expected that averaging over the trees found would lead to better predictions than single tree models.

6.4.3 Improving the generated 'sample'

In the previous subsection we highlighted the fact that sampling from the full posterior distribution is not feasible with current MCMC methodology. However, two strategies have been proposed which try to combat some of the problems associated with the sampling. They do not attempt to combat inefficiencies in the sampler but just try and ensure that trees that fit the data well are found.

Firstly, Denison *et al.* (1998b) try and ensure that the sampling algorithm initially finds a good local mode. They do this by restricting the tree's growth at the beginning of the algorithm as during this time nearly all *BIRTH* steps are accepted as the tree is small and does not fit the data well. Allowing unrestricted growth of the tree initially can lead to only very large trees being able to fit the data well as the poor initial splits which are unlikely to be changed need to be compensated for lower down the tree. The idea that Denison *et al.* (1998b) propose to counter this is to restrict the tree at the beginning of the burn-in period to prevent the sampler going down the route of having poor initial splits. In their paper they restrict the trees from having more than six terminal nodes for the first few hundred iterations and then allow the tree to grow indiscriminantly. This strategy works well and further experimentation is needed to determine good restriction regimes (e.g. at most four nodes for first 500 iterations, then eight for next 500, then unrestricted). Although, in our experience, the presence

Table 6.1 Summary of the Pima Indian dataset.

Predictor name	Description
`preg`	Number of times pregnant
`plasma`	Plasma glucose concentration
`bp`	Diastolic blood pressure (mmHg)
`skin`	Triceps skin thickness (mm)
`insulin`	Two-hour serum insulin (μU ml^{-1})
`bmi`	Body mass index (kg m^{-2})
`ped`	Diabetes pedigree function
`age`	Age (years)

of a restricted phase of growth, or lack of it, is far more significant to the overall outcome than the actual regime employed.

Another method of improving the sample, proposed by Chipman *et al.* (1998a), is to restart the tree from the root node many times during the run. So, instead of one long run of the sampler from the initial tree with no splits, many short runs are undertaken, all started at the tree with a single terminal node. Although some of these short runs may become trapped in regions of the posterior where only poor trees exist, many of these short runs will find good trees. Restriction at the beginning is not required as it is not so vital that each run finds good trees as a restart will occur soon. It is only when one long run takes place that the initial tree is of vital importance.

The multiple-run method of Chipman *et al.* (1998a) has one particular advantage over the restricted growth method. Trees with very different initial splits are encountered regularly, whereas with one long run this is much less likely to be the case. However, many of these diverse trees are quickly rejected as they have little support from the data.

6.5 Classification Using Bayesian Trees

6.5.1 The Pima Indian dataset

We demonstrate the basic tree methodology using a binary classification example. We choose to analyse the Pima Indian diabetes dataset available from the University of California, Irvine machine learning database repository (`ftp:// ftp.ics.uci.edu/pub/machine-learning-databases`). The aim is to classify the 768 female patients into those with diabetes and those without on the basis of the eight predictor variables which are described in Table 6.1. Note that out of the total of 768 cases 268 were classified as having diabetes and we shall denote these as class 1 with the rest assigned to class 0.

To demonstrate the differences between the two sampling approaches outlined earlier we analysed the Pima Indian dataset in both ways. Firstly we ran the algorithm similarly to Denison *et al.* (1998b) for one long run of 250 000 iterations. Note that in this run we did restrict the tree from growing past seven terminal nodes for the first 600 iterations. We also followed the approach suggested by Chipman *et al.* (1998a) by generating a sample from 50 short runs of 5000 iterations. Both algorithms were run for the same total number of iterations so computation times were very similar.

As we are assuming that the sampler does not converge in the time we have available to run it, we cannot make probability statements about features in the model (e.g. the posterior distribution of the number of terminal nodes). This gives us the freedom to store every single model visited by the sampling algorithm, as independence between the models is no longer an issue, and we do not discard the initial trees visited as there is no need for a burn-in period.

Note that for both of these runs we employed the sampling approach of Denison *et al.* (1998b) and only ran the algorithm using the four move types *BIRTH*, *DEATH*, *CHANGE-RULE* and *CHANGE-SPLIT* and did not incorporate the *SWAP* step of Chipman *et al.* (1998a).

6.5.2 *Selecting trees from the sample*

One of the hardest things to do when analysing the output of the Bayesian tree algorithm is to summarise the wealth of information it provides. It would appear that there are many competing ways to select 'good' trees from the generated sample, whether that sample was collected from one long run, or multiple restarts at the root node. The most obvious would be to evaluate $p(\mathcal{D} \mid \boldsymbol{\theta})p(\boldsymbol{\theta})$ and then approximate the posterior $p(\boldsymbol{\theta} \mid \mathcal{D})$ by normalizing this quantity over the generated sample. However, Chipman *et al.* (1998a) point out a problem with this approach. Using their example, consider all possible trees with a single splitting node. Suppose there are two predictor variables, one binary and one with 100 possible split points. If the marginal likelihoods for two splitting questions are identical, one using the binary predictor and one using the one with many possible splits, the relative posterior weight for the split with the binary predictor will be much greater because small weight is associated with each of the 100 splits. This is due to the neighbourhood around the splits not being taken into account by the posterior.

To avoid this problem we use the marginal likelihood to compare tree models. Chipman *et al.* (1998a) suggest using this along with the misclassification rate to determine good models. However, the marginal likelihood is a good criterion as it does contain a natural dimension penalty, whereas the misclassification rate (or sum of squares for regression problems) is well-known to give an over-optimistic guide to a tree's true predictive capabilities (see, for example, Breiman *et al.* 1984).

6.5.3 *Summarising the output*

In Figure 6.6 we display a summary of the output from running the chain once and restricting its initial growth. We see that although the chain does not move quickly about the space, over the long run reasonable mixing takes place and many local

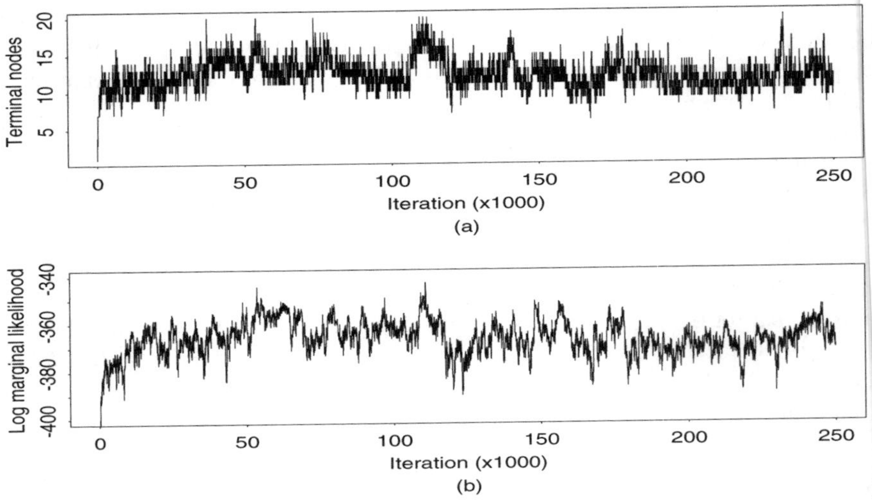

Figure 6.6 Results from one long run of the chain: (a) the number of terminal nodes, and (b) the log marginal likelihood, for each tree found by the algorithm.

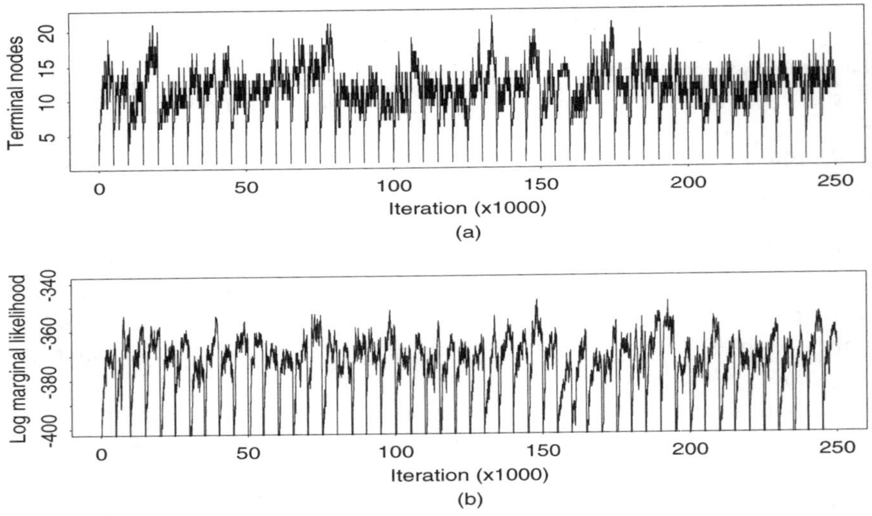

Figure 6.7 Same as Figure 6.6 but for 50 restarts of the chain every 5000 iterations.

modes are visited. For instance, at around 110 000 iterations the algorithm finds a clump of good trees with around 18 terminal nodes. However, around 230 000 iterations the sampler again visits trees with about 18 nodes but finds that these have low marginal likelihood so quickly moves away. However, this figure also shows the lack of convergence of the sampling algorithm. The marginal likelihood never settles down and appears to drift slowly from poor to good models and back again.

Figure 6.7 is the corresponding plot to Figure 6.6 but for the chain that was restarted 50 times at the root node. We see how sometimes a block of 5000 iterations yields a tree with a high log marginal likelihood but at other times it gets stuck in a poor local mode. This behaviour motivated the restart strategy of Chipman *et al.* (1998a). They give an example where the true tree had only five nodes for which the restart strategy works well, but for this particular dataset it seems that a much larger tree is required so restarts must only occur after a substantial number of iterations. Choosing how many iterations before a restart then becomes vital but it is better to err on the side of having too many than too few. However, having many iterations before a restart means that the method requires a lot of computational power, so it may only be possible to restart the algorithm a few times.

6.5.4 Identifying good trees

To stay within the Bayesian philosophy of this book we shall select trees according to their marginal likelihood rather than any other criterion, even though Chipman *et al.* (1998b) discuss ways to identify locally modal trees by defining between-tree distance metrics. Now, in Table 6.2 we give the highest log marginal likelihood found using each of the above methods for some of the more common numbers of terminal nodes. These results are typical and suggest little overall difference in the two sampling strategies although, with unlimited computing power, restarting with very long blocks may be expected to outperform the single long run method. It seems that the restart strategy certainly visits many more modes, some good and some bad but can fail to find good models with many nodes. The single run, although it gets trapped in modes for long periods of time (sometimes for about 100 iterations), given enough time, it does jump between other models with high marginal likelihoods and can be better at finding good large trees. The main drawback of the single-run strategy is the difficulty it has in changing the top nodes of the tree. However, we found that over the entire run the question in the top node changed 400 times so some mixing over very different trees is achieved.

In Figure 6.8 we display the tree with the highest marginal likelihood. It has 19 nodes and was found by the single-run strategy. Some aspects of the tree methodology are nicely demonstrated here. For instance the `skin` predictor is not split on so has no influence on the final classification rule produced by the tree. We can also get some idea of the relative importance of the predictors by seeing how many data points reach a node containing a split on each of the predictors. Using this idea we see that all the data points meet the first split on `plasma`, 679 reach one with a split on `bmi` and 543 reach one containing `age`. In some loose sense, this tree suggests that `plasma` is the most important predictor, followed by `bmi`, then `age` and so on.

Another tree of interest is displayed in Figure 6.9. This tree could be worth further investigation as it has the lowest misclassification rate of all those trees with the highest marginal likelihood for their number of nodes. Chipman *et al.* (1998a) suggest using misclassification rate as an indicator of good trees, but having two criteria to choose trees makes it unclear how to perform routine analysis. Having said this, for this

Table 6.2 The highest log marginal likelihoods (LMLs) found for the trees with between 11 and 20 terminal nodes with their corresponding number of misclassified points.

Terminal nodes	Highest LML		Misclassified points	
	No restarts	50 restarts	No restarts	50 restarts
11	−352.8	−351.6	143	149
12	−349.9	−353.1	151	156
13	−349.2	−348.5	150	162
14	−349.6	−349.2	151	162
15	−348.5	−353.1	148	157
16	−347.1	−349.5	151	154
17	−345.4	−348.2	145	153
18	−343.6	−347.8	147	154
19	−343.1	−347.7	147	152
20	−344.6	−348.9	146	152

example we find that the tree with the lowest misclassification rate is parsimonious (it has only 11 terminal nodes) and, as such, is unlikely to have overfitted the data. With this smaller tree it is also easier to identify the relevant variables as now we find that three of the predictors have been eliminated by the structure (`skin` again, as well as `bp` and `ped`). The other important predictors are similar to those found for the 19-node tree, although the effect of `insulin` is far more pronounced with this tree.

The trees in Figures 6.8 and 6.9 were both found using a single long run of the algorithm, so it is worth noticing how different they are at the top few nodes of the tree. The initial split on the 11-node tree immediately yields a terminal node while the 19-node tree has six terminal nodes to the right of the root node. However, similarities do remain, most notably the left-hand side branch of both trees which split on `plasma`, `age`, `preg` and `bmi`.

6.6 Discussion

We have seen how a stochastic search based on the MCMC sampling algorithm can be used to grow trees. From experience, we have found that a completely stochastic search may not be the best method for uncovering good trees. Especially when the tree required to explain the data are of reasonable size (greater than 10 terminal nodes), starting the algorithm from a good tree with only a few nodes can produce better results.

We found that better trees were discovered when we reanalysed the Pima Indian dataset starting each run from the four-node tree grown using a greedy search strategy. This tree is displayed in Figure 6.10. This tree, in common with those found by the completely stochastic searches, also suggests that `plasma` is a good variable to split on at the root node. We list the results found by starting from Figure 6.10 in Table 6.3. These are much more competitive with the classical tree growing algorithm also given.

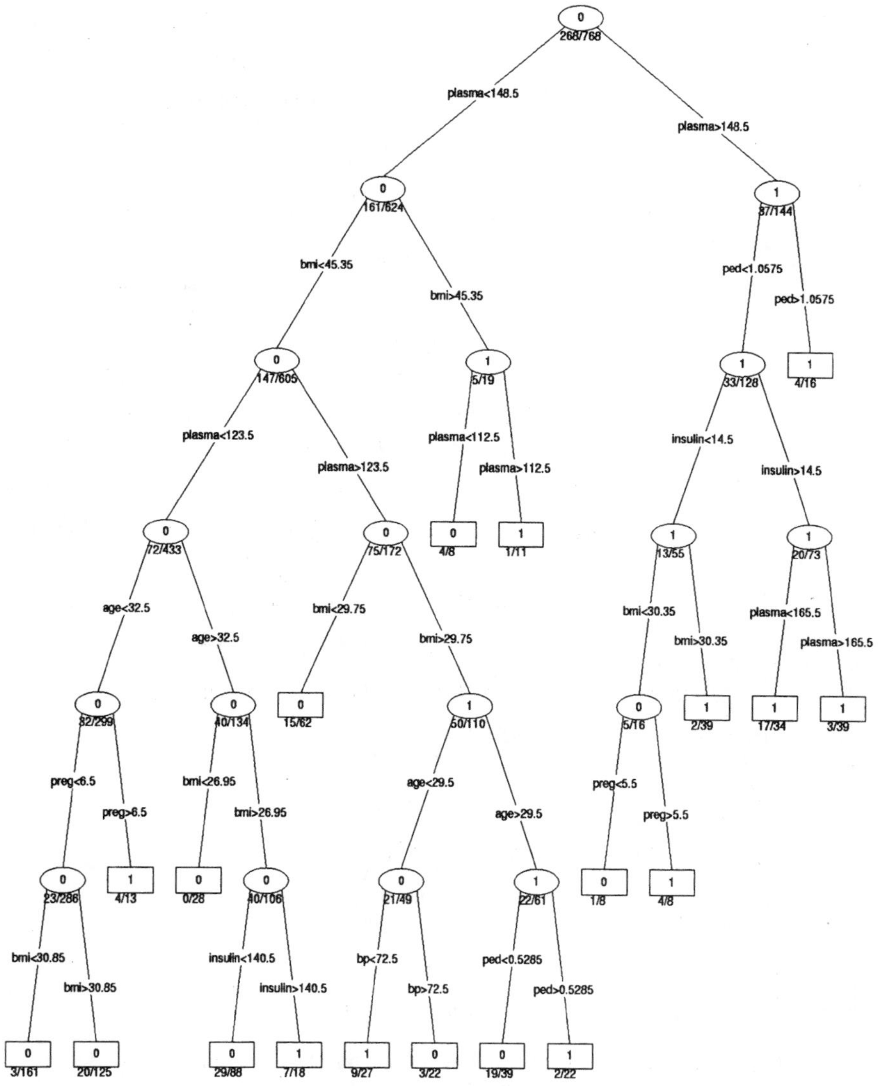

Figure 6.8 The best tree found using both methods and comparing trees with the log marginal likelihood. The numbers within each node give the most prevalent class at that node while the numbers below the nodes give the fraction of misclassified points.

Note that we used the `tree()` function in *S*+ (Clark and Pregibon 1992; Venables and Ripley 1997) to perform the greedy search for the four-node tree as well as the greedy classification tree results given in Table 6.3. Note that this strategy of starting from a good small tree was originally proposed by Chipman *et al.* (1998a).

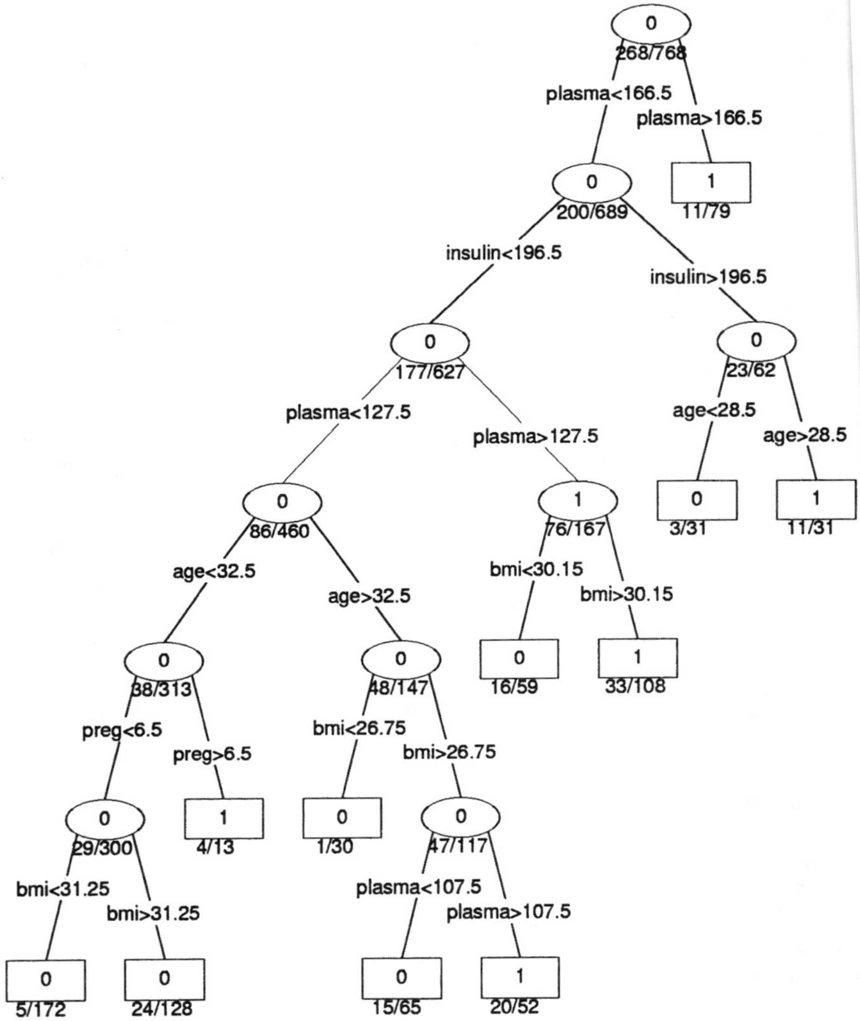

Figure 6.9 The tree found with the lowest misclassification rate of those referred to in Table 6.2.

The single long run, somewhat surprisingly, again appears to outperform the restart strategy. Even by doubling the block length we found that the restart strategy did not spend long enough in the local modes it found to uncover the best trees.

The results for the long run are comparable with the greedy search strategy for this dataset. In both Chipman *et al.* (1998a) and Denison *et al.* (1998b) they show how the Bayesian search strategy performs better than the usual greedy search in slightly easier problems. However, in examples such as this where a large tree is needed to fit the data

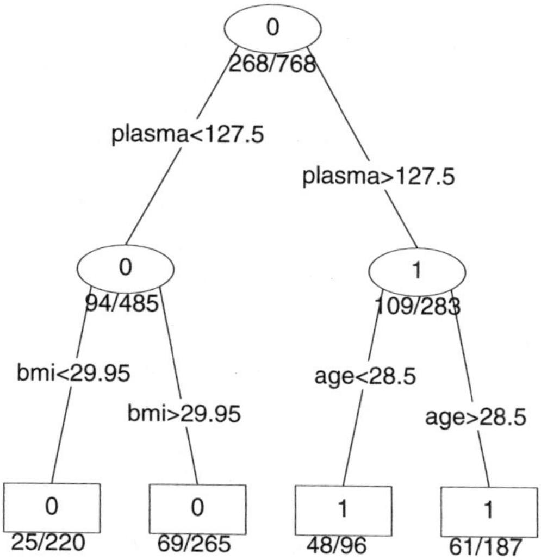

Figure 6.10 The four-node tree found by a greedy search, which was used to start the algorithm off to give the results in Table 6.3.

Table 6.3 The highest log marginal likelihoods (LMLs) found for the trees with between 10 and 20 terminal nodes with their corresponding number of misclassified points when starting the algorithms off at the greedy four-node tree. We also display the results from the classical greedy search strategy in the columns marked GCT (greedy) classification tree).

Terminal nodes	Highest LML			Misclassified points		
	0 restarts	50 restarts	GCT	0 restarts	50 restarts	GCT
11	−347.5	−355.6	−348.0	141	157	160
12	−342.5	−351.3	−344.1	135	157	144
13	−342.7	−351.3	−341.1	135	157	144
14	−339.9	−348.6	−338.4	146	146	144
15	−338.4	−346.5	−336.4	134	153	144
16	−336.9	−346.3	−334.5	143	154	144
17	−336.9	−347.1	NA	142	145	NA
18	−336.4	−334.9	NA	142	144	NA
19	−334.7	−331.2	NA	135	143	NA
20	−333.7	−330.6	−328.8	135	142	144

the stochastic search does not necessarily work better than the greedy strategy. For most of the other models in the book the Bayesian method outperforms the greedy search in nearly every instance so this potential deficiency is another unique, and undesirable, characteristic of the Bayesian tree method.

6.7 Further Reading

The use of stochastic search techniques to find trees is not a new one. Different search techniques have been suggested by many authors (see, for example, Isik and Ammar 1992; Lovasz *et al.* 1995; Nikolaev and Slavov 1997; Safavian and Landgrebe 1991) ever since the original search method was suggested by Breiman *et al.* (1984). A good example is the work of Bucy and Diesposti (1993), who use simulated annealing to find good trees. However, there remain many unanswered questions on how to perform effective searches of the posterior tree space. Running parallel tree searches and using the ensemble of trees to make updates seems a promising idea. Simpler methods to improve mixing that need further work are restarting at different trees and picking the trees from which to start the algorithm.

Making sense of the generated sample and picking the 'best' trees is also problematic. Using the marginal likelihood to select trees has been favoured by Bayesian practitioners but as the method is itself not strictly Bayesian maybe other non-Bayesian measures of fidelity will determine trees with better predictive power.

As sampling from the complete model space is infeasible using the simulation techniques we have described, averaging amongst sampled trees (see, for example, Oliver and Hand 1996; Shannon and Banks 1999) does not seem appealing in a Bayesian context. Trees are motivated as easy to interpret models rather than as great prediction tools. So, if prediction is the aim, averaging over models designed to predict well, rather than be interpretable, is a potentially superior strategy. However, when simple trees are involved (e.g. stumps, as in Oliver and Hand 1994) averages can be determined analytically and these can provide good, and reasonably fast, predictions.

Although Bayesian model averaging can be adopted to combine the predictions made by stumps another method of combining stumps, known as boosting, has been developed in the machine learning community by Schapire (1990), Freund (1995) and Schapire and Singer (1999). As with model averaging, predictions are made with a weighted sum of stumps. The difference is that the weights and form of each stump is found using weighted versions of the data. Empirically, this has been shown to be a very good, and computationally efficient, method for classifying data (see, for example, Bauer and Kohavi 1999; Dietterich 2000); however, the statistical properties of the approach are the subject of current research (Friedman *et al.* 2000; Schapire *et al.* 1998). In particular the relationship, if any, between boosting and Bayesian model averaging is being investigated (Mertens and Hand 1999) and Denison (2001) demonstrates a way to combine the two approaches. Boosting has also been applied to regression problems (see, for example, Avnimelech and Intrator 1999; Drucker 1997) but in this context it has not, so far, proved so popular.

Extensions to standard tree models include linear splitting questions (see, for example, Lim *et al.* 2000; Murthy *et al.* 1994) and fitting other distributions in the terminal nodes. Rather than fitting constant levels in the bottom nodes Chipman *et al.* (2000b) suggest performing linear regression in each segment of the predictor space. Other work involves fitting survival trees which assume that data in each terminal node has the same survivor function (Ahn 1996; Leblanc and Crowley 1993; Nelson *et al.*

1998). The next chapter brings together these sort of ideas as we shall see how to fit other distributions to the terminal nodes in tree-like structures which allow for more flexible partitioning schemes.

6.8 Problems

6.1 Prove that, in all binary-tree structures, there exists one more terminal node than splitting node.

Use this result to prove that there is at least one site in any binary-tree structure where two terminal nodes have a common parent splitting node as a direct ascendant (i.e. at least one site as shown in Figure 6.5 exists).

6.2 Suppose we are interested in performing binary or 0–1 classification with a stump and the probability of being in class j ($j = 0, 1$) when in node i ($i = 1, 2$) is denoted by ϕ_{ij}. If we choose to use the prior

$$p(\phi_{10}, \phi_{11}, \phi_{20}, \phi_{21}) = \mathrm{Di}_1(\phi_{10}, \phi_{11} \mid \alpha_{10}, \alpha_{11})\ \mathrm{Di}_1(\phi_{20}, \phi_{21} \mid \alpha_{20}, \alpha_{21}),$$

where the α_{ij}s are strictly positive integers, show that the marginal likelihood of the data given a stump structure that partitions the data so that there are m_{ij} points of class j in node i is given by

$$\prod_{i=1}^{2}\left\{\frac{\alpha_{i\bullet}-1}{m_{i\bullet}+\alpha_{i\bullet}-1}\binom{\alpha_{i\bullet}-2}{\alpha_{i0}-1}\binom{m_{i\bullet}+\alpha_{i\bullet}-2}{m_{i0}+\alpha_{i0}-1}^{-1}\right\},$$

where $\alpha_{i\bullet} = \alpha_{i0} + \alpha_{i1}$ and similarly for $m_{i\bullet}$.

$$\left[\textit{Note that } \Gamma(n+1) = n! \textit{ and } \binom{n}{r} = \frac{n!}{r!(n-r)!}.\right]$$

Suppose that the m_{ij} terms are large in relation to the prior terms α_{ij}. Use Stirling's approximation ($n! \approx \sqrt{2\pi} n^{n+1/2} \exp(-n)$) to show that the log of the marginal likelihood derived above is approximately equal to

$$m_{i0} \log m_{i0} + m_{i1} \log m_{i1} - m_{i\bullet} \log m_{i\bullet}.$$

Note that the final term in the expression above is a constant for all stumps and the first two terms correspond to minus Quinlan's information gain splitting criterion (Quinlan 1986), which is also known as the entropy function.

6.3 If we allow N possible splitting questions at each splitting node of a tree, find the total number of trees with four or fewer terminal nodes.

Suppose we are interested in classification and wish to determine analytically the posterior predictive distribution for the class assignment of a point with predictor values $\boldsymbol{x}$, i.e.

$$p(y = i \mid \boldsymbol{x}, \mathcal{D}) = \sum_{\boldsymbol{\theta} \in \boldsymbol{\Theta}} p(y = i \mid \boldsymbol{x}, \mathcal{D}, \boldsymbol{\theta}) p(\boldsymbol{\theta} \mid \mathcal{D}).$$

If we allow ourselves to perform no more than one million evaluations of the marginal likelihood (which is proportional to the posterior $p(\boldsymbol{\theta} \mid \mathcal{D})$), what is the largest value that N can take?

7

Partition Models

7.1 Introduction

The C&RT models of Chapter 6 demonstrate the general idea of partitioning the product space into disjoint regions, where the distributions of the points in each region are taken to be independent. However, as we noted at the time, there are difficulties with using the standard tree set-up, most notably the hierarchy in the basis functions leading to poor mixing in the generated posterior sample, and the axis-parallel nature of the partitions. The chapter introduces partition models which aim to overcome some of these difficulties.

First of all we shall define what we mean by a partition models, defining them similarly to the product partition models of Barry and Hartigan (1993).

Definition 7.1. A partition model is made up of a number of disjoint regions $R_1, \ldots, R_k$ whose union is the domain of interest, so that $R_i \cap R_j = \emptyset$ for $i \neq j$ and $\bigcup_1^k R_i = \mathcal{X}$. The responses in each region, given the parameters $\boldsymbol{\phi}$, are taken to be exchangeable and to come from the same class of distribution f. Thus the likelihood of the data given the model parameters $\boldsymbol{\theta}$ (which define the regions) and $\boldsymbol{\phi}$ is given by $\prod_1^n f(y_i \mid \phi_{j(i)})$, where $j(i) = \{j : \boldsymbol{x}_i \in R_j\}$.

Thus, the predictor space $\mathcal{X}$ is split into a number of disjoint regions, where the data contained in each region is assumed to come from a common density indexed by some region-specific parameters given in $\boldsymbol{\phi}$. The motivation behind this model is the reasonable premise that *points nearby in predictor space come from the same local distribution.*

For instance, when performing regression we may assume that points within a given region come from a distribution which has a fixed mean level. With this definition trees are special cases of partition models. However, our definition also allows for terminal node distributions that do not necessarily correspond to either classification or regression trees. Also, the partitioning structure has not been explicitly defined so can take any form.

Even though trees are special cases of partition models we need to consider this extra class of models due to the inherent difficulties associated with the Bayesian analysis of standard trees. In the previous chapter we saw that the biggest difficulty with fitting trees in a Bayesian framework was in simulating from the full posterior distribution of the structure. Also, only allowing axis-parallel splits hampers the model from finding

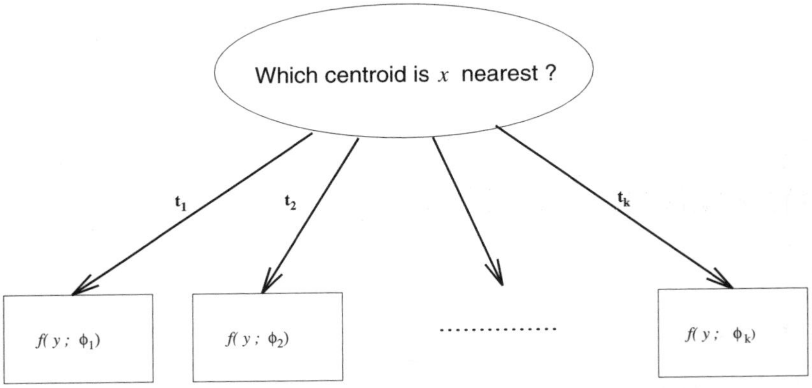

Figure 7.1 The Bayesian partition model represented as a tree structure. A new point with predictors $\boldsymbol{x}$ is assigned to the terminal node corresponding to whichever centroid $\boldsymbol{x}$ is closest to. The distribution of $y \mid \boldsymbol{x}$ within each terminal node is from the same family but indexed with different parameters ϕ_i.

parsimonious representations of the data. Thus, the more general class of Bayesian partition models (BPM) are designed to allow us to easily sample from the target posterior and to have more flexible boundaries between regions (or terminal nodes in the tree context). The advantages of this are immense as now a complete Bayesian analysis of the model can be performed. The main drawback comes in interpretability of the model. However, throughout this book prediction has been our main aim so the lack of interpretability is not a great issue.

Partition models remove much of the hierarchy in the basis functions by allowing only, at most, a few splits in any route down the tree. However, doing this and only allowing binary splits would lead to poor fits to the data due to the small number of terminal nodes that would be possible. However, by allowing more than just two-way splits at each splitting node we can overcome this problem.

The general partition model which we shall consider in this chapter can be represented as the tree structure shown in Figure 7.1. As with usual tree models the structure is grown using a set of training data and then predictions are made on a new point with predictor values $\boldsymbol{x}$ dependent on what terminal node $\boldsymbol{x}$ is assigned to. Remember that each of the terminal nodes is assigned a distribution indexed with parameters ϕ_i, with the complete set of terminal node parameters given by $\boldsymbol{\phi} = (\phi_1, \ldots, \phi_k)$.

The first thing to note about the partition model is that there is only one splitting node, located at the root node. Further, the tree allows non-binary splits at this node to compensate for the lack of extra splitting nodes. We define the partition of the predictor space via the random parameters $\boldsymbol{t}_1, \ldots, \boldsymbol{t}_k$. In essence, these give us the question at the splitting node with which we assign data points to regions (or, equivalently, terminal nodes). Following the convention of previous chapter we denote the complete vector of model parameters by $\boldsymbol{\theta} = (\boldsymbol{t}_1, \ldots, \boldsymbol{t}_k)$.

7.2 One-Dimensional Partition Models

We saw back in Section 3.2.1 how to use Bayesian methods to detect changepoints when we assumed that data were generated with normal errors. This model assumed one (or more) shifts in the mean level at some unknown locations. However, this model could also be used to smoothly fit the data by posterior averaging over the models with different changepoint locations. We find that the posterior predictive distribution found in this way varies smoothly with the predictor location.

However, when we encounter situations in which the data are not generated normally, we found that it was more difficult to sample from the required posterior to estimate the predictive distribution. In these cases we used Bayesian generalised nonlinear models to make inference by fitting flexible functions to the mean of the response so that

$$E(y \mid \boldsymbol{x}) = \beta_0 + \sum_{1}^{k} \beta_i B_i(\boldsymbol{x}).$$

However, the coefficients $\boldsymbol{\beta} = (\beta_0, \ldots, \beta_k)$ cannot be integrated out when the response is assumed to come from an exponential family distribution, but not the Gaussian one.

Partition models are useful as they give us a general framework for making inference using 'changepoint' models, and they also allow us to use conjugate models to ease posterior inference. They do this by taking

$$E(y \mid \boldsymbol{x} \in R_i) = \beta_i,$$

for $i = 1, \ldots, k$, so that the responses in each region R_i are assumed to come from some distribution with a fixed first moment. Let us see first how we can implement such a model when we only have one-dimensional data.

Figures 7.2 and 7.3 give examples of non-normal datasets that can be analysed using partition models. Figure 7.2 relates to the number of coal-mining disasters there were in Britain for every year from 1851 to 1962. In the early years it appears that there were many more disasters but after the turn of the century the disaster rate appears to be much less. This has been studied by various authors as a changepoint dataset (Green 1995; Jandhyala and Fotopoulos 1999; Lee 1998) and was introduced by Raftery and Akman (1986). As the observations relate to the number of random events occurring in a fixed interval of time, the obvious model assumes that each response is drawn from a Poisson distribution with some unknown mean parameter.

The other dataset, displayed in Figure 7.3, was first studied by Kitagawa (1987) and concerns the rainfall in Tokyo, Japan, in 1983 and 1984. For each day of the year in both 1983 and 1984 it was recorded whether or not over one millimetre of rainfall fell in Tokyo. Thus the data consist of a time series of counts, where each one can be modelled as a binomial experiment with two independent trials and an unknown probability of success which depends on the day of the year, except for the 29 February for which only one trial took place.

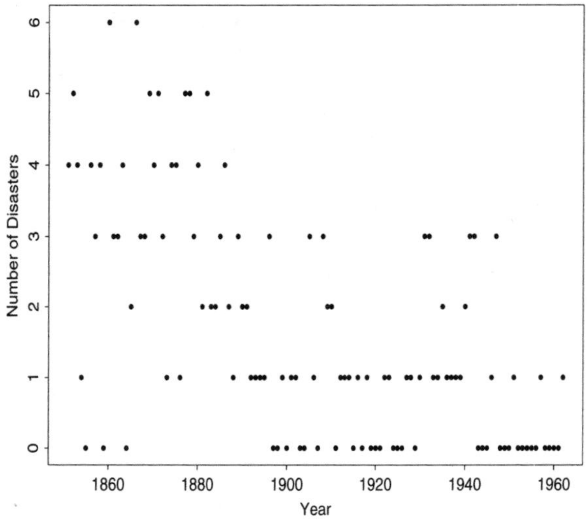

Figure 7.2 Scatterplot of the coal-mining disasters dataset giving the number of disasters in Britain each year from 1851 to 1962.

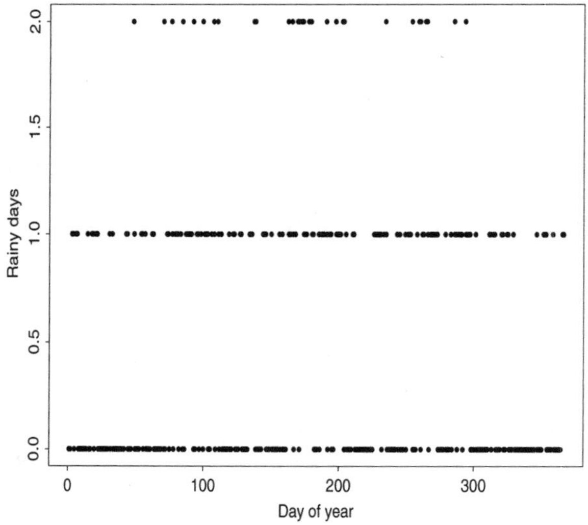

Figure 7.3 The Tokyo rainfall dataset. This was collected from 1983 to 1984 and gives whether, on each day of the year, no rain was detected in 1983 or 1984, it rained in one of the years, or it rained in both of the years.

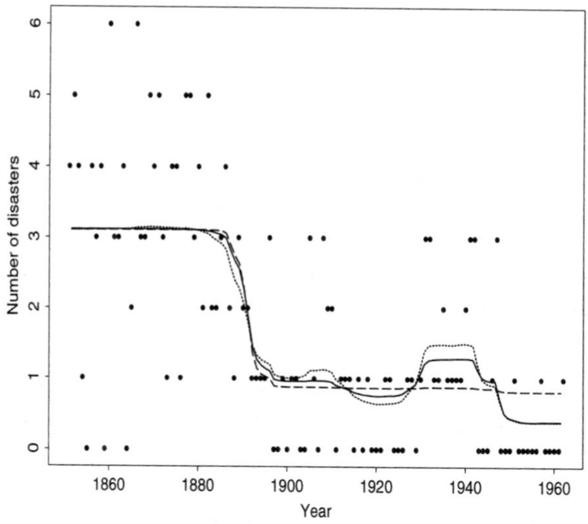

Figure 7.4 Results for the coal-mining data example using three different prior specifications on the Poisson mean of the response: Ga(0.0167, 0.01) (dashed line); Ga(0.167, 0.1) (solid line) and Ga(1.67, 1) (dotted line).

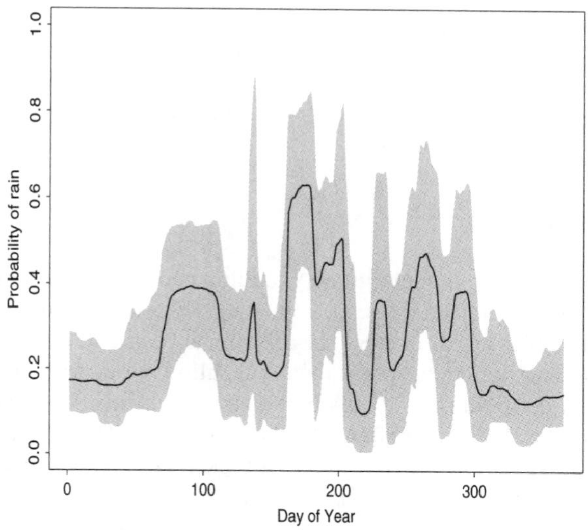

Figure 7.5 Results for the Tokyo rainfall dataset using a uniform prior on the probability of rain. The shading gives the 95% credible intervals for the mean.

We are going to use a Bayesian partition model to analyse both of these datasets. The unknown mean parameter of the Poisson distribution is estimated with a BPM for the coal-mining example, whereas for the Tokyo rainfall the unknown probability of rainfall is modelled with a BPM. We shall go into more details into how we actually do this later on, but for now we content ourselves with displaying and interpreting the results.

We ran the coal-mining example assigning a conjugate gamma prior to the number of accidents in a year, taking its prior mean to equal the empirical mean in the data (1.67 accidents per year). This prior ensures that the BPM can model deviations from 'average' behaviour.

In Figure 7.4 we display the posterior mean estimate to the mean number of accidents using the BPM. This is made up of an average over piecewise constant models such as that displayed in Figure 7.6, which happens to be the model found with the highest marginal likelihood. We display the results using three different priors, all with mean 1.67 but with different variances. We see that the results are quite stable to the choice of prior although the most diffuse one, given by the dashed line, does appear to be underfitting.

In Figure 7.5 we give the results found using a BPM for the Tokyo rainfall dataset. We give the posterior mean estimate to the probability of rainfall on a particular day of the year together with 95% credible intervals for this mean. In this example we assigned a uniform prior (i.e. a $\text{Be}(1, 1)$) to this probability rather than choosing it empirically as in the coal-mining example. We see the high variability associated with the probability of rain because the prior has a significant effect on the posterior for this dataset. In fact, here the prior has an influence on the posterior equal to the data, as the sum of the beta parameters of the prior is equal to the number of trials performed at each day of the year. This case demonstrates how influential a uniform specification can be. In this case another prior with both beta parameters less than one might be more appropriate as this lets the data dictate the form of the posterior much more.

7.2.1 Changepoint models

In one dimension, partition models are easy to think about. They are just changepoint models where, between adjacent changepoints, the responses come from a common distribution with different parameters. In a Bayesian context, all we need to do is to make inference on the changepoint locations, allowing us to then determine the required predictive distributions. The disjoint regions are easily defined by the location of these changepoints, which are often taken to be ordered. This was the general idea behind Bayesian partitioning when it was first looked at by Yao (1984) and Hartigan and co-workers (see, for example, Barry and Hartigan 1992, 1993; Hartigan 1990). Other notable work includes Carlin *et al.* (1992) and Stephens (1994). These papers were driven by understanding the changing nature of responses with only one predictor and Barry and Hartigan (1993) even suggested that partition models are unsuitable methods for approximating smooth functions. Although the curves in Figures 7.4 and

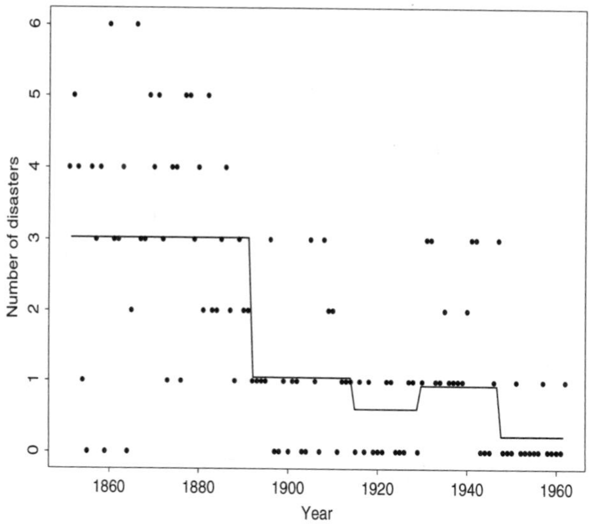

Figure 7.6 The partitioning found for the coal-mining dataset with the highest marginal likelihood with a Ga(0.167, 0.1) prior.

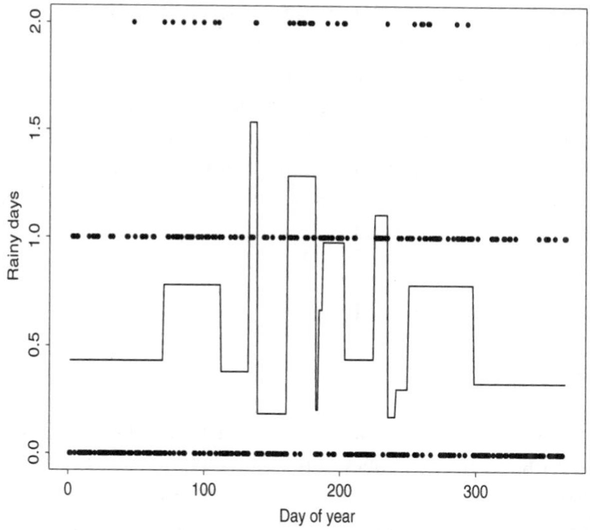

Figure 7.7 The partitioning found for the Tokyo rainfall dataset with the highest marginal likelihood.

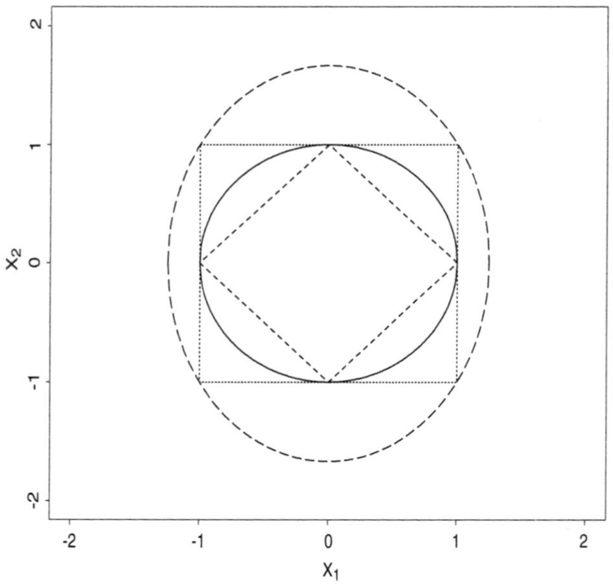

Figure 7.8 The sets of points which are a unit distance from the origin using the metrics in Table 7.1: Euclidean (solid line); absolute (short-dashed); city blocks (dotted) and Mahannobolis with $(w_1, w_2) = (\frac{16}{25}, \frac{9}{25})$ (long-dashed).

7.5 are not completely smooth, they again demonstrate how the mean of the posterior predictive distribution takes into account model misspecification.

Remember that our focus is still on good predictions to the data. This is in contrast to general changepoint methodology that wants to make inference about the location and number of changepoints. However, we can easily produce our best estimates to this by taking the maximum *a posteriori* sample or the one with the highest marginal likelihood. In Figures 7.6 and 7.7 we display the models with the highest marginal likelihood for the changepoint datasets.

7.3 Multidimensional Partition Models

7.3.1 Tessellations

Partition models in one dimension are easy to formulate because there is a natural ordering of the elements in that space. However, as soon as we move beyond one dimension we lose all ideas of ordering. Even defining the median of a bivariate distribution is not easy. Thus, regions for partition models in more than one dimension are not obvious to describe. Using a tree representation is one way to formulate disjoint regions that span the predictor space but we want to move away from such a restrictive partitioning regime.

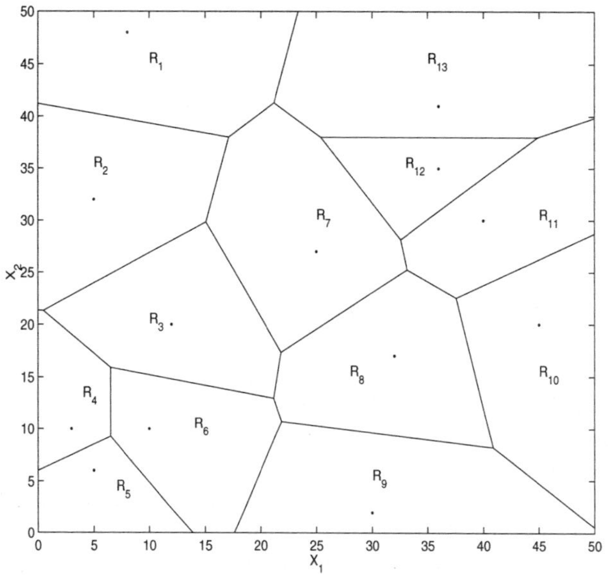

Figure 7.9 An example of a Voronoi tessellation found using the Euclidean distance metric. The dots mark the generating points of the tessellations with the lines giving the boundaries of the regions.

As mentioned in the introduction we wish to define regions so that points close together have the same distribution. We now describe one way that we can do this when the data are not univariate. We wish to define the splits via the centres of the regions, which we shall denote by $\boldsymbol{t}_1, \ldots, \boldsymbol{t}_k$. We then assign all the points nearest to $\boldsymbol{t}_i$ as being in region R_i so that

$$R_i = \{\boldsymbol{x} \in \mathcal{X} : D(\boldsymbol{x}, \boldsymbol{t}_i) < D(\boldsymbol{x}, \boldsymbol{t}_j) \text{ for all } j \neq i\},$$

where $D(\boldsymbol{x}_1, \boldsymbol{x}_2)$ is a prespecified distance metric defined for all points $\boldsymbol{x}_1, \boldsymbol{x}_2 \in \mathcal{X}$. We call the set of regions a *tessellation* as they form disjoint regions that span the predictor space. More specifically this particular structure is known as a Voronoi (or Dirichlet) tessellation (Green and Sibson 1978; Okabe *et al.* 2000; Voronoi 1908).

There are a number of ways to define the *distance* $D(\boldsymbol{x}_1, \boldsymbol{x}_2)$ between two points in $\boldsymbol{x}_1, \boldsymbol{x}_2 \in \mathcal{X}$ in a Voronoi tessellation. Some of the more common metrics are given in Table 7.1, and which one is chosen will obviously affect the regions produced. Figure 7.8 shows how the distance metrics given alter the set of points which are a unit distance from the origin, i.e. $\{\boldsymbol{x} : D(\boldsymbol{x}, \boldsymbol{0}) = 1\}$.

In Figure 7.9 we display an example of a Voronoi tessellation of a two-dimensional predictor space, where $\mathcal{X} = [0, 50] \times [0, 50]$ using the simple Euclidean metric. This is, in fact, equivalent to the special case of the Mahannobolis distance when $w_1 = w_2 = 0.5$.

Table 7.1 Commonly used distance metrics.

Distance metric	$D(\boldsymbol{x}_1, \boldsymbol{x}_2)$
Euclidean	$\sum_i (x_{1i} - x_{2i})^2$
Absolute	$\sum_i \lvert x_{1i} - x_{2i} \rvert$
City blocks	$\arg\max_i \lvert x_{1i} - x_{2i} \rvert$
Mahannobolis	$\sum_i w_i (x_{1i} - x_{2i})^2, \quad \sum_i w_i = 1$

7.3.2 Marginal likelihoods for partition models

For classification problems the marginal likelihood of the data given the parameters is identical to that given for classification trees in (6.2) using the same prior specification for $\boldsymbol{\phi}$ conditioned on the model. Further, using the prior specification given in (6.4), when we just want to fit a constant level to the data in each region we use (6.5).

In some cases it may be more appropriate to fit a regression model within each partition. For the regression tree this is usually taken to be just a constant level in each terminal node, but for the partition model we fit a linear plane in each region. This means that the resulting model is a combination of discontinuous planes (this is in contrast to the piecewise linear model of Section 4.4.4).

We assume that the responses in region R_i, denoted $Y_i = (y_{i,1}, \ldots, y_{i,n_i})'$, follow a multivariate normal distribution with mean $\boldsymbol{B}_i \boldsymbol{\beta}_i$ and variance $\sigma^2 \boldsymbol{I}_{n_i}$, where $\boldsymbol{B}_i$ and $\boldsymbol{\beta}_i$ are the basis matrix and corresponding coefficients in region R_i, respectively. To each set of coefficients $\boldsymbol{\beta}_i$ we assign the standard conjugate normal prior, i.e.

$$p(\boldsymbol{\beta}_i \mid \boldsymbol{\theta}, \sigma^2) = N(\boldsymbol{0}, \sigma^2 \boldsymbol{V}_i),$$

where $\boldsymbol{V}_i$ is the prior variance matrix for $\boldsymbol{\beta}_i$, which we shall take as $\boldsymbol{V}_i = v\boldsymbol{I}$. Note that we use a single overall regression variance, rather than one for each region, so take $p(\sigma^2) = \mathrm{Ga}(a, b)$.

Using these priors we find that the marginal likelihood of the data is given by

$$p(\mathcal{D} \mid \boldsymbol{\theta}) = \frac{(b)^a \Gamma(a^*)}{\pi^{n/2} \Gamma(a)} (b^*)^{-a^*} \prod_{i=1}^{M} \frac{|\boldsymbol{V}_i^*|^{1/2}}{|\boldsymbol{V}_i|^{1/2}},$$

where $\boldsymbol{V}_i^* = (\boldsymbol{B}_i' \boldsymbol{B}_i + \boldsymbol{V}_i^{-1})^{-1}$, the posterior variance in the ith partition, and a^*, b^* are found using (2.19) and (2.20). Perhaps unsurprisingly, we find again that we can write this partition model as a standard Bayesian linear model (Section 2.4) as we can write

$$Y = \boldsymbol{B}\boldsymbol{\beta} + \boldsymbol{\epsilon}, \tag{7.1}$$

where $\boldsymbol{B} = \mathrm{diag}(\boldsymbol{B}_1, \ldots, \boldsymbol{B}_k)$, $\boldsymbol{\beta} = (\boldsymbol{\beta}_1, \ldots, \boldsymbol{\beta}_k)'$ and $\boldsymbol{\epsilon}$ is the error vector. The prior we use in this representation is a NIG($\boldsymbol{0}$, $\boldsymbol{V}$, a, b), where $\boldsymbol{V} = \mathrm{diag}(\boldsymbol{V}_1, \ldots \boldsymbol{V}_k)$. The reason that we shall not stress this more compact notation is that it does not illustrate that we can evaluate the model much more efficiently by performing k separate linear regressions (k inversions of $p \times p$ matrices) rather than one large one (one $(kp) \times (kp)$ inversion).

7.3.3 *Prior on the model structure*

We use the usual prior specification for $\boldsymbol{\theta}$, which assumes every model in a single dimension is equally likely, and each dimension is equally probable, *a priori*. Hence

$$p(\boldsymbol{\theta}) = \binom{T}{k}^{-1} \frac{1}{K+1},$$

where T is the number of positions where we can place the centroids, $\boldsymbol{t}_i$, and K is the maximum number of generating points allowed. We can choose the T possible locations of the generating points either on a grid that covers the predictor space or at the data points themselves. In low dimensions it is often better to work on a grid, but this becomes impossible in high dimensions, where we prefer having candidate sites at the data points, or even the marginal predictors values.

The prior specification given above is completely general and allows us to specify partition models in any dimension. However, as we have already discussed, the original work on partition models focused mainly on problems in one dimension. In this case other priors were suggested which took advantage of the natural ordering in one-dimensional spaces. To explain these prior specifications it is easier to think of the model parameter $\boldsymbol{\theta}$ being made up of a fixed n-dimensional vector of zeros and ones. We use a one in the ith position to indicate a changepoint (i.e. the start of a new region) at the ith ordered X_1 value in the dataset. This is, in fact, identical to the curve fitting approach of Smith and Kohn (1996) using basis functions of the form $I(X_1 > x_{(i)})$ instead of cubic splines.

Yao (1984) follow Smith and Kohn (1996) and assign a constant prior probability p to the chances of a changepoint at every location and choose the prior on each coefficient as an independent normal distribution with common mean μ_0 and variance σ_0^2. However, Barry and Hartigan (1992) try an alternative approach and, assuming changepoints at $i_0 = 0 < i_1 < \cdots < i_k = n$, take the model prior as

$$p(\boldsymbol{\theta}) \propto (i_1 - i_0)^{-2} \left\{ \prod_{j=2}^{k-1} (i_j - i_{j-1})^{-3} \right\} (i_k - i_{k-1})^{-2}.$$

They motivate this prior specification by the desirable consistency properties it induces in the posterior partition distribution.

The methods of Yao (1984) and Barry and Hartigan (1992) can be generalised to higher dimensions but this is difficult. As noted during the discussion on the similar method of Smith and Kohn (1996) we find that this type of set-up will involve a parameter vector that grows exponentially with the dimension of the predictor space. Further, the prior used by Barry and Hartigan (1992) suggests that we use penalise large regions more than small ones. It is simple to work out the areas of the regions in one dimension but is often problematic in higher dimension if a flexible tessellation structure is allowed such as a Voronoi tessellation, although Heikkinen and Arjas (1998, 1999) demonstrate how to work out areas for regions in a two-dimensional Voronoi tessellation.

7.3.4 *Computational strategy*

Again, we suggest a simple form for the sampling algorithm consisting of three steps that are each chosen with equal probability.

BIRTH. Add a new generating point to the tessellation drawn uniformly from those allowable.

DEATH. Remove a randomly chosen generating point from the tessellation.

MOVE. Move a generating point by redrawing its position randomly from the allowable locations.

Thus the basic algorithm again uses the reversible jump sampler to move about a posterior space that consists of an unknown number of knot locations. All we do is sample from the posterior of these knot points, given the data. However, when using the Mahannobolis distance metric an extra move type is required to draw the weights $\boldsymbol{w} = (w_1, \ldots, w_p)$ as these too are then considered random. This move type is chosen with the same probability as the others and is described by the following.

ALTER. Alter the distance weighting vector $\boldsymbol{w}$ using a normal proposal tuned during the burn-in phase to give acceptance of around 30% (see Holmes *et al.* (1999b) for more details).

7.4 Classification with Partition Models

We now demonstrate the use of partition models for classification problems. We feel that they have better properties in this context as definite boundaries are often preferred in classification problems so the lack of smoothness between partitions is not a significant issue. It is for the same reason that classification trees are much more popular than regression trees.

7.4.1 *Speech recognition dataset*

We consider a classification task using the speech recognition dataset analysed by Peng *et al.* (1996). The data are shown in Figure 7.10 and represent 10 categories (or classes) of words uttered by 75 speakers with each speaker repeating each word twice. There are 1494 data points as three speakers for the seventh category are missing. The two predictors are the first and second formants taken from the spectral analysis of the

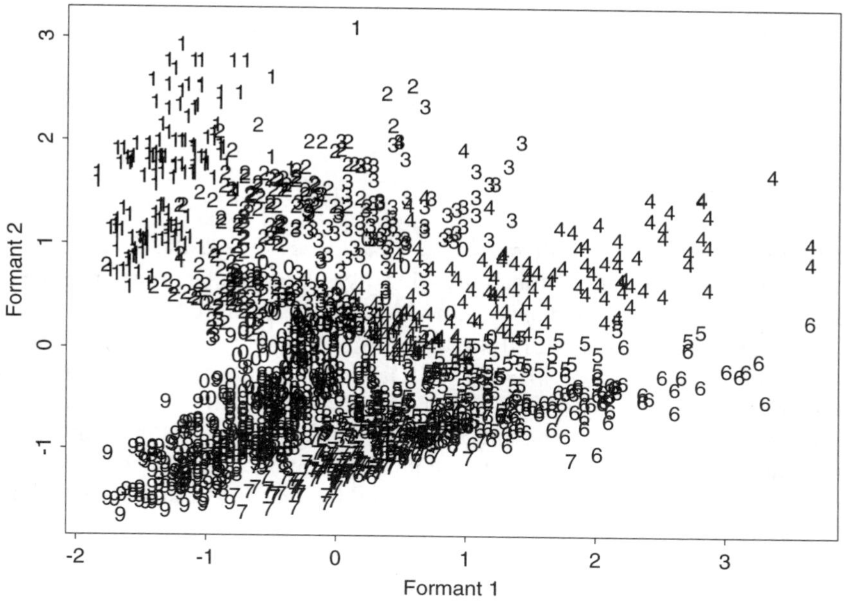

Figure 7.10 Speech recognition dataset. See Table 7.2 for labels.

Table 7.2 Words spoken which correspond to the classes in the speech recognition dataset.

Word	Class	Word	Class
Heard	0	Hud	5
Heed	1	Hod	6
Hid	2	Hawed	7
Head	3	Hood	8
Had	4	Who'd	9

recorded utterances and have been normalised to have zero mean and unit variance (as in Peng *et al.* 1996). The formants are the vocal tract's resonant frequencies. The particular words and their labels are listed in Table 7.2. We immediately notice that the class boundaries are not parallel to the axes and some appear very complicated so usual classification trees will be poor for this dataset. Further, there are some regions where there is little doubt of the dominating class (e.g. the scattering of 1s in the top left), while other areas are far more unsure (e.g. around the origin).

We analyse this dataset using a partition model with two different prior specifications and refer to the resulting models as BPM-N and BPM-I. BPM-N uses a uniform

Table 7.3 Average misclassification errors, over three runs with different seeds, for the 1345 point test sets for each training dataset for both the BPM-N and BPM-I models.

	Dataset				
	1	2	3	4	5
BPM-N	426	387	457	359	393
BPM-I	416	421	348	360	388

prior for the $\boldsymbol{\phi}_i$, hence we choose $\boldsymbol{\alpha}_i = (1, \ldots, 1)$. This suggests that in each region each class has the same prior probability. However, in many classification contexts we believe that each class dominates in at least one region in the predictor space. Hence for BPM-I we assign an informative prior on $\boldsymbol{\phi}$ which reflects this, taking $\boldsymbol{\alpha}_i = (1 + n\delta_{i1}, \ldots, 1 + n\delta_{iC})$ $(i = 1, \ldots, k)$, where $\delta_{ij} = 1$ if $i = j$ else $\delta_{ij} = 0$. This prior leads to region R_j $(j \leqslant C)$ finding an area in $\mathcal{X}$ where class j dominates. Thus, the generating points for these regions tend to place themselves in the middle of their allocated class clusters. Note that for both of these models we also take $p(k) = 0$ for $k < C$, the total number of classes, to ensure that the BPM-I prior makes sense for all possible k.

To demonstrate the relative predictive power of the partition model with the two different prior specifications, we randomly generated five splits of the data into 149 point training sets and 1345 point test sets. Again we found that after 50 000 burn-in iterations the marginal likelihood of the models had settled down. We then took every 100th model in the next one million iterations to be in our generated sample. This ensured that the dependencies between all the parameters in the model (including k) were small, although the stable predictions could be found using many fewer iterations. In Table 7.3 we give the results for the two different prior specifications.

In this example it appears that the informative prior aids in the classification performance of the BPM as on average it predicts better over these few splits of the data (an average misclassification rate of 387 against 404). This is because, for multiclass (i.e. more than two-class) problems the BPM-N occasionally fails to be the most probable class in any of the partitions which leads to a 100% classification error on this missing class. However, using the BPM-I prior does not provide us with information about the variability of class assignment over the predictor space as it produces a posterior which makes almost certain class assignments around the first C generating points even if the data do not provide such certainty.

In Figure 7.11 we show again how posterior averaging ensures that the predictions are a smooth function of $\boldsymbol{x}$, even though the single models themselves consist of hard boundaries. We display contour plots for the class conditional probabilities, $p(y = C_k \mid \boldsymbol{x}, \mathcal{D})$, for $k = 1, \ldots, 4$. The difficult decision boundaries are captured well by the BPM-N model and some notion of variability is present. One particularly nice feature is how, as we move further away from the observed data, we start to predict with the prior.

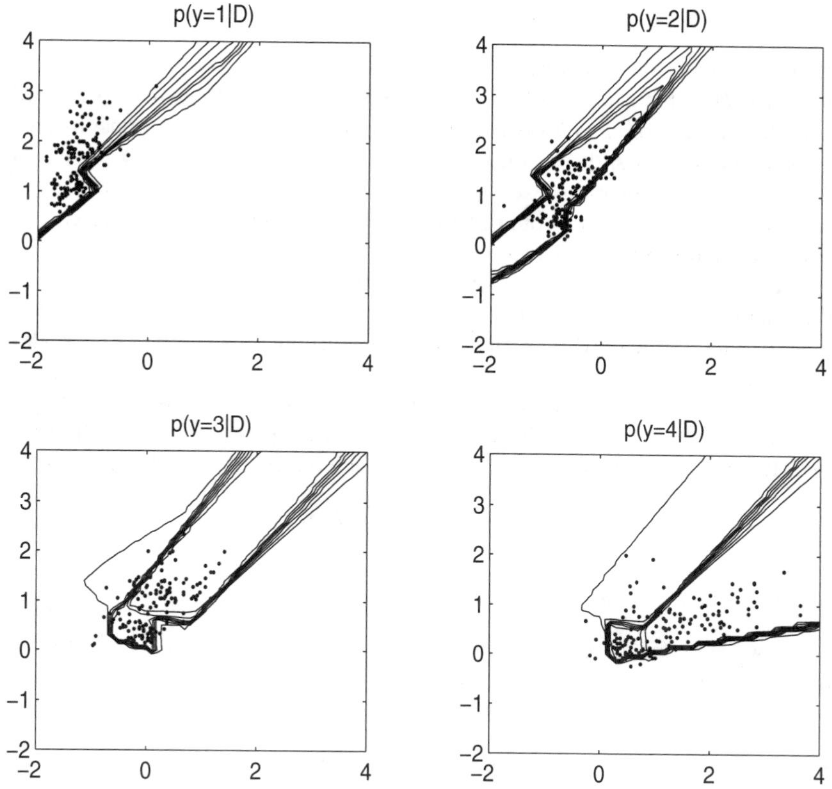

Figure 7.11 The contours of posterior probability for the speech recognition dataset for classes 1 to 4 using BPM-N.

7.5 Disease Mapping with Partition Models

7.5.1 Introduction

This section concerns the use of partition models for a specific spatial statistics problem. Spatial data analysis (for a review see Cressie 1993) involves determining the effect the location $\boldsymbol{x}$ has on the response y. Sometimes, each data point includes, in addition to the location information, further covariates which summarise other characteristics of the individual datum.

The partition model is best suited to analysing data of relatively few ($\leqslant 5$) dimensions so spatial data are ideal as the location information is most commonly given in two dimensions. In high dimensions points become more spread out due to the curse of dimensionality so partitions based on distance can behave poorly. Also, regions in high dimensions are no longer local so local modelling, as in the partition model, is not so attractive.

7.5.2 The disease mapping problem

Here we focus on the well-researched disease mapping problem (e.g. Clayton and Kaldor (1987), Schlattmann and Bohning (1993), Bernardinelli *et al.* (1995), Lawson *et al.* (1999), Pascutto *et al.* (2000)). This concerns estimating the relationship between location and disease risk based on a set of n observations $\mathcal{D} = (y_i, N_i, \boldsymbol{x}_i)_1^n$, where each $\boldsymbol{x}_i$ is a spatial location, each y_i is a count of the incidences of the disease at the location and each N_i is a count of the population susceptible to the disease at $\boldsymbol{x}_i$. Often the data are aggregated at some level and the (y_i, N_i) related to a geographical region (e.g. electoral ward, county, state) and the $\boldsymbol{x}_i$ are the centroids of these regions.

We can think of every member of the population as having an unknown probability (or risk) of contracting the disease in question. When concerned with spatial modelling we assume that the probability of contracting the disease at location $\boldsymbol{x}$ is given by $\xi(\boldsymbol{x})$. Disease mapping then concerns modelling this unknown function and here we choose to do this using a BPM. This involves taking the probability of contracting the disease to be a constant, ϕ_i $(i = 1, \ldots, k)$, in each region, R_i and define

$$\xi(\boldsymbol{x}) = \{\phi_i : \boldsymbol{x} \in R_i\}.$$

7.5.3 The binomial model for disease risk

Suppose we are concerned with the response y at a general location $\boldsymbol{x}$ where there is a susceptible population N. Assuming that individuals contract the disease independently given $\boldsymbol{x} \in R_i$ and ϕ_i we know that

$$\begin{aligned} p(y \mid \boldsymbol{x} \in R_i, N, \phi_i) &= \text{Bi}(N, \phi_i) \\ &= \binom{N}{y} \phi_i^y (1 - \phi_i)^{N-y}, \end{aligned} \tag{7.2}$$

for $y = 0, 1, \ldots, N$. By assigning the conjugate priors to this likelihood, namely taking

$$\begin{aligned} p(\phi_i) &= \text{Be}(\gamma, \delta) \\ &= \frac{\Gamma(\gamma + \delta)}{\Gamma(\gamma)\Gamma(\delta)} \phi_i^{\gamma-1} (1 - \phi_i)^{\delta-1}, \end{aligned}$$

for $\phi \in (0, 1)$ and $i = 1, \ldots, k$. We find that we can determine both the conditional posterior of ϕ_i (see Problem 7.8) and the marginal likelihood given the tessellation structure $\boldsymbol{\theta}$, which is

$$p(\mathcal{D} \mid \boldsymbol{\theta}) = \prod_{i=1}^{k} \left\{ \frac{\Gamma(\gamma + \delta)}{\Gamma(\gamma)\Gamma(\delta)} \prod_j \binom{N_{ij}}{y_{ij}} \times \frac{\Gamma(\gamma + y_{i\bullet})\Gamma(\delta + N_{i\bullet} - y_{i\bullet})}{\Gamma(\gamma + \delta + N_{i\bullet})} \right\}, \tag{7.3}$$

where y_{ij} and N_{ij} are the disease count and susceptible population for the jth data point in region R_i. Further, we take $y_{i\bullet} = \sum_j y_{ij}$ and define $N_{i\bullet}$ similarly.

Note that we used this model to analyse the Tokyo rainfall data of Section 7.2. Here we took the prior to be uniform which corresponds to setting $\gamma = \delta = 1$.

7.5.4 *The Poisson model for disease risk*

For diseases with high incidence rates (e.g. childhood asthma) the binomial model is appropriate, but more often N is large relative to y. In these cases the Poisson approximation to the binomial is nearly always used so instead we model

$$p(y \mid \boldsymbol{x} \in R_i, N, \phi_i) = \text{Poi}(N\phi_i) = \frac{(N\phi_i)^y \exp(-N\phi_i)}{y!}, \tag{7.4}$$

for $y = 0, 1, 2, \ldots$. The gamma distribution is conjugate to this Poisson likelihood so it is convenient to assume that

$$p(\phi_i) = \text{Ga}(\gamma, \delta) = \frac{\delta^\gamma}{\Gamma(\gamma)} \phi_i^{\gamma-1} \exp(-\delta\phi_i),$$

for $\phi_i > 0$ and $i = 1, \ldots, k$. Using this conjugate Poisson-gamma model we can again determine the marginal likelihood of the data given the parameters. This is given by

$$p(\mathcal{D} \mid \boldsymbol{\theta}) = \prod_{i=1}^{k} \left\{ \frac{\delta^\gamma \prod_j N_{ij}^{y_{ij}}}{\Gamma(\gamma) \prod_j y_{ij}!} \times \frac{\Gamma(\gamma + y_{i\bullet})}{(\delta + N_{i\bullet})^{\gamma + y_{i\bullet}}} \right\}. \tag{7.5}$$

Further, we can work out the predictive density for the disease rate in each partition for both this Poisson-gamma model and the binomial-beta model described before (see Problem 7.8).

7.5.5 *Example: leukaemia incidence data*

The example we present here closely follows the work of Denison and Holmes (2001). The dataset we shall use comes from Waller *et al.* (1994) and is available from Statlib at `http://lib.stat.cmu.edu/datasets/csb/`. The data concern the incidence of cancer in an eight-county region in upstate New York and Waller *et al.* (1994) used the dataset to locate clusters of cancer cases. Specific attention was paid to 11 hazardous waste sites in the study area as it was of interest to determine whether or not these waste sites were associated with the higher than normal disease rate. The waste sites were picked as they produced the chemical trichloroethylene (TCE) as effluent and this compound had previously been linked with leukaemia incidence.

This dataset was available as a set of centroids of regions, denoted by $\boldsymbol{x}_1, \ldots, \boldsymbol{x}_n$, together with the number of leukaemia cases in each region, y_i, and the population in the region, N_i. There were 790 data points in total with their location, and those of the

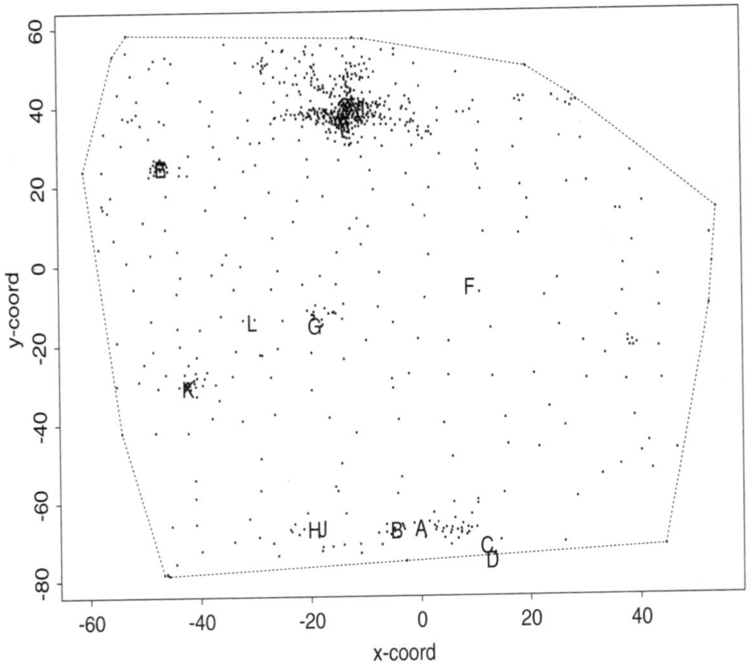

Figure 7.12 A scatterplot of the locations of the data points in the leukaemia incidence dataset. At each point a susceptible number of cases, N_i, and a number of leukaemia cases, y_i, is available. (Reproduced by permission of the International Biometric Society.)

waste sites, given in Figure 7.12 (note that the large cluster of points near the top of the map is the city of Syracuse, NY). Many of the responses y_i are not integers, as some incidences had undetermined locations. In these cases the incidence is population weighted among the regions where it could possibly have occurred. This causes us no problems as we can use the continuous version of the Poisson distribution which just replaces the factorial in (7.4) with a gamma function.

As the number of cases of leukaemia is small compared to the region populations we can use the BPM together with a Poisson distribution to model the risk $\xi(\boldsymbol{x})$ associated with location $\boldsymbol{x}$. We follow many other studies (see, for example, Diggle *et al.* 1997) and assume that each response y_i follows a Poisson distribution with mean $N_i\xi(\boldsymbol{x}_i)$. This model slightly alters the marginal likelihood in (7.5) with the factorials being replaced by gamma functions.

The dataset contains 592 cases of leukaemia amongst the total population of the 790 regions of just over one million people with the empirical mean risk being $\xi_0 = 5.60 \times 10^{-4}$, or 0.56 cases per thousand people. We adopt a data-based prior for each ϕ_i by shrinking the expected risk to the empirical mean as suggested by Clayton and

Kaldor (1987) and used for the coal-mining dataset example earlier in this chapter. To aid us in setting this prior we reparametrise it by taking

$$p(\phi_i) = \mathrm{Ga}(\xi_0 c, c).$$

Thus the prior mean for the disease risk is the empirical mean, ξ_0, and the prior variance is ξ_0/c. The smoothness of the prior is then completely determined by the choice of the only user-set parameter, c.

We could choose c by setting it empirically as well, or just by adopting a sensible value. Its effect is similar to that of the prior variance of the coefficients, v, for the linear model. As we saw in Section 3.3.5, we can remove much of the effect of such a parameter by allowing it to be random and assigning a hyperprior to it. Denison and Holmes (2001) favoured this approach and they suggested using

$$p(\log c) = U(\log c_{\mathrm{L}}, \log c_{\mathrm{U}}),$$

or, equivalently,

$$p(c) = \frac{1}{c(c_{\mathrm{U}} - c_{\mathrm{L}})}, \qquad c \in (c_{\mathrm{L}}, c_{\mathrm{U}}),$$

for $i = 1, \ldots, k$ and some prior constants c_{L} and c_{U}. Denison and Holmes (2001) choose $c_{\mathrm{L}} = 600$ and $c_{\mathrm{U}} = 50\,000$ so that the prior probability of the disease risk lying between $(\frac{1}{5}\xi_0, 5\xi_0)$ is between 0.5 ($c_{\mathrm{L}} = 600$) and 1 ($c_{\mathrm{U}} = 50\,000$) (within some acceptable tolerance). This reflects the premise that we have a better than even chance, at the very least, of the disease risk in any partition being within a fifth and five times the overall empirical risk.

Drawing c just requires an extra step in the sampler. We add a Metropolis–Hastings step and use the prior as the proposal density. In this case this proposal distribution gave adequate acceptance rates but in other cases a normal proposal density on the logarithm of the disease risk may be more appropriate.

Although good results can be found with fewer iterations the sampling algorithm was run for 50 000 burn-in iterations with every 1000th iteration in the next 10 million iterations taken as the independent sample from the posterior distribution. The acceptance rates for each move type were between 25 and 60% suggesting that the sampler moves adequately around the posterior, even with such simple move proposals.

7.5.6 *Convergence assessment*

We now describe one method of assessing convergence of the sampling algorithm for this particular model. However, these ideas can also be used for convergence assessment when using any of the other models outlined in this book.

Parameters for models that change dimension have little interpretation from one model to the next. For example, when using the partition model the location of the 'tenth' centre, $\boldsymbol{t}_{10}$, does not have the same meaning across all the models in the generated sample. When there are less than 10 partitions it is not even present in the model and it is likely to be in a very different position depending on the location and number of the other centres. This makes tracking the position of this particular centre throughout the sample meaningless. However, conventional convergence diagnostic checks (see, for example, Brooks 1998; Brooks and Roberts 1999; Cowles and Carlin 1996; Gelman and Rubin 1992; Robert 1995a; Roberts and Tweedie 1996) rely on showing that deviations from stationarity are not present in individual parameters throughout the run of the sampler. Because of the varying dimension of the posterior distributions we work with throughout the book the most widely used convergence diagnostics are not applicable. However, this is a promising area of future research (Brooks and Giudici 1999). For the time being we have chosen to come up with a practical and well-motivated rule-of-thumb for assessing convergence, or more accurately, diagnosing non-convergence.

As prediction is our focus we choose to look for convergence, or otherwise, in the predictions made by the models in the generated sample. If different subsamples of models do not predict similarly we have evidence of non-convergence. We now demonstrate how we can do this for the partition model for disease mapping.

We split the generated sample of 10 000 models into five sequential blocks and checked that the predictions at the waste sites using each separate block were not unduly different from the overall predictions made by the complete sample. This can be undertaken by comparing the predictions in each block with the appropriate binomial distribution, where the true predictive probabilities are estimated by those found with the entire sample.

In Table 7.4 we give the probabilities that the relative risk, defined by $r(\boldsymbol{x}) = \xi_0^{-1} E\{\xi(\boldsymbol{x}) \mid \boldsymbol{x}, \mathcal{D}\}$, was less than 1.2 for each subset of the sample at the 11 waste sites. We also give the overall posterior estimate to this probability found using the complete sample of 10 000 models. We can then compare the probabilities found for each subsample with the distribution of the probability found using the entire sample. For instance, look at the column relating to Site A. Using the whole of the generated sample our estimate is $P(r(\boldsymbol{x}) < 1.2) = 0.062$. Hence, if this were the true model we can assume that the number of times, in a sample of 2000 models, we would predict a relative risk value below 1.2 at Site A follows a $\mathrm{Bi}(2000, 0.062)$ distribution. By inspection of this distribution we see that 95% of the time we would expect this number to lie between 103 and 146. So, estimates to the predicted probability for each sample are expected to lie between 0.0515 and 0.0730, the 2.5% and 97.5% points indicated in the table. For Site A we see that in the third subset the predicted probability was outside this range.

Overall we find that in the 55 subsets looked at in Table 7.4, the posterior probabilities were within the 95% limits in all but five cases. The 97.5% percentile point for a $\mathrm{Bi}(55, 0.05)$ distribution is 6, so this provides no evidence to reject the idea that the predictions are not stationary. That is, we can assume that the algorithm has converged to an acceptable degree.

We could choose any points in the predictor space to monitor the predictions but in this example the waste sites are a natural choice, as it is at these locations that we are most interested in. In other models these test locations could be scattered randomly over the predictor space or possibly on a grid.

Table 7.4 Predictions for the probability that the relative risk is less than 1.2 made by five subsets of 2000 models at each of the 11 hazardous waste sites. We also give the overall mean prediction found from the complete generated sample of 10 000 models. The 2.5% and 97.5% values are the quantiles of the binomial distribution with $n = 2000$ and p set to be the overall mean for that site.

	Site number										
	A	B	C	D	E	F	G	H	I	J	K
Subset 1	0.061	0.074	0.221	0.217	0.345	0.661	0.110	0.781	0.789	0.984	0.571
Subset 2	0.056	0.072	0.232	0.225	0.361	0.651	0.112	0.792	0.804	0.981	0.581
Subset 3	0.074	0.084	0.251	0.245	0.355	0.686	0.124	0.782	0.785	0.982	0.562
Subset 4	0.068	0.083	0.209	0.209	0.370	0.691	0.116	0.775	0.787	0.983	0.570
Subset 5	0.052	0.070	0.235	0.230	0.348	0.685	0.128	0.777	0.789	0.981	0.589
Mean	0.062	0.076	0.230	0.225	0.356	0.675	0.118	0.782	0.791	0.982	0.575
2.5%	0.052	0.065	0.212	0.207	0.335	0.655	0.104	0.764	0.773	0.976	0.554
97.5%	0.073	0.088	0.249	0.244	0.377	0.696	0.133	0.800	0.809	0.988	0.597

We have argued that for the models we have already presented in this book, model parameters have little meaning across the generated sample. Nevertheless, Brooks and Giudici (1999) demonstrate a clever method that assesses convergence for graphical models (Lauritzen 1996), when using a reversible jump sampler. However, in general, it does not appear straightforward to parametrise all varying dimensional models in the way suggested although this is a topic of current research interest.

7.5.7 Posterior inference for the leukaemia data

In Figure 7.13 we display, as a contour plot, the posterior mean relative risk surface, $r(\boldsymbol{x}) = \xi_0^{-1} E(\xi(\boldsymbol{x}) \mid \boldsymbol{x}, Y, \boldsymbol{X})$. Again we see the smooth look to the posterior mean surface and we can see which regions have higher than average relative risk. However, just using this plot to make inference can be misleading as the posterior variances of these risks are not represented even though these can be readily determined.

In Figure 7.15 we plot the densities (as histograms) of the relative risk of leukaemia incidence at the 11 waste site locations. Note that in this plot as each of the bins is of width 0.25 the density is four times the posterior probability of being in each bin. It is only by viewing this plot that we can make inferences about the possible effect, or otherwise, of each of the sites.

From these results we see that there is considerable evidence that the relative risk at Sites A and B is above average as they both have less than 1% of their posterior probability associated with a relative risk of less than one. Of the other sites none have less than 5% of their posterior probability less than one. Site G has a mean relative risk of 1.55 yet, as it is located in a location with relatively few nearby data points (see Figure 7.12), the variance associated with this estimate is so high that a relative risk of 1 falls within its 95% credible interval. Hence there is no evidence to

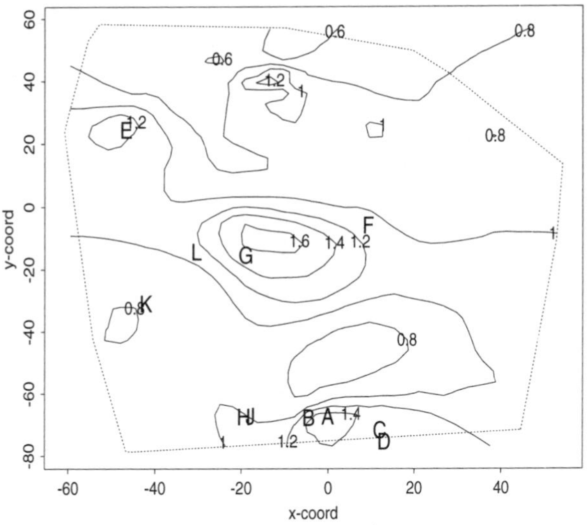

Figure 7.13 A contour plot of the posterior mean relative risk surface. The dotted lines give the convex hull of the dataset and the letters are the 11 waste sites where we are interested in the predicting the risk. (Reproduced by permission of the International Biometric Society.)

assume that there is a higher than average risk at G. This would have been missed by only examining the contour plot of the mean risk surface and highlights the need to determine posterior variances of things which we wish to make inference about.

After their analysis, Waller *et al.* (1994) discovered that Site A was near a well that operated during the time of the study (1978–1982) but was taken out of service in 1983 due to a New York Department of Environmental Conservation study which classified it as being a significant threat to public health due to TCE contamination. The well provided drinking water for the population south of the Susquehanna River, which runs just south of Site A. Thus, one reason the model suggested a high relative risk just south of A may be due to the effects of the contaminated drinking water provided by the well. The partition model is particularly well suited to picking up this sort of non-stationary nature in the underlying risk because it can allow for discontinuities in the risk function, which can often occur at natural boundaries like rivers. Other possible reasons for such discontinuities are mountain ranges, national boundaries or even city limits.

Just to show what a possible partition of the predictor space can look like, in Figure 7.14 we display the tessellation structure that gave the highest marginal likelihood value. It had 48 tessellations and we can see that there are a high number of tessellations around Syracuse, where there were a lot of data points. This is typical of such a model which tries to ensure that regions have similar numbers of cases and susceptible populations.

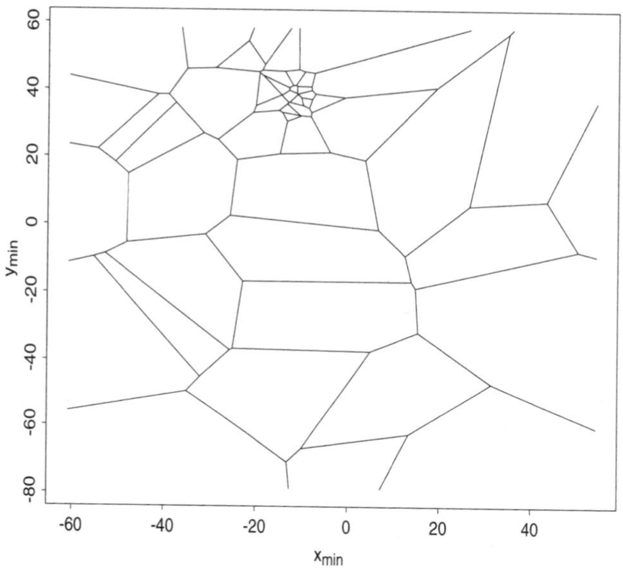

Figure 7.14 The regions for the model found with the highest marginal likelihood. It contains 48 regions.

7.6 Discussion

Partition models are proposed for their predictive power but they throw up some particular philosophical issues about Bayesian modelling. The partition structure, leading to discontinuous regression functions, are proposed as they allow us to still make use of conjugate models. However, one criticism of the models is that they do not actually represent the modeller's beliefs about the true underlying regression function. It is only through posterior smoothing that we obtain functions that mimic the truth more closely as the posterior expectation is much smoother than any single model.

We adopt an $\mathcal{M}_{\text{open}}$ perspective to Bayesian modelling (Bernardo and Smith 1994) so we never believe that the true model lies within the set of possible models. So, partition models are like every other model suggested in this book in that they are not expected to, and cannot, exactly mimic the truth. Instead, they are only used as ways of representing our beliefs about the relationship between y and $\boldsymbol{x}$. We find that, *a priori*, each predictor location $\boldsymbol{x}$ has an identical prior distribution on y, for example a $\text{Ga}(\xi_0 c, c)$ one. Hence marginally, y has a perfectly reasonable prior distribution despite the introduction of the model parameters $\boldsymbol{\theta}$.

Modellers would rarely know how many partitions to fit but are more likely to have some idea that y varies slowly with $\boldsymbol{x}$. Hence, although individual partition models

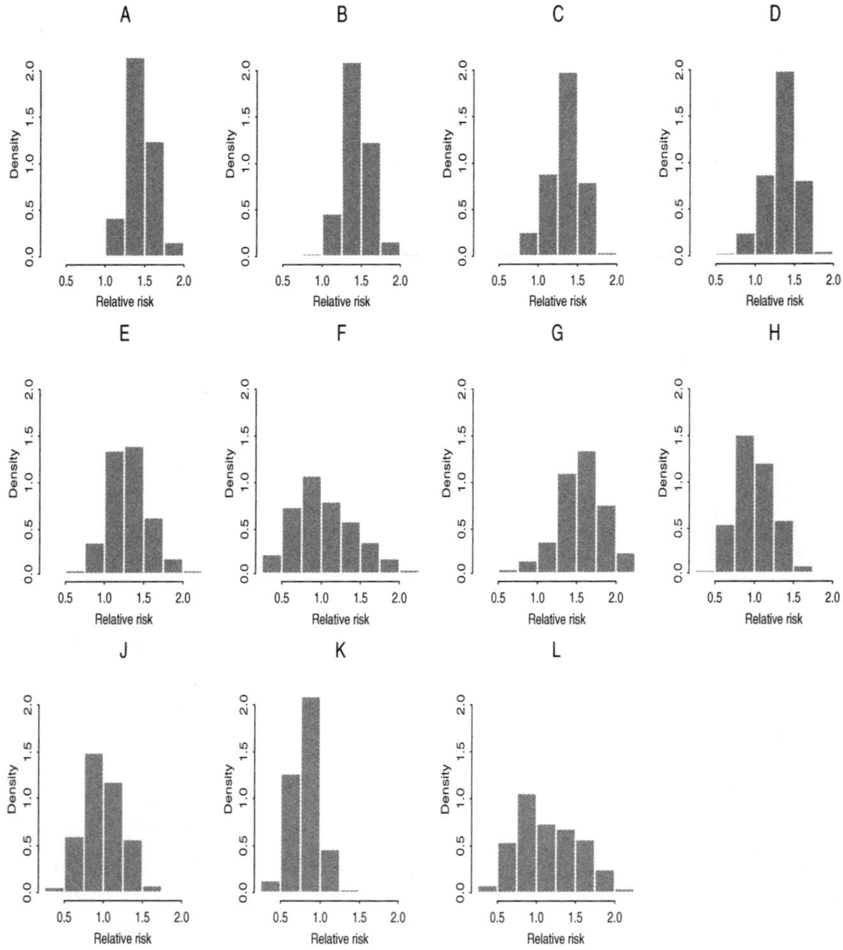

Figure 7.15 The densities of the relative risk at each of the 11 waste sites. (Reproduced by permission of the International Biometric Society.)

may appear unlikely when viewed singly, the prior predictive distribution,

$$p(y \mid \boldsymbol{x}) = \int_{\boldsymbol{\Theta}} p(y \mid \boldsymbol{x}, \boldsymbol{\theta}) p(\boldsymbol{\theta}) \, \mathrm{d}\boldsymbol{\theta}$$

$$\approx \frac{1}{T} \sum_{t=1}^{T} p(y \mid \boldsymbol{x}, \boldsymbol{\theta}^{(t)}),$$

where $\boldsymbol{\theta}^{(1)}, \ldots, \boldsymbol{\theta}^{(T)}$ is an approximate sample from the prior distribution, $p(\boldsymbol{\theta})$, and may be investigated to see if its characteristics reflect our beliefs.

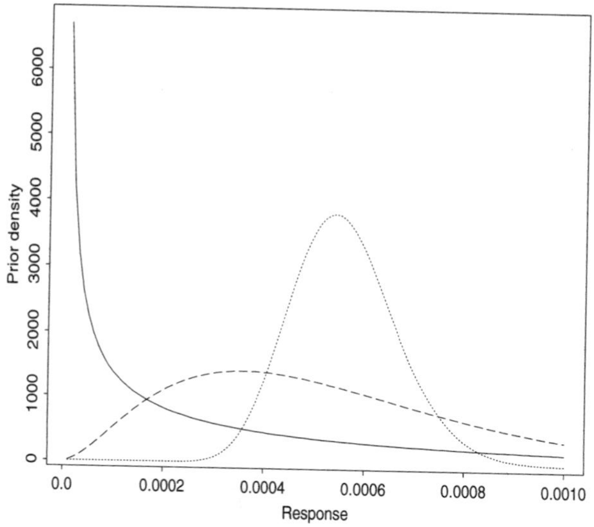

Figure 7.16 The prior predictive distribution (i.e. a Ga($\xi_0 c$, c)) for the responses for the disease mapping partition model given various values of c: $c = 600$, solid line; $c = 5000$, dashed line; $c = 50\,000$, dotted line.

To draw from $p(y \mid \boldsymbol{x})$ we just pick one of the generated sample at random and then draw from the distribution in the region that $\boldsymbol{x}$ lies within. However, for the prior model, each region is identically distributed with a Ga($\xi_0 c$, c) distribution, given c. We plot some of these distributions in Figure 7.16. The main point to note is that this prior density remains constant across the predictor space so *a priori* we are saying that y comes from some fixed mixture density, independent of $\boldsymbol{x}$. Thus, the partition model implies a reasonable dependence structure between y and $\boldsymbol{x}$ when $\boldsymbol{\theta}$ is marginalised out, even though it appears unreasonable conditional on $\boldsymbol{\theta}$.

In a similar spirit to the $\mathcal{M}_{\text{open}}$ perspective in the rejoinder to the discussion of Raftery *et al.* (1997) it is suggested that models cannot be used to represent prior beliefs. Instead, priors should be thought of as weights of evidence amongst the set of possible models in $\boldsymbol{\Theta}$.

Regression models allow us to view the effect of posterior smoothing particularly easily. As with many of the other regression models we have met previously we know from (7.1) that the partition model can be written as a linear model. So, by comparing the results found using the partition model and the spline models of Chapter 3 we could compare the relative merits of the basis functions given in (3.2) and those defined by the partition model.

In Figure 7.17 we plot a true curve together with data generated by the addition of normal errors. This function happens to be taken from a simulated data example given by Fan and Gijbels (1995). In Figure 7.18 we display the posterior mean estimate to

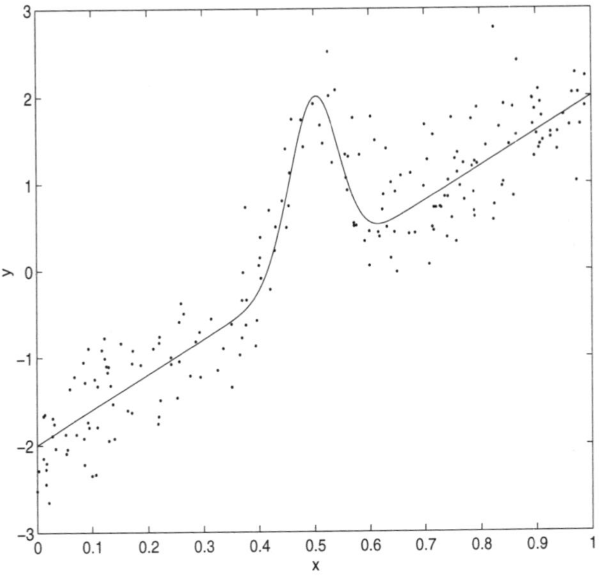

Figure 7.17 A simulated dataset, together with the true function, from an example of Fan and Gijbels (1995).

the truth and again see how the marginalisation induces smoothness in $p(y \mid \boldsymbol{x}, \mathcal{D})$. In addition, for linear models, we can write

$$E(Y \mid \mathcal{D}) = \boldsymbol{B}(\boldsymbol{B}'\boldsymbol{B} + \boldsymbol{V}^{-1})^{-1}\boldsymbol{B}'Y := \boldsymbol{S}Y,$$

where $\boldsymbol{S}$ is known as the smoothing (or sometimes the 'hat') matrix. It is an $n \times n$ matrix that gives us the relative weight each observed response has on each expected response value. By plotting the rows of $\boldsymbol{S}$ that relate to certain predictor values we can see how the partition model automatically incorporates adaptive smoothness into the posterior distribution.

Figure 7.19 gives the rows of the smoothing matrix at $x = 0$, 0.5 and 0.8. These are drawn as lines but they are actually discrete sets of points. We see at $x = 0.5$, when the true curve is varying most, the fitted value is a weighted mean of the observed responses between 0.4 and 0.6. As would be expected, the weight decays off from its peak at 0.5. However, at 0.8 the fitted value takes into account the observed responses between 0.6 and 1. A much wider range is used here as at this point the curve is just linear and all the points between 0.6 and 1 are effectively from the same linear model and can help in determining a good estimate at 0.8. Obviously, in areas where the weights are more spread out, the resulting posterior estimate is smoother.

We can also see how the partition model prior imposes smoothness on the observations. By simulating from the prior in a region of dimensions $[0, 100] \times [0, 200]$ we find that the smoothing matrix for the partition model, using the Euclidean distance

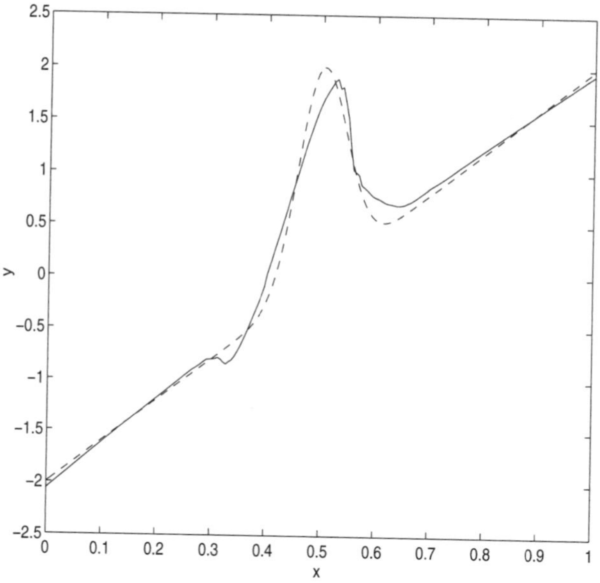

Figure 7.18 The posterior mean prediction using the partition model (solid line), together with the true curve (dashed line).

metric, can be represented by Figure 7.20. Here we see how, away from the boundary, the partition prior suggests that the correlation between points decreases with increasing distance, very similarly to a bivariate Gaussian distribution. Hence, the partition model prior imposes a well-motivated prior for $p(y_1, \ldots, y_n)$, even though the model itself appears to have some undesirable properties.

Recall that the representation theorem was introduced as a way of allowing us to assign suitable priors on the observed responses $y_1, \ldots, y_n$. In fact, one way to judge how good a parametric representation of the data is, is in terms of how well it mimics our beliefs about the density $y_1, \ldots, y_n$. Further, if two models give similar prior representations of $p(y_1, \ldots, y_n)$ it would seem natural to favour the model with which it is easier to make posterior inference. Partition models do well on both accounts. They mimic our beliefs about the correlations between data points in $\mathcal{X}$ decreasing with increasing distance *a priori* (Figure 7.20), as well as being particularly straightforward to make posterior inference with.

7.7 Further Reading

The main reading for further discussion of the work in this chapter is Holmes *et al.* (1999b), Denison and Holmes (2001) and Ferreira *et al.* (2002). However, the usefulness of partition models derives from their similarity with the tree models of Chapter 6 and the references given there. The work of Chipman *et al.* (2000a) is

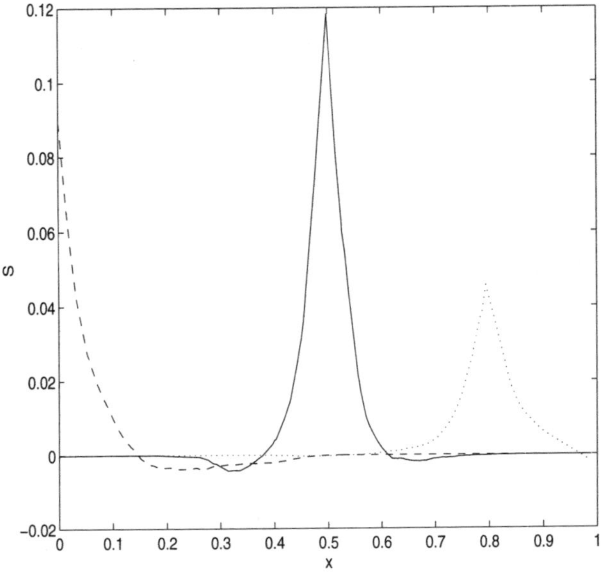

Figure 7.19 The posterior mean smoothing matrix at $x = 0$ (dashed line), $x = 0.5$ (solid line) and $x = 0.8$ (dotted line).

especially relevant and this too proposes partition models which fit independent linear regressions in each of the regions. However, they choose to still partition using a tree structure.

The early Bayesian literature on partition models was mainly concerned with changepoint analysis, nearly always in one dimension. Notable work includes Yao (1984), Carlin *et al.* (1992), Barry and Hartigan (1993) and Stephens (1994). These involve splitting up the one-dimensional predictor space into independent blocks. More recent work on changepoint problems is given in Crowley (1997), Lee (1997, 1998) and Quintana and Iglesias (2001).

The restoration of ion channels signals involves the determination of an unknown number of changepoints between a channel being 'open' or 'closed'. This application has attracted much interest in the statistical community (for an overview see Ball and Rice 1992) as it is a particularly useful outlet for (one-dimensional) changepoint methodology. Work on changepoint detection driven by this application includes Ball and Davies (1995), Rosales *et al.* (1998), Hodgson (1999), Hodgson and Green (1999) and Stark *et al.* (1999).

As well as the one-dimensional work on changepoint analysis, partition models in two dimensions are well known in image and spatial data analysis. The main difference between these disciplines is that images are usually made up of a collection of responses at a regular grid of locations whereas the data locations in most spatial data analyses are less structured. Both disciplines use similar techniques and often require the splitting up of the predictor space into regions of similar colour/texture

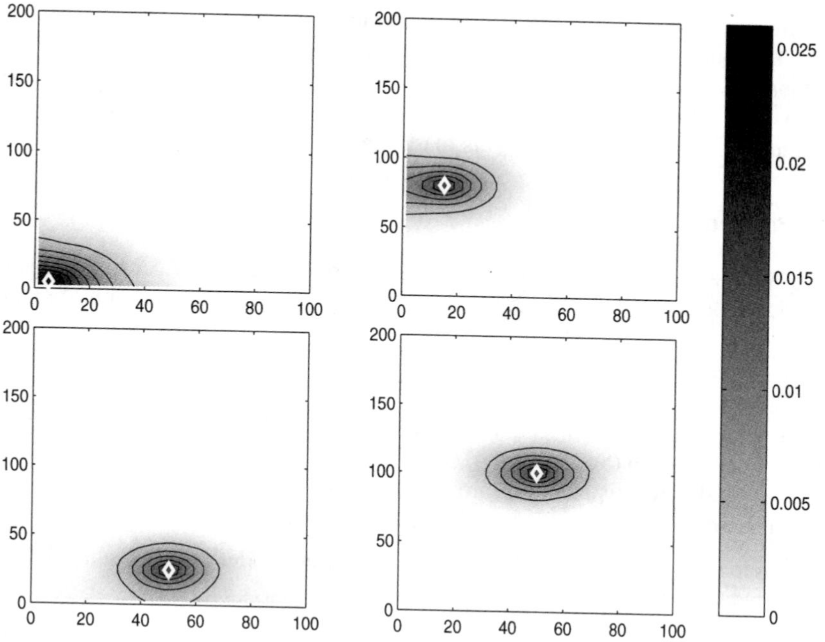

Figure 7.20 A two-dimensional representation of the smoothing matrix associated with the prior partition model with Euclidean distance on a rectangle of dimensions $[0, 100] \times [0, 200]$. The four points where we have determined the prior smoothing matrix are (5,5), (15,80), (50,25) and (50,100). All the image plots are given on the same colour scale.

(image analysis) or disease risk/intensity (spatial data analysis). Interesting examples of image analysis include Silverman and Cooper (1988), Phillips and Smith (1994), Aykroyd (1995), Rue and Hurn (1999) and the review of template models in Jain *et al.* (1998). Also, Byers (1992), Ghosh *et al.* (1997), Knorr-Held and Besag (1998), Nicholls (1997, 1998), Heikkinen and Arjas (1998, 1999), Sambridge (1999a, 1999b) and Møller and Skare (2001) give examples of partition modelling for various spatial data applications. Especially relevant to the disease mapping example is the work of Knorr-Held and Rasser (2000), who analyse similar data to that presented in this chapter, except that the data are already given in geographical areas. So the problem is to find which of these areas can be joined together and combined to form one partition region. Thus, the main difference is just in the way the model that defines the partitions is defined. Further work by Giudici *et al.* (2000) extends this model to allow for categorical predictor variables.

An area for future research is to apply partitioning ideas to high-dimensional data, especially for data-mining applications where there are typically many data points and many predictors. The main difficulty with applying partition models formed by Voronoi tessellations here is the time it takes to assign points to regions as the distance metric becomes expensive to compute. Also, as many of the variables may be irrelevant we need a good way to exclude inputs that have little effect. Early work in this area includes Denison *et al.* (2002), who partition each variable individually and then determine a sensible way to combine these results, and Hoggart and Griffin (2001), who adopt a partitioning scheme involving hyperplanes which allows more efficient determination of the regions than standard Voronoi tessellations.

7.8 Problems

7.1 For the binomial model for disease mapping (Section 7.5.3) show that the conditional posterior for the risk in the ith region, ϕ_i, using the notation of equation (7.3) is given by

$$p(\phi_i \mid \mathcal{D}, \boldsymbol{\theta}) = \text{Be}(\gamma + y_{i\bullet}, \delta + N_{i\bullet} - y_{i\bullet}).$$

Now determine the same distribution when we use the Poisson approximation to the binomial distribution (Section 7.5.4).

Show that the two distributions are approximately equal when the number of cases of the disease is small relative to the number of susceptibles.

7.2 Suppose that we wish to fit a Bayesian partition model to some data. We believe that the responses, y_i, given the predictors $\boldsymbol{x}_i$ ($i = 1, \ldots, n$), follow

$$p(y_i \mid \boldsymbol{x}_i \in R_j, \boldsymbol{\theta}, \boldsymbol{\phi}) = U(0, \phi_j),$$

where $\boldsymbol{\theta}$ is a vector of model parameters that define the partitions $R_1, \ldots, R_k$ and $\boldsymbol{\phi} = (\phi_1, \ldots, \phi_k)$. If we assume that

$$p(\phi_j) = \gamma\delta^{\gamma}\phi_j^{-(\gamma+1)}, \qquad \phi_j > \delta,$$

for $j = 1, \ldots, k$ and some fixed constants γ and δ, find the marginal likelihood of the data given $\boldsymbol{\theta}$.

Now show that the predictive density, $p(\phi_j \mid \mathcal{D}, \boldsymbol{\theta})$, is given by the same distribution as the prior but with updated parameters

$$\gamma^* = \gamma + n_j \quad \text{and} \quad \delta^* = \max(\delta, \max_{i \in R_j} y_i),$$

where n_j is the number of points in R_j.

7.3 Smith (1980) considered data (Y_{1i}, Y_{2i}), $i = 1, \ldots, 13$, which represented the number of occurrences of two types of pronoun-endings observed in 13 chronologically ordered medieval manuscripts. Stephens (1994) assumes an independent binomial model with N_i trials for $Y_1 = (Y_{11}, \ldots, Y_{1,13})$ with two changepoints, so that

$$p(Y_{1i} \mid \theta_1, \theta_2, \theta_3, N_i) \propto \begin{cases} \theta_1^{Y_{1i}}(1-\theta_1)^{Y_{2i}}, & i = 1, \ldots, r_1, \\ \theta_2^{Y_{1i}}(1-\theta_2)^{Y_{2i}}, & i = r_1 + 1, \ldots, r_2, \\ \theta_1^{Y_{1i}}(1-\theta_1)^{Y_{2i}}, & i = r_2 + 1, \ldots, 13, \end{cases}$$

for $j = 1, 2$ and where $1 \leqslant r_1 < r_2 < 13$ and $Y_{2i} = N_i - Y_{1i}$. Assume independent beta priors for each of the unknowns $(\theta_1, \theta_2, \theta_3)$ so that

$$p(\theta_1, \theta_2, \theta_3) \propto \prod_{j=1}^{3} \theta_j^{-1}(1 - \theta_j)^{-1},$$

and also assume a uniform prior on all the possible pairs of values for the other unknowns (r_1, r_2). Show that the joint posterior distribution for r_1 and r_2 is given by

$$p(r_1, r_2 \mid Y, N) \propto \prod_{j=1}^{3} \frac{\Gamma(S_{1j})\Gamma(S_{2j})}{\Gamma(S_{2j} + S_{2j})},$$

where

$$S_{j1} = \sum_{i=1}^{r_1} Y_{ji}, \qquad S_{j2} = \sum_{i=r_1+1}^{r_2} Y_{ji}, \qquad S_{j3} = \sum_{i=r_2+1}^{13} Y_{ji},$$

for $j = 1, 2$.

7.4 Consider again Problem 7.8 and now suppose that you want to use a Gibbs sampling algorithm to obtain the marginal posterior distributions of r_1 and r_2. Firstly, obtain the full conditional distributions of r_1 and r_2 and show that the conditional distribution of θ_j is

$$p(\theta_j \mid r_1, r_2, Y_j) = \text{Be}(S_{j1} + 1, S_{j2} + 1),$$

for $j = 1, 2, 3$.

Extend the analysis when the number of changepoints is assumed to be unknown, rather than fixed to be two. Describe a Gibbs sampling scheme to simulate from the posterior of interest.

8

Nearest-Neighbour Models

8.1 Introduction

The partition models of the previous chapter broke up the predictor space $\mathcal{X}$ into regions within which the response followed an identical distribution. In the regression context this had the effect that the mean value of the responses within each region were constant, while in the classification context the responses in a region were assumed to be draws from a single probability distribution. In this way the partition model utilised 'nearby' points to guide the predictions and so could be thought of as a local model. In a similar way, nearest-neighbour models also use local regions in predictor space to make predictions but define the regions in a very different manner.

In this chapter we consider nearest-neighbour models and, in particular, describe a method that allows us to put the simple idea behind the approach into a probabilistic model. One great advantage of this is that in a modelling framework we can average out over the uncertainty in the actual model specification in much the same way that the posterior mean prediction using an average of partition models is preferred over one made using just a single model (not least because the average is motivated by decision-theoretic results using squared-error loss). We find that conventional nearest-neighbour is solely an algorithm for predicting a class label and gives no idea of the uncertainty associated with the label chosen. This makes the Bayesian approach, proposed by Holmes and Adams (2002) particularly attractive here.

In this chapter we start in Section 8.2 by introducing the nearest-neighbour classification algorithm and then Section 8.3 describes how to set the algorithm in a probabilistic modelling framework and make inference from the posterior of the model parameters. Section 8.4 demonstrates the effectiveness of the Bayesian nearest-neighbour method compared to the standard nearest-neighbour algorithm. Finally, in Sections 8.5 and 8.6 we give a discussion of the approach as well as further reading material.

8.2 Nearest-Neighbour Classification

The standard nearest-neighbour approach is so straightforward that it can be defined in just a couple of lines. Suppose you are required to predict the class label, y, at a given location $\boldsymbol{x} \in \mathcal{X}$. On the basis of the observed dataset, $\mathcal{D} = \{y_i, \boldsymbol{x}_i\}_{i=1}^n$, the standard k-nearest-neighbour algorithm proceeds as follows.

1. Let $d_i = \|\boldsymbol{x} - \boldsymbol{x}_i\|$, the distance (according to some predefined metric) between the ith predictor vector and the location where we wish to predict the class label.

2. Let $d_{(i)}$ represent the ith ordered distance so that $d_{(1)}$ is the minimum distance observed and $d_{(n)}$ the maximum. Further, let $y_{(1)}, \ldots, y_{(n)}$ be the class labels ordered in the same way, i.e. according to the $d_{(i)}$.

3. Predict y using the most frequently occurring class in $y_{(1)}, \ldots, y_{(k)}$. In the event of a tie predict y by randomly choosing between the tied classes.

Simply put, the nearest-neighbour algorithm finds the k points closest to $\boldsymbol{x}$ and predicts y by choosing the mode of the class labels relating to the k nearest points. In reality, not all of the sorting suggested above needs to be carried out as we only need to know which are the k closest points. The distance metric that is chosen is usually the Euclidean one but others are also possible (see Section 7.3.1).

Figure 8.1 illustrates the approach using three nearest neighbours. Here the points and circles represent data points of two classes in predictor space. The two crosses denote points at which we wish to make a prediction. The algorithm finds the three nearest neighbours to each prediction point, illustrated by the arrows, and then makes a prediction using the most common class (crosses or circles) of these neighbours. The proportion of each class found is offered as a measure of confidence in the forecast; so that the lower point is predicted to be of class '$\cdot$' with a (2/3) vote, while the upper point is declared a circle with a (3/3) vote.

The k-nearest-neighbour (k-nn) algorithm is amongst the most popular methods used in statistical pattern recognition. This is, in part, due to its simplicity and ease of implementation but also there is strong empirical evidence to suggest that the method performs well in a variety of situations compared to other classification algorithms (see, for example, Michie *et al.* 1994). The method is attributed to Fix and Hodges (1951) in an unpublished report and Dasarathy (1991) provides a comprehensive collection of around 140 key papers.

The standard k-nn method outlined above is an algorithm which can only be used to provide predictions at a new location, rather than to determine a suitable model for the data. Hence, the procedure gives forecasts, yet no probabilistic interpretation can be attached to these predictions. This lack of a formal modelling framework, while on the surface attractively simple, gives rise to a number of drawbacks. Most notably the choice of k, a critical parameter in determining the overall 'smoothness' (as a function of location in $\mathcal{X}$) of the predictions, cannot be integrated over to take into account the uncertainty in it. For instance, with $k = 1$ the algorithm partitions the predictor space $\boldsymbol{x}$ with a Voronoi tessellation with centres at the data points $\{\boldsymbol{x}_i\}_1^n$ and within each region the prediction is labelled as the associated $\{y_i\}_1^n$. Yet for $k = n$ the algorithm reports the most common class within $\mathcal{D}$ as the class label of all predictions. In reality, k lies somewhere between these two extremes.

Theoretical results exist that suggest that optimally k should tend to ∞ and $k/n \to 0$ as $n \to \infty$ (Devroye *et al.* 1996). However, this asymptotic result is of little practical use for moderate n (Fukunaga and Hostetler 1973; McLachlan 1992). Rough guidelines based on simulation studies do exist, such as in Enas and Choi (1986), where they suggest scaling k as $n^{2/8}$ or $n^{3/8}$. However, in practice most people either simply set $k = 1$ or choose k via cross-validation on misclassification rate (Ripley 1996). In contrast, we choose to embed the nearest-neighbour classification procedure in a Bayes framework so that the predictions made by the algorithm lie on a coherent, rather than just heuristic, foundation. In addition, we can model average over the plausible values of the number of neighbours, k, to use. As we have seen many times earlier in the book, this allows the final predictions to incorporate the uncertainty associated with this crucial parameter, the only user-set one in standard nearest-neighbour algorithms.

8.3 Probabilistic Nearest Neighbour

8.3.1 *Formulation*

To be able to fit nearest-neighbour methods into a probabilistic structure we need to define a prior distribution over the unknown parameters and a likelihood, and then we make inference using the implied posterior distribution.

We take the likelihood of the data given the parameters β and k to be given by

$$p(\mathcal{D} \mid \beta, k) = \prod_{i=1}^{n} \frac{\exp((\beta/k)\sum_{j \overset{k}{\sim} i} \delta_{y_i y_j})}{\sum_{q=1}^{Q} \exp((\beta/k)\sum_{j \overset{k}{\sim} i} \delta_{q y_j})}, \tag{8.1}$$

where k is the number of neighbours that influence each prediction, β is a parameter that controls the strength of association between the neighbouring y_is, δ_{ab} is a Kronecker delta (i.e. equals one if $a = b$, zero otherwise) and

$$\sum_{j \overset{k}{\sim} i}$$

denotes the summation over the k nearest neighbours of $\boldsymbol{x}_i$ in the set of the observed predictor locations not including the ith one. The term

$$\frac{1}{k} \sum_{j \overset{k}{\sim} i} \delta_{q y_j}$$

records the proportion of points of class q in the k nearest neighbours of $\boldsymbol{x}_i$.

Holmes and Adams (2002) assume some temporal ordering of the data points which leads to the predictive distribution for the response at $\boldsymbol{x}$ following from conditional

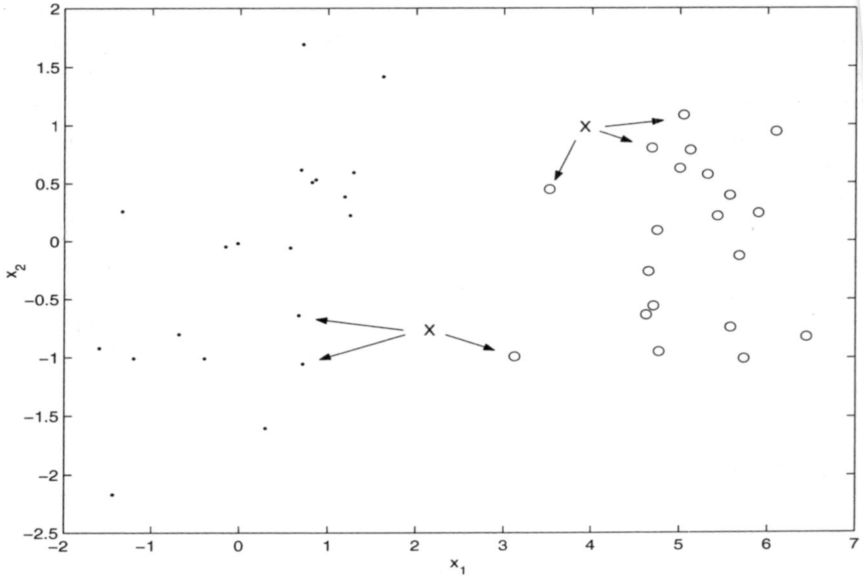

Figure 8.1 Illustration of the k-nearest-neighbour algorithm using $k = 3$. The points and circles mark the predictor values of 20 data points from two classes. A prediction is required for the two points denoted by crosses. The three nearest neighbours to each point are found, illustrated by the arrows, and a prediction is made using the most common class, points or circles. For the lower prediction point the forecast is that the point is from the left-hand class, as we have two points and one circle. For the upper point we have three circles as the nearest neighbour and hence the forecast is for this class.

probability and being given by

$$p(y \mid \boldsymbol{x}, \beta, k, \mathcal{D}) = \frac{\exp((\beta/k)\sum_{j \overset{k}{\sim} n+1} \delta_{y_{n+1}y_j})}{\sum_{q=1}^{Q} \exp((\beta/k)\sum_{j \overset{k}{\sim} n+1} \delta_{qy_j})}, \tag{8.2}$$

so that the modal class for y is given by the most common class found amongst its k nearest neighbours. The interaction parameter β acts like a regression coefficient and both (8.1) and (8.2) have the form of a local logistic regression in the number of neighbouring classes.

The construction of the neighbourhood in (8.1) gives equal weight to each of the k neighbours in the prediction in (8.2). However, it would be equally valid to use a weighted distance metric such as a tricube kernel (Fan and Gijbels 1996) leading to the alternative form of (8.1) given by

$$p(\mathcal{D} \mid \beta, k) = \prod_{i=1}^{n} \frac{\exp(\beta \sum_{j \overset{k}{\sim} i} w(\|x_i - x_j\|)\delta_{y_i y_j})}{\sum_{q=1}^{Q} \exp(\beta \sum_{j \overset{k}{\sim} i} w(\|x_i - x_j\|)\delta_{qy_j})}, \tag{8.3}$$

where the weight function $w(r)$, defined on $r \geqslant 0$, is a monotonically decreasing function of distance r. Using this weighted version of the likelihood is consistent with the belief that points closest to the prediction location $\boldsymbol{x}$ should have the most influence on the predicted response distribution.

The joint distributions of the data in (8.1) and (8.3) are reminiscent of the priors found in Markov random field models used in spatial statistics. This is not surprising as the Markov random field priors are also motivated by the local conditional distributions that they induce (Besag and Kooperburg 1995; Besag *et al.* 1991). One reason for using (8.1) and (8.3) is that it is a normalized distribution whose normalizing constant is independent of β and k. This normalization greatly aids the analysis when we come to consider β and k as random.

Treating β and k as known and fixed *a priori*, as in standard nearest-neighbour algorithms, is unrealistic and fails to account for a key component of uncertainty in the model. To accommodate this uncertainty in a Bayesian setting requires the specification of a prior over the joint distribution of β and k. We have little prior knowledge as to the likely values of k and β, other than the fact that β should be positive. Hence, we suggest adopting the following independent default prior densities:

$$\begin{aligned} p(k) &= U\{1, \ldots, k_{\max}\}, \qquad k_{\max} = \min(n, 200), \\ p(\beta) &= U(0, \infty). \end{aligned} \tag{8.4}$$

The prior on β is improper so if a proper prior is to be preferred we recommend taking $p(\beta) = 2N(0, c)I(\beta > 0)$ with c set as large as possible.

The flat priors in (8.4) do not lead to predictive overfitting. This is apparent from (8.1) and (8.3), which both have the form of a cross-validation probability measure as the ith response is not considered a neighbour of itself. Hence, the prediction of y_i is not influenced by the observed class label for y_i. This is similar to the predictive approach of Geisser and Eddy (1979), who defined a quasi-Bayesian cross-validated likelihood (see also Geisser 1975; Laud and Ibrahim 1995).

8.3.2 *Implementation*

As we cannot form the marginal likelihood analytically through evaluation of

$$\begin{aligned} p(\mathcal{D}) &= \sum_{k=1}^{k_{\max}} p(k) \int_0^{\infty} p(\mathcal{D} \mid \beta, k) p(\beta)\, \mathrm{d}\beta \\ &= \frac{1}{k_{\max}} \sum_{k=1}^{k_{\max}} \int_0^{\infty} p(\mathcal{D} \mid \beta, k)\, \mathrm{d}\beta, \end{aligned} \tag{8.5}$$

we need to resort to some approximation method to make posterior inference. However, in contrast to many of the methods presented in this book the vector of unknown parameters is fixed and, even more importantly, low-dimensional (there are only two unknowns). So, although Markov chain Monte Carlo algorithms could still be used to approximate both the integral over β and the summation over k we find that in some circumstances it is easier to use an alternative approach.

The summation in (8.5) required to evaluate the integral is always technically possible to perform, given enough computing time. The reason that we cannot evaluate $p(\mathcal{D})$ is because of the analytically intractable integral over β. However, there exist a whole host of approximation algorithms for one-dimensional integrals including trapezium rule, quadrature and Monte Carlo integration (see Press (1992) for discussion and computer code to perform these one-dimensional integration routines). Thus, if the number of points where predictions are required is small Holmes and Adams (2002) suggest using one of these (they choose quadrature) to approximate the $k_{\max}$ integrals required to determine $p(\mathcal{D})$.

When many predictions are required standard numerical analysis algorithms for integration may be prohibitively slow. In these cases we can revert back to a Markov chain Monte Carlo algorithm. With this simple model this is straightforward as we have a fixed two-dimensional posterior distribution to sample from, $p(\beta, k \mid \mathcal{D})$, and a standard Metropolis–Hastings sampler can be used. As we would expect k and β to have high posterior correlation it is advisable to jointly update them. One way is to propose the move from (β, k) to new values (β', k'), where

$$\left.\begin{aligned} k' &\sim U\{k-3, k-2, k-1, k, k+1, k+2, k+3\}, \\ \beta' &\sim N(\beta, v), \end{aligned}\right\} \qquad (8.6)$$

where v is a user-set constant that can be chosen during the burn-in period to achieve a reasonable acceptance rate of proposed models. Further, we impose a reflection condition at the boundary of the range of β so that if $\beta' < 0$ we reset β' as $-\beta'$.

Using the uniform priors in (8.4) and the symmetric proposals in (8.6) leads us to accept the proposed move to parameters (β', k') if, u, a draw from a $U[0, 1]$ distribution, is less than

$$\min\left\{1, \frac{p(\mathcal{D} \mid \beta', k')}{p(\mathcal{D} \mid \beta, k)}\right\},$$

otherwise the current values of k and β are retained.

8.4 Examples

We now demonstrate some of the features of the probabilistic nearest-neighbour method on some simulated data, the arm tremor data seen earlier and a spatial data analysis problem relating to the distribution of trees.

8.4.1 Ripley's simulated data

First of all we look at a simulated dataset taken from Ripley (1994). The dataset is a two-class classification problem where each population is an equal mixture of two bivariate normal distributions. A training set of 250 points is used and the model is

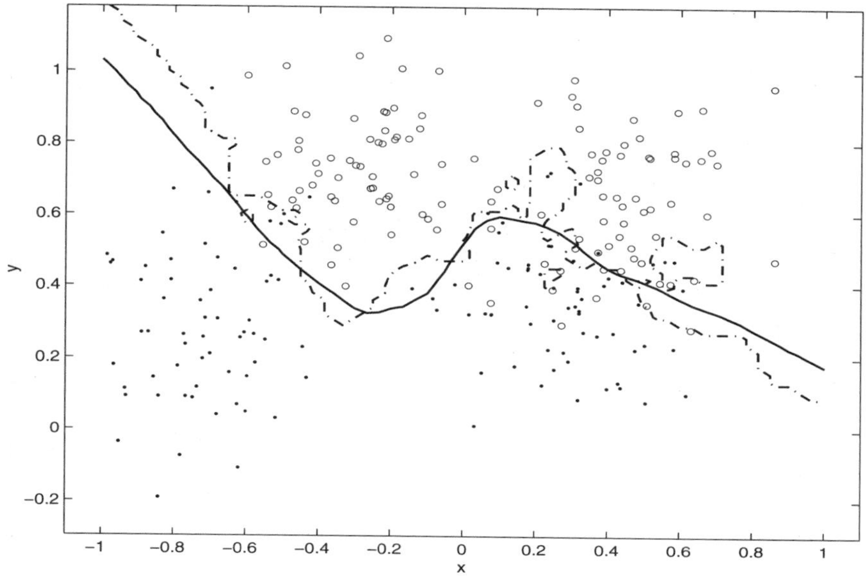

Figure 8.2 Training set with decision boundary $p(y = 1 \mid \mathcal{D}) = p(y = 0 \mid \mathcal{D})$ for p-nn (solid) and the five-nearest-neighbour method (dot-dashed), taken from Ripley (1994).

tested on a set of 1000 points. We use the same data as in Ripley (1994) for ease of comparison. The data can be obtained from the web site that accompanies Ripley (1994) at `www.stats.ox.ac.uk/pub/PRNN/`.

In Figure 8.2 we reproduce the results given in Holmes and Adams (2002) which compare the standard nearest-neighbour approach with the Bayesian one. The lines give the contours of equal probability for each class, i.e. the line where $p(y = 1 \mid \mathcal{D}) = p(y = 0 \mid \mathcal{D})$. The solid line is for the probabilistic nearest-neighbour method, whereas the dot-dash line was found using a standard five-nearest-neighbour algorithm. The probabilistic method is much smoother as it accounts for the uncertainty in the number of neighbours to use and, not surprisingly, we also find that it classifies new data more accurately than the standard method which appears to overfit.

Figure 8.3 gives a complete contour plot for the probabilistic nearest-neighbour results. The plot is given on a different scale so that the features of the contours can be more easily seen. The contours in Figure 8.3 are seen to vary smoothly over the predictor space $\mathcal{X}$ and spread out in regions of low data density where there is greater uncertainty in the predictions.

It is interesting to examine the effect of adopting a uniform weighting for each point within the k nearest neighbours, as opposed to one where the weights decrease for points further from the prediction location. In Figure 8.4 we display the results found by reanalysing the simulated dataset but this time using the probabilistic method with weighted predictions. We show the predictive contours found by adopting the tricube

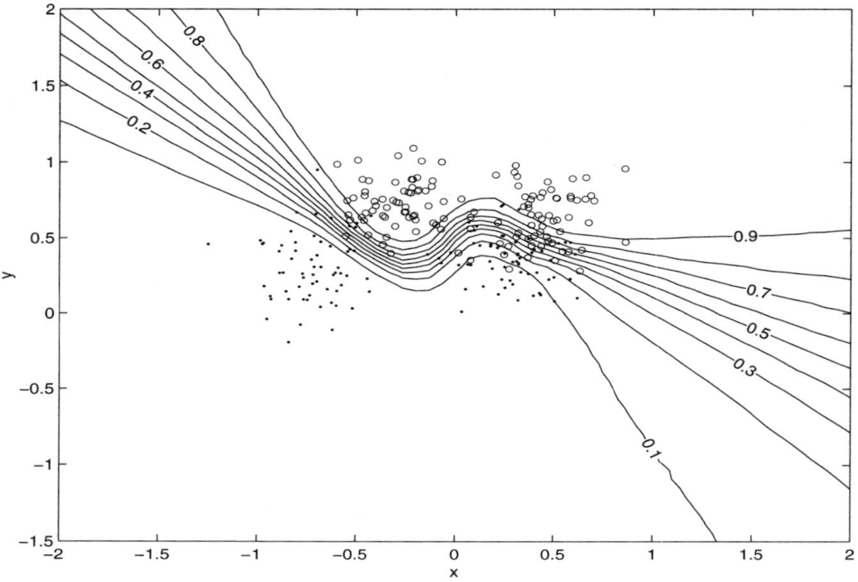

Figure 8.3 Contour plot for $p(y = 1 \mid \mathcal{D})$ in the simulated example of Section 8.4.1.

weighting function

$$w(x_i, x_j) = (1 - \|x_i - x_j\|^3 / d_{i,(k)})^3$$

and using (8.3), where $d_{i,(k)}$ is the Euclidean distance from x_i to the kth nearest neighbour.

We note how the weighting function leads to predictions coming from the prior mean when predicting far away from the observed data. However, within the convex hull of the dataset the method predicts similarly to the unweighted version. For low-dimensional examples the weighting makes little difference but as the number of dimensions increases it is likely to have much more impact. Firstly, as distance is not robust in high dimensions so the weighting function may be difficult to define. Also, in high dimensions many of the locations where we might want to predict lie outside the convex hull of the data so it might be better to predict with more uncertainty (as in the weighted version) rather than making too strong predictions as the unweighted version is liable to do.

8.4.2 Arm tremor data

For comparison, in Figure 8.5 we display the contour plot found by applying the probabilistic nearest-neighbour method to the arm tremor dataset seen earlier. If we compare this with Figure 5.1 we see that the nearest neighbour contours are more complex than those found using a Bayesian MARS for classification. This is because

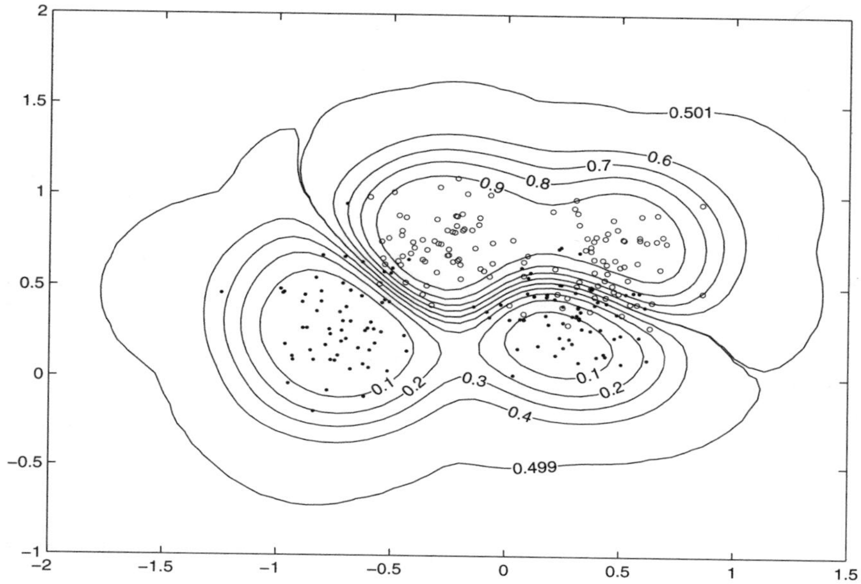

Figure 8.4 Contour plot for $p(y = 1 \mid \mathcal{D})$ for the simulated example of Section 8.4.1 using the tricube weighting function.

the BMARS basis functions naturally favour splits that are axis parallel whereas the nearest-neighbour model does not contain this structure as it makes no underlying structural assumptions about the form of the probability contours.

Figure 8.6 plots the points found in the sample generated to make posterior inference with the arm tremor dataset. We see that the joint posterior of β and k is strongly related in that as k gets larger, β tends to become smaller. Further, we see how the algorithm visits many different values of k (from 7 to 21).

8.4.3 Lancing Woods data

Our final example is chosen to illustrate the ability of the method to fit complicated class conditional fields, as well as the simple ones shown in Figure 8.3. For this purpose we use the Lancing Woods data described in Diggle (1983). The data originate from a study by Gerrard (1969), who provides the locations of three major species of trees (hickories, maples and oaks) in a 19.6 acre plot in Lancing Woods, Clinton County, MI, USA. The data are plotted in the left-hand column of Figure 8.7.

Diggle (1983) was concerned with investigating spatial dependence between the patterns. From our perspective, the data are interesting because of the strongly overlapping class conditional distributions. Our goal is to provide predictive distributions for the three classes and answer questions of the type 'given that I have observed a tree at location $\boldsymbol{x}$, what is the probability that it is a hickory, maple or oak?'. The posterior

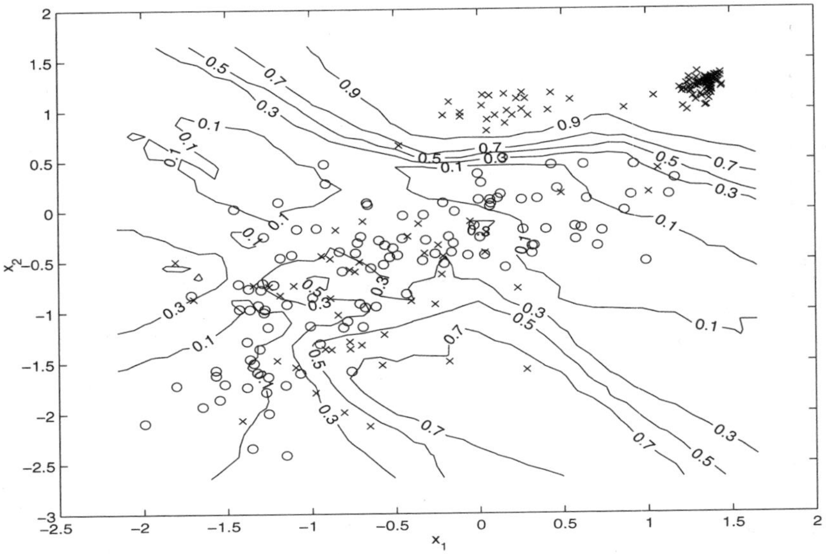

Figure 8.5 The contour plot for the arm tremor dataset using the weighted version of the probabilistic nearest-neighbour algorithm.

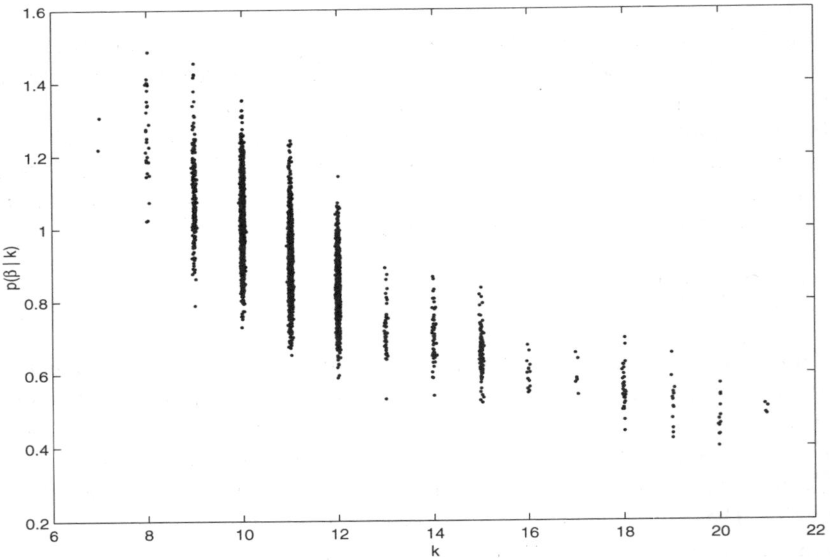

Figure 8.6 The joint posterior distribution, $p(\beta, k \mid \mathcal{D})$, found from the generated sample for arm tremor dataset. Note that some random jitter has been applied to the k values to allow the spread of β for each value of k to be judged better.

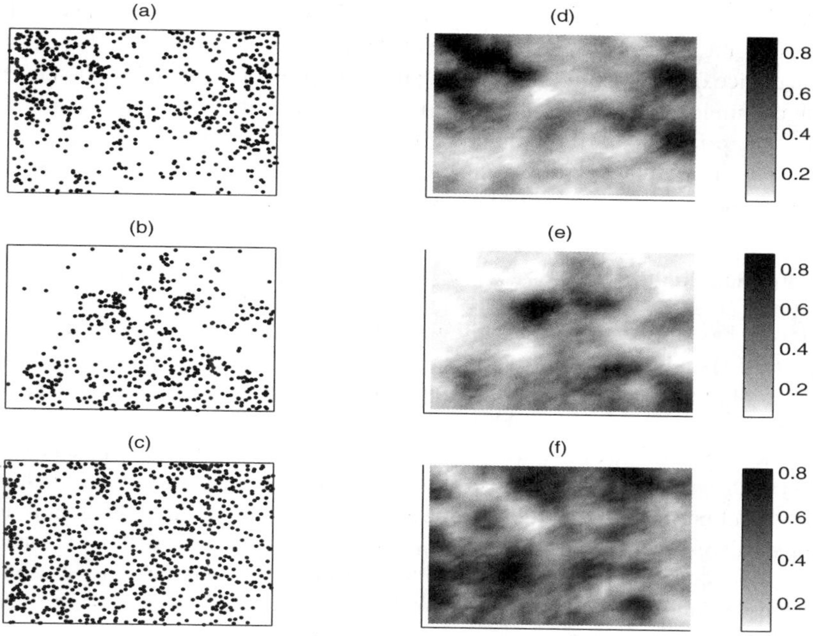

Figure 8.7 Location of each species of tree in Lancing Woods: (a) 703 hickories; (b) 514 maples; (c) 929 oaks. Also shown are the posterior probabilities of each tree over the space of interest estimated using the probabilistic nearest-neighbour algorithm: (d) $p(\text{tree} = \text{hickory})$; (e) $p(\text{tree} = \text{maple})$; (f) $p(\text{tree} = \text{oak})$. The posterior expectation of k was 34 for this example.

class conditional probability fields produced by the probabilistic nearest-neighbour with a uniform kernel (8.1) are shown in the right-hand column of Figure 8.7. The predictive fields in Figure 8.7 should be compared with that of Figure 8.3. This illustrates that the probabilistic nearest-neighbour, with the same default prior, is capable of fitting very complex probability fields if the classes are strongly overlapping, and simpler ones when the classes are reasonably separated.

8.5 Discussion

The probabilistic nearest-neighbour method accounts for a key component of uncertainty that is neglected in the standard k-nn algorithm. The marginalization over the neighbourhood size and the interaction level between neighbours was shown to produce smooth probability fields. The method can be considered as nonparametric in that it makes no assumptions about the underlying class conditional distributions $p(\boldsymbol{x} \mid y)$. The method is also fully automatic with no user set parameters and we believe it to be one of the few off-the-shelf Bayesian nonparametric classification methods available. It was shown that the default prior can fit both simple and complex class conditional probability fields depending on the separability of the underlying classes. For many applications we envisage that the MCMC simulation will be unnecessary with all integrals being performed by quadrature methods.

The choice of whether or not to use the uniform (8.1) or weighted version (8.3) of the algorithm is problem-specific. In theory there is nothing to stop one sampling the form of the prior as part of the model. That is, we could propose to switch from the uniform to the weighted and back again during the MCMC simulation. This would produce a model that mixes over the two priors.

8.6 Further Reading

The model has strong similarities with the Markov random field models described in Besag (1974) and subsequently in Besag (1986) and Tjelmeland and Besag (1998). They have found extensive use in epidemiology and spatial statistics and image analysis, for example Kato *et al.* (2001) and Best *et al.* (2001). Prequential analysis has been discussed in Dawid and Vovk (1999), Dawid (1997) and Skouras and Dawid (1998) with extensions by Arjas and Gasbarra (1997) and its use as a method of model selection was presented in Modha and Masry (1998). Nearest-neighbour forecasting is also known as ‘lazy learning’ or ‘case-based reasoning’ within the machine learning and computer science communities, where it has found extensive use (see, for example, Aha 1997; Watson and Marir 1994; Zheng and Webb 2000). The case of nearest neighbours for regression is reviewed in Atkeson *et al.* (1997). The nearest-neighbour algorithm has been used in a wide variety of applications (see, for example, Ben-Dor *et al.* 2000; Liu and Nakagawa 2001). Some of the theoretical properties of the nearest neighbour classifier are reported in Snapp and Venkatesh (1998).

9

Multiple Response Models

9.1 Introduction

In some situations the observed response (or output) may not be just a single value, but a collection of values. We call such situations *multiple response models* and they introduce their own difficulties when we wish to model dependence across the responses. Multiple response data are commonly available but often the multivariate structure is ignored. This might be due to the fact that if the response variable is Gaussian and a classical linear model is used, then the mean prediction for the two or more responses are unaltered no matter what the correlation between them is. That is, the mean prediction is the same as that when the responses are modelled separately and the full marginal distributions are also unaltered. However, this is not the case when we come to consider nonlinear models and, in particular, basis function models.

In multiple response situations the Bayesian linear model can be generalised to the multivariate linear model (see, for example, Brown 1993; Gelman *et al.* 1995) and we use this to aid us in analysing multiple response data. We start off this chapter in Section 9.2 with a description of the multiple response model, giving the notation that will be used. Section 9.3 describes the conjugate Bayesian multivariate linear regression, and when this model can be utilised. We find that when we allow different basis functions to be used for different responses, we can no longer write down a fully conjugate model for both the coefficients and regression variance. Hence, Section 9.4 introduces the seemingly unrelated regression (SUR) model which introduces correlation in the responses by modelling the joint distribution of the residuals of the fits in each response.

9.2 The Multiple Response Model

In the multivariate regression problem we have measurements on two or more response variables which we shall denote $\boldsymbol{Y} = \{Y_1, \ldots, Y_q\}$, together with a set of potential basis functions $\boldsymbol{B} = \{B_1, \ldots, B_k\}$, where $Y_j = \{y_{1j}, \ldots, y_{nj}\}$, $y_{ij} \in \mathbb{R}$. We assume *a priori* that there is dependence between the regressions, i.e. the Y_is. The basis functions that we could use to model each response could again be any of the ones considered earlier this book.

A useful way to accommodate interactions between the regressions is to assume that their residuals are correlated. In this manner the q regressions for the ith observation take the form,

$$\left.\begin{aligned} y_{i1} &= \boldsymbol{b}_i^{(1)\prime}\boldsymbol{\beta}_1 + \epsilon_{i1}, \\ &\vdots \\ y_{iq} &= \boldsymbol{b}_i^{(q)\prime}\boldsymbol{\beta}_q + \epsilon_{iq}, \end{aligned}\right\} \tag{9.1}$$

for $i = 1, \ldots, n$ and where $\boldsymbol{B}_j = \{\boldsymbol{b}_1^{(j)}, \ldots, \boldsymbol{b}_n^{(j)}\}'$ and $\boldsymbol{b}_i^{(j)}$ represents the ith row of the basis design matrix $\boldsymbol{B}^{(j)}$ for the jth regression. The noise process $\boldsymbol{\epsilon}_i = \{\epsilon_{i1}, \ldots, \epsilon_{iq}\}$ is correlated via the distribution,

$$\boldsymbol{\epsilon}_i \sim N_q(\boldsymbol{0}, \boldsymbol{\Sigma}), \tag{9.2}$$

where $N_q(\boldsymbol{0}, \boldsymbol{\Sigma})$ denotes the q-dimensional multivariate normal density with zero mean vector and variance matrix $\boldsymbol{\Sigma}$. The correlation in the noise process instills correlation in the joint density $p(\boldsymbol{y}_i \mid \boldsymbol{b}_i)$, where $\boldsymbol{y}_i = \{y_{i1}, \ldots, y_{iq}\}$. Note that if $\boldsymbol{\Sigma}$ is diagonal, then the q regressions are independent and the analysis on each model can be performed separately.

9.3 Conjugate Multivariate Linear Regression

If we use the same set of basis functions for each response, then a conjugate analysis is available (Brown *et al.* 1998) allowing us to adapt many of the results discussed in Chapter 2. In particular, we can derive analytic results for the marginal likelihood of any basis set given a set of data. Our workings here follow those presented in Brown (1993) and Brown *et al.* (1998) and in particular their use of matrix variate distributions described in Dawid (1981). The matrix distributions allow for simple symbolic Bayesian manipulation as shown here.

Let $\boldsymbol{U}(p \times q)$ be a random matrix where each element is mean zero. Then $\boldsymbol{U}$ is said to have a matrix normal distribution, $\mathcal{N}(\boldsymbol{P}, \boldsymbol{Q})$ if $p_{ii}\boldsymbol{Q}$ and $q_{jj}\boldsymbol{P}$ define the covariance matrices of the ith row and jth column of $\boldsymbol{U}$ for positive definite matrices $\boldsymbol{P}$ and $\boldsymbol{Q}$.

The matrix normal may be represented by the density,

$$p(\boldsymbol{U}) = (2\pi)^{-pq/2}|\boldsymbol{P}|^{-q/2}|\boldsymbol{Q}|^{-p/2}\exp\{(-1/2)\,\mathrm{tr}[\boldsymbol{P}^{-1}\boldsymbol{U}\boldsymbol{Q}^{-1}\boldsymbol{U}']\}, \tag{9.3}$$

and $\mathrm{tr}(\boldsymbol{A})$ denotes the trace of (square) matrix $\boldsymbol{A}$, i.e. the sum of the diagonal terms of $\boldsymbol{A}$.

Now for the multivariate Bayesian model we assume

$$\boldsymbol{Y} - \boldsymbol{B}\boldsymbol{\beta} \sim \mathcal{N}(\boldsymbol{I}_n, \boldsymbol{\Sigma}),$$

where as before $\boldsymbol{Y}$ is $n \times q$ and $\boldsymbol{I}_n$ denotes the $n \times n$ identity matrix. Using the matrix normal notation we can specify conjugate priors for $\boldsymbol{\beta}$ and $\boldsymbol{\Sigma}$ as

$$\left.\begin{aligned} p(\boldsymbol{\beta} \mid \boldsymbol{\Sigma}) &= \mathcal{N}(\lambda^{-1}\boldsymbol{I}_p, \boldsymbol{\Sigma}), \\ p(\boldsymbol{\Sigma}) &= \mathcal{IW}(\delta, \boldsymbol{Q}), \end{aligned}\right\} \tag{9.4}$$

where λ is a prior constant and we have implicitly defined $\boldsymbol{\beta}$ to be mean zero *a priori*. Further, $\mathcal{IW}(\cdot)$ denotes the inverse-Wishart distribution with δ degrees of freedom and $q \times q$ matrix $\boldsymbol{Q}$, given by

$$\begin{aligned} p(\boldsymbol{\Sigma}) &= \mathcal{IW}(\delta, \boldsymbol{Q}) \\ &= c(q, \delta)|\boldsymbol{Q}|^{(\delta+q-1)/2}|\boldsymbol{\Sigma}|^{-(\delta+2q)/2} \exp\{-\tfrac{1}{2}\operatorname{tr}(\boldsymbol{\Sigma}^{-1}\boldsymbol{Q})\}, \end{aligned}$$

with

$$c(q, \delta) = \frac{2^{-q(\delta+q-1)/2}}{\Gamma\{(\delta+q-1)/2\}}.$$

We can then derive the marginal likelihood $p(\mathcal{D} \mid \boldsymbol{\theta})$ for a given basis function matrix $\boldsymbol{B}$, completely defined by the parameter vector $\boldsymbol{\theta}$. Under a flat prior for $p(\boldsymbol{\theta})$ this turns out to be

$$p(\mathcal{D} \mid \boldsymbol{\theta}) \propto (\lambda^{-k}|\boldsymbol{P}|)^{-q/2}|\boldsymbol{Q}^*|^{-(n+\delta+q-1)/2},$$

where k is the number of basis functions in $\boldsymbol{B}$ and

$$\begin{aligned} \boldsymbol{Q}^* &= \boldsymbol{Q} + \boldsymbol{Y}'\boldsymbol{Y} - \hat{\boldsymbol{\beta}}'\boldsymbol{P}\hat{\boldsymbol{\beta}}, \\ \boldsymbol{P} &= \boldsymbol{B}'\boldsymbol{B} + \lambda\boldsymbol{I}_k, \\ \hat{\boldsymbol{\beta}} &= \boldsymbol{P}^{-1}\boldsymbol{B}'\boldsymbol{Y}. \end{aligned}$$

The marginal likelihood can then be used within an MCMC sampler constructed to sample from $p(\boldsymbol{\theta} \mid \mathcal{D})$ in much the same way as in the univariate case of Chapter 3.

9.4 Seemingly Unrelated Regressions

The results in the previous section are applicable when we can assume that all the responses share the same basis function matrix. For the rest of this chapter we will concentrate on the more general case of (9.1) known as the seemingly unrelated regressions (SUR) (Zellner 1962). The SUR model is distinguished by the fact that the basis set is not fixed across regressions so that $\boldsymbol{B}^{(j)} \neq \boldsymbol{B}^{(k)}$ for $j \neq k$. Hence, the regressions are 'seemingly unrelated' due to differing $\boldsymbol{B}$, though they are actually related though the noise process $\boldsymbol{\epsilon}_i$. This might appear a rather limited construction, however, it has found widespread use especially in econometrics (see, for example, Bauwens and Lubrano 1996; Bauwens and Richard 1985; Geweke 1989).

One of the most common, and most illustrative, applications of SUR models is in the study of related autoregressive (AR) time series. Consider two AR time series

(Y_1, Y_2), where

$$y_{i1} = \alpha_1 y_{i-1,1} + \cdots + \alpha_{d_1} y_{i-d_1,1} + \epsilon_{i1},$$
$$y_{i2} = \gamma_1 y_{i-1,2} + \cdots + \gamma_{d_2} y_{i-d_2,2} + \epsilon_{i2},$$

where d_j is the order of the jth series. The series appear unrelated as they are driven solely by their own lagged values. However, if we impose a noise process on $\{\epsilon_{i1}, \epsilon_{i2}\}$ as in (9.2), then this is a SUR model. This special case of the SUR model is known as vector autoregression (VAR) in the time series literature (Harvey 1993).

In our Bayesian analysis of the SUR model, prior distributions are placed on all unknown parameters. Traditionally, the Bayesian analysis of the SUR model adopted prior densities on the coefficients $\boldsymbol{\beta}$ and the covariance matrix $\boldsymbol{\Sigma}$ (Percy 1992, 1996). Using the methods introduced in Chapters 3 and 4 we can extend the SUR model to take into account uncertainty in the basis set $\boldsymbol{B}$ as well.

The analysis of the SUR model is aided considerably by noting that the multivariate regression can be expressed equivalently as a univariate regression with a suitably redefined error structure (see Gelman *et al.* 1995, Section 15.2). The univariate representation concatenates the responses $Y_1, \ldots, Y_q$ into a single $N \times 1$ column vector $\boldsymbol{Y} = \{\boldsymbol{y}_1, \boldsymbol{y}_2, \ldots, \boldsymbol{y}_n\}$, where now $\boldsymbol{y}_i = \{y_{i1}, \ldots, y_{iq}\}$. In matrix form the univariate regression is written as

$$\boldsymbol{Y} = \boldsymbol{B}\boldsymbol{\beta} + \boldsymbol{\epsilon}, \tag{9.5}$$

where $\boldsymbol{\beta}$ is the column matrix of combined regression coefficients $\boldsymbol{\beta} = (\boldsymbol{\beta}'_1, \ldots, \boldsymbol{\beta}'_q)'$ and $\boldsymbol{B}$ is the univariate design matrix for which we have

$$\boldsymbol{B}\boldsymbol{\beta} = \begin{pmatrix} \boldsymbol{b}_1^{(1)} & \boldsymbol{0}' & \cdots & \boldsymbol{0}' \\ \boldsymbol{0}' & \boldsymbol{b}_1^{(2)} & \cdots & \boldsymbol{0}' \\ \vdots & \vdots & \ddots & \vdots \\ \boldsymbol{0}' & \boldsymbol{0}' & \cdots & \boldsymbol{b}_1^{(q)} \\ \boldsymbol{b}_2^{(1)} & \boldsymbol{0}' & \cdots & \boldsymbol{0}' \\ \vdots & \vdots & \vdots & \vdots \\ \boldsymbol{0}' & \boldsymbol{0}' & \cdots & \boldsymbol{b}_n^{(q)} \end{pmatrix} \begin{pmatrix} \boldsymbol{\beta}_1 \\ \boldsymbol{\beta}_2 \\ \vdots \\ \boldsymbol{\beta}_q \end{pmatrix}, \tag{9.6}$$

where there are k_j predictors associated with the jth regression so that $\boldsymbol{b}_i^{(j)}$ is a $1 \times k_j$ vector, hence $\boldsymbol{B}$ is an $N \times k$ matrix, where $N = nq$ and $k = \sum_{j=1}^{q} k_j$ and the univariate noise process follows,

$$\boldsymbol{\epsilon} \sim \mathrm{N}_N(\boldsymbol{0}, \boldsymbol{\Phi}^{-1}), \tag{9.7}$$

with precision matrix

$$\boldsymbol{\Phi} = [\boldsymbol{I}_n \otimes \boldsymbol{\Sigma}^{-1}] = \begin{pmatrix} \boldsymbol{\Sigma}^{-1} & \boldsymbol{0} & \cdots & \boldsymbol{0} \\ \boldsymbol{0} & \boldsymbol{\Sigma}^{-1} & \cdots & \boldsymbol{0} \\ \vdots & \vdots & \ddots & \vdots \\ \boldsymbol{0} & \boldsymbol{0} & \cdots & \boldsymbol{\Sigma}^{-1} \end{pmatrix}, \tag{9.8}$$

where $\otimes$ denotes the Kronecker product and clearly $\boldsymbol{\Phi}$ is an $N \times N$ matrix.

The form of the model (9.5)–(9.7) may appear complicated, but it is equivalent to the somewhat simpler looking model in (9.1) and (9.2). The rationale for using the univariate form is that the computation and parameter distribution updating becomes easier.

The assumption of Gaussian noise made in (9.7), determines the likelihood

$$\begin{aligned} p(\mathcal{D} \mid \boldsymbol{\beta}, \boldsymbol{\Phi}, \boldsymbol{\theta}) &= |\boldsymbol{\Phi}|^{1/2}(2\pi)^{-N/2} \exp\{-\tfrac{1}{2}(\boldsymbol{Y} - \boldsymbol{B\beta})'\boldsymbol{\Phi}(\boldsymbol{Y} - \boldsymbol{B\beta})\} \\ &= |\boldsymbol{\Sigma}|^{n/2}(2\pi)^{-N/2} \exp\{-\tfrac{1}{2}(\boldsymbol{Y} - \boldsymbol{B\beta})'\boldsymbol{\Phi}(\boldsymbol{Y} - \boldsymbol{B\beta})\}, \end{aligned} \tag{9.9}$$

where $|\cdot|$ indicates the determinant of a matrix.

In conventional Bayesian linear regression models conjugate priors are usually adopted for the parameters. Conjugacy considerably aids the computational aspects of the modelling. In the Bayesian SUR model with different basis sets for each regression, there is no natural conjugate prior for $\boldsymbol{\beta}$ and $\boldsymbol{\Sigma}$. Instead we suggest independent priors for $\boldsymbol{\beta}$ and $\boldsymbol{\Sigma}$ of the form,

$$\begin{aligned} p(\boldsymbol{\beta}, \boldsymbol{\Sigma}) &= p(\boldsymbol{\beta})p(\boldsymbol{\Sigma}) \\ &= N(\mathbf{0}, \boldsymbol{\Omega}^{-\mathbf{1}})\,\mathrm{Wi}(\alpha, \boldsymbol{\Theta}) \\ &= C|\boldsymbol{\Omega}|^{1/2}|\boldsymbol{\Sigma}|^{\alpha-(q+1)/2} \exp[-\tfrac{1}{2}\{\mathrm{tr}(\boldsymbol{\Theta\Sigma}) + \boldsymbol{\beta}'\boldsymbol{\Omega\beta}\}], \end{aligned} \tag{9.10}$$

where $\boldsymbol{\Omega}$ is a prior precision (inverse variance) matrix, $\mathrm{Wi}(\alpha, \boldsymbol{\Theta})$ denotes the Wishart density, with parameters $\boldsymbol{\Theta}$ a $q \times q$ positive-definite matrix and scalar α, and C is a normalizing constant given by

$$C = |\boldsymbol{\Theta}|^{\alpha}(2\pi)^{-k/2}\pi^{-q(q-1)/4} \prod_{i=1}^{q} \Gamma[\tfrac{1}{2}(2\alpha + 1 - i)]^{-1}. \tag{9.11}$$

The Wishart distribution places positive probability on the space of symmetric positive-definite matrices and is the multivariate extension of the gamma density. The parameters α and $\boldsymbol{\Theta}$ play a similar role to the corresponding parameters in a gamma density, for example, the mean of the Wishart is $\alpha\boldsymbol{\Theta}^{-1}$ (see Press (1972) for further details).

The joint density follows by combining the likelihood (9.9) and prior (9.10) to give

$$\begin{aligned} p(\boldsymbol{\beta}, \boldsymbol{\Sigma} \mid \mathcal{D}, \boldsymbol{\theta}) = {} & C|\boldsymbol{\Omega}|^{1/2}|\boldsymbol{\Sigma}|^{\alpha+(n-q-1)/2} \\ & \times \exp[-\{\mathrm{tr}(\boldsymbol{\Theta\Sigma}) + \boldsymbol{\beta}'\boldsymbol{\Omega\beta} + (\boldsymbol{Y} - \boldsymbol{B\beta})'\boldsymbol{\Phi}(\boldsymbol{Y} - \boldsymbol{B\beta})\}/2], \end{aligned} \tag{9.12}$$

which is not of standard distributional form.

However, the choice of prior distributions allows for partial conjugacy. Specifically, the conditional posterior density $p(\boldsymbol{\beta} \mid \boldsymbol{\Sigma}, Y, \boldsymbol{B})$ is normal and $p(\boldsymbol{\Sigma} \mid \boldsymbol{\beta}, Y, \boldsymbol{B})$ is Wishart (although the joint density $p(\boldsymbol{\beta}, \boldsymbol{\Sigma} \mid Y, \boldsymbol{B})$ is analytically intractable). This partial conjugacy aids in the computation as we shall demonstrate later on. After a little algebra, we find

$$p(\boldsymbol{\beta} \mid \boldsymbol{\Sigma}, \boldsymbol{Y}, \boldsymbol{B}) = A \exp(-\tfrac{1}{2}\{(\boldsymbol{\beta} - \hat{\boldsymbol{\beta}})'\boldsymbol{\Lambda}^{-1}(\boldsymbol{\beta} - \hat{\boldsymbol{\beta}}) + a\}), \tag{9.13}$$

where A is a constant that does not depend on $\boldsymbol{\beta}$ and

$$\boldsymbol{\Lambda} = (\boldsymbol{\Omega} + \boldsymbol{B}'\boldsymbol{\Phi}\boldsymbol{B})^{-1}, \tag{9.14}$$

$$\hat{\boldsymbol{\beta}} = \boldsymbol{\Lambda}\boldsymbol{B}'\boldsymbol{\Phi}\boldsymbol{Y}, \tag{9.15}$$

$$a = \operatorname{tr}(\boldsymbol{\Theta}\boldsymbol{\Sigma}) + \boldsymbol{Y}'\boldsymbol{\Phi}\boldsymbol{Y} - \hat{\boldsymbol{\beta}}'\boldsymbol{\Lambda}^{-1}\hat{\boldsymbol{\beta}}, \tag{9.16}$$

and the marginal likelihood $p(\mathcal{D} \mid \boldsymbol{\Sigma}, \boldsymbol{\theta}) = \int p(\mathcal{D} \mid \boldsymbol{\beta}, \boldsymbol{\Sigma}, \boldsymbol{\theta}) p(\boldsymbol{\beta} \mid \boldsymbol{\Sigma}, \boldsymbol{\theta})\, \mathrm{d}\boldsymbol{\beta}$ can be obtained analytically, yielding

$$p(\mathcal{D} \mid \boldsymbol{\Sigma}, \boldsymbol{\theta}) = CA|\boldsymbol{\Lambda}|^{-1/2}(2\pi)^{k/2}\exp(-\tfrac{1}{2}a). \tag{9.17}$$

To derive the conditional distribution of $\boldsymbol{\Sigma}$ conditioned on the regression coefficients $\boldsymbol{\beta}$ and basis set $\boldsymbol{B}$, it helps to define the multivariate matrix of residuals. Let $\boldsymbol{R}$ be the $n \times q$ matrix of errors defined as

$$\boldsymbol{R} = \begin{pmatrix} \epsilon_{11} & \epsilon_{12} & \cdots & \epsilon_{1q} \\ \epsilon_{21} & \epsilon_{22} & \cdots & \epsilon_{2q} \\ \vdots & \vdots & \ddots & \vdots \\ \epsilon_{n1} & \epsilon_{n2} & \cdots & \epsilon_{nq} \end{pmatrix}, \tag{9.18}$$

where ϵ_{ij} is the residual fit associated with the ith data point for the jth response, $\epsilon_{ij} = y_{ij} - \boldsymbol{b}_i^{(j)}\boldsymbol{\beta}_j$. The likelihood (9.9) can then be rewritten in multivariate format as

$$p(\mathcal{D} \mid \boldsymbol{\beta}, \boldsymbol{\Sigma}, \boldsymbol{\theta}) = |\boldsymbol{\Sigma}|^{n/2}(2\pi)^{-N/2}\exp\{-\tfrac{1}{2}\operatorname{tr}(\boldsymbol{\Theta}^*\boldsymbol{\Sigma})\}, \tag{9.19}$$

where $\boldsymbol{\Theta}^* = \boldsymbol{R}'\boldsymbol{R}$ and we have made use of the equivalence

$$\begin{aligned}(\boldsymbol{Y} - \boldsymbol{B}\boldsymbol{\beta})'\boldsymbol{\Phi}(\boldsymbol{Y} - \boldsymbol{B}\boldsymbol{\beta}) &= (\boldsymbol{Y} - \boldsymbol{B}\boldsymbol{\beta})'[\boldsymbol{I}_n \otimes \boldsymbol{\Sigma}](\boldsymbol{Y} - \boldsymbol{B}\boldsymbol{\beta}) \\ &= \operatorname{tr}(\boldsymbol{\Theta}^*\boldsymbol{\Sigma}).\end{aligned}$$

The use of the multivariate form for the likelihood is conjugate to the Wishart prior for $\boldsymbol{\Sigma}$ to leave the posterior probability density for $\boldsymbol{\Sigma}$ as Wishart,

$$p(\boldsymbol{\Sigma} \mid \mathcal{D}, \boldsymbol{\beta}, \boldsymbol{\theta}) = \operatorname{Wi}(\alpha + \tfrac{1}{2}n, \boldsymbol{\Theta} + \boldsymbol{R}'\boldsymbol{R}). \tag{9.20}$$

The conditional and marginal distributions defined in (9.13), (9.17), (9.20) are used in our simulations to make inference statements about features of the model space.

9.4.1 *Prior on the basis function matrix*

In the previous section, when deriving prior distributions and posterior forms for $\boldsymbol{\beta}$ and $\boldsymbol{\Sigma}$, we conditioned on the set of parameters $\boldsymbol{\theta}$ that defined the design matrix $\boldsymbol{B}$. We now look at priors on $\boldsymbol{\theta}$ over the space of potential basis sets. This space might contain the original predictors but equally we could consider arbitrary transformations of the original explanatory variables.

Smith and Kohn (2000) studied the Bayesian SUR model and used a data-based prior for the regression coefficients $\boldsymbol{\beta}$, which leads to a nonstandard conditional distribution for $\boldsymbol{\Sigma}$. Smith and Kohn (2000) use binary indicator variables on a set of potential basis functions to determine which are included in the model. The approach of Smith and Kohn (2000) therefore requires the elicitation of a potential set before the analysis taking place.

In this section we describe a more general procedure which includes the case when the number of potential predictors is very large or even uncountable, as in the examples discussed later in Section 9.6. In addition we advocate a prior which allows direct simulation from the full conditional distribution of $\boldsymbol{\Sigma}$, given in (9.20). Note that much of the work in this chapter is based on the technical report by Holmes *et al.* (1999a).

The space of basis sets will invariably be determined by the modelling task under consideration. For example, in the AR models, $p(\boldsymbol{\theta})$ is over the space of lagged response variables up to some maximum order defined by the user. Alternatively, if we are curve fitting, then $p(\boldsymbol{\theta})$ might be the space of all regression spline models with different numbers and positions of knots. However, regardless of the modelling task, we recommend that $p(\boldsymbol{\theta})$ be taken to be an uninformative proper flat density over the basis space, arguing that Bayesian methods naturally guard against overfitting.

9.5 Computational Details

The posterior distribution of the model can be explored using a Markov chain Monte Carlo (MCMC) simulation algorithm. The method we suggest utilises a hybrid approach that iterates between updating $\boldsymbol{\beta}$, $\boldsymbol{\Sigma}$ and $\boldsymbol{\theta}$ with each update conditional on the other parameters being fixed as well as the data. The method starts off from an initial state $\{\boldsymbol{\beta}^{(0)}, \boldsymbol{\Sigma}^{(0)}, \boldsymbol{\theta}^{(0)}\}$. Then the following three move proposals are performed in turn.

1. Update $\boldsymbol{\theta} \mid \mathcal{D}, \boldsymbol{\Sigma}$ using a Metropolis–Hastings-type kernel described below.
2. Update $\boldsymbol{\beta}$ by drawing a new value from the density $p(\boldsymbol{\beta} \mid \mathcal{D}, \boldsymbol{B}, \boldsymbol{\Sigma})$ in (9.13).
3. Update $\boldsymbol{\Sigma}$ by drawing a new value from $p(\boldsymbol{\Sigma} \mid \mathcal{D}, \boldsymbol{\beta}, \boldsymbol{B})$ in (9.20).

In the above algorithm, the updates to $\boldsymbol{\beta}$ and $\boldsymbol{\Sigma}$ are simple and are just draws from multivariate normal and Wishart densities defined by (9.13) and (9.20): standard methods exist for sampling from these distributions (Press 1972). However, the updating step for $\boldsymbol{\theta}$ is a little more complex so we now go on to describe it in more detail.

9.5.1 *Updating the parameter vector* $\boldsymbol{\theta}$

The sampling step of updating $\boldsymbol{\theta} \mid \mathcal{D}, \boldsymbol{\Sigma}$, uses a reversible Metropolis step (Green 1995) that attempts one of three move proposal types with equal probability.

BIRTH. Add a predictor (column) in $\boldsymbol{B}$. First choose one of the q regressions at random and then choose to add a potential predictor associated with that regression that is not currently in the model.

DEATH. Remove a predictor (column) of $\boldsymbol{B}$. First choose one of the q regressions at random and then choose to remove one of the predictors associated with that regression.

MOVE. Alter a predictor in $\boldsymbol{\theta}$. First choose one of the q regressions at random and then swap a current predictor in the regression with a potential predictor that is currently not in the regression. Note that $\dim(\boldsymbol{\theta})$ is unchanged for this move.

These steps are undertaken with the restriction that at least one term is retained in each regression and that a predetermined maximum number of predictors for each regression is not exceeded. Any move that violates these restrictions is rejected and another draw made.

We are free to define any model space of predictors; however, our choice will invariably be guided by the task under consideration. For vector AR models, $\boldsymbol{B}$ is a matrix of lagged responses taken from the original time series, up to some maximum lag. For regression spline models $\boldsymbol{B}$ could relate to the original design matrix plus nonlinear transformations of the design using spline basis functions with varying knot points (Eubank 1999).

In accordance with MCMC theory, the proposed changes to $\boldsymbol{\theta}$ (which define the basis function matrix $\boldsymbol{B}$) are not automatically accepted. Having proposed a change from $\boldsymbol{\theta}$ to, say, $\boldsymbol{\theta}'$, the proposal is accepted with a certain probability Q given by

$$Q = \min\left\{1,\ R\frac{p(\mathcal{D} \mid \boldsymbol{\Sigma}, \boldsymbol{\theta}')}{p(\mathcal{D} \mid \boldsymbol{\Sigma}, \boldsymbol{\theta})}\right\}, \tag{9.21}$$

when we assume a flat prior for $p(\boldsymbol{\theta})$, $p(\mathcal{D} \mid \boldsymbol{\Sigma}, \boldsymbol{\theta})$ is the marginal likelihood defined in (9.17), and R is the ratio of move probabilities,

$$\frac{p(\mathcal{D} \mid \boldsymbol{\Sigma}, \boldsymbol{\theta}')}{p(\mathcal{D} \mid \boldsymbol{\Sigma}, \boldsymbol{\theta})} = \frac{|\boldsymbol{\Omega}'|^{1/2}|\boldsymbol{\Lambda}'|^{-1/2}\exp(-\frac{1}{2}a')}{|\boldsymbol{\Omega}|^{1/2}|\boldsymbol{\Lambda}|^{-1/2}\exp(-\frac{1}{2}a)}, \tag{9.22}$$

where $\boldsymbol{\Omega}$, $\boldsymbol{\Lambda}$ and a are defined in (9.14)–(9.16).

9.6 Examples

In this section we analyse a number of simple examples. The examples are artificial but chosen to illustrate the method. First we consider a vector autoregressive time series of unknown order where stationarity constraints are not enforced. This is a standard Bayes linear regression model with multiple responses and unknown number of predictors. The set of potential predictors is taken as the set of lagged response variables up to a maximum of 20 lags, $k_{\max} = 20$. In the second example we examine a multiple curve fitting problem using linear splines with an unknown number of knot points.

In all the simulations we adopted the same independent vague priors on the regression coefficients $\boldsymbol{\Omega}^{-1} = 100\boldsymbol{I}_k$ and an improper vague prior on $\boldsymbol{\Sigma}$ with $\boldsymbol{\Theta} = 0.01\boldsymbol{I}_q$, and $\alpha = 0.01$. Note that the posterior density of $\boldsymbol{\Sigma}$ will be proper provided we have more than $(q+1)$ data observations. The use of identical and independent priors on $\boldsymbol{\beta}$ is appropriate as for both the spline and AR examples the predictors are measured on the same scale and the variance of 100 is uninformative. All results are taken using a burn-in of 50 000 iterations after which every fifth sample was retained.

9.6.1 *Vector autoregressive processes*

Time series analysis has seen the most applications of the SUR model, perhaps due to the common occurrence of cross-correlation in physical processes that are recorded jointly in time. In this section we examine the effect that failing to account for correlation has on model order determination when correlation actually exists. Specifically, we examine correlated vector autoregression models (see, for example, Ghatak 1998; Harvey 1993).

To perform the comparison we simulated data from a VAR with additive correlated noise and then performed a Bayesian analysis using both a univariate approach, where each series is modelled independently, and then the multivariate one laid out above. In the special case of VAR models we require only two move proposals when sampling $\boldsymbol{\theta}$. These relate to an increase or decrease in the AR order of one of the series.

The following stationary VAR model was used in the analysis,

$$\begin{pmatrix} y_{1,t} \\ y_{2,t} \end{pmatrix} = \begin{pmatrix} 0.8y_{1,t-1} \\ 0.7y_{2,t-1} - 0.4y_{2,t-2} \end{pmatrix} + \begin{pmatrix} \epsilon_{1,t} \\ \epsilon_{2,t} \end{pmatrix}, \tag{9.23}$$

where $\boldsymbol{\epsilon}_t = (\epsilon_{1,t}, \epsilon_{2,t})$ is taken from a zero mean multivariate normal $\boldsymbol{\epsilon}_t \sim N_2(0, \boldsymbol{\Sigma})$ with covariance,

$$\boldsymbol{\Sigma} = \begin{pmatrix} 1.0 & 0.5 \\ 0.5 & 1.0 \end{pmatrix}.$$

To test our methodology we ran the MCMC sampler on a time series of length 100 generated from (9.23). We then repeated the analysis 20 times using different draws of $\boldsymbol{\epsilon}$. Figure 9.1 shows the marginal posterior model order probabilities averaged over the 20 simulations for both methods. The left-hand column of histograms relates to the multivariate analysis and the right-hand to the univariate approach. Not surprisingly, the multivariate method places, on average, much more probability mass on the true model order. The averaged mass assigned to the true model orders in Figure 9.1 are 0.97 and 0.96 for the multivariate approach and only 0.76 and 0.91 for the univariate one.

This simple example demonstrates the marked effect that correlation has on model order inference. Marginal predictive distributions summarised using the MCMC samples clearly indicate the correlation between series. This is shown in Figure 9.2 for the point $(y_{1,20}, y_{2,20})$.

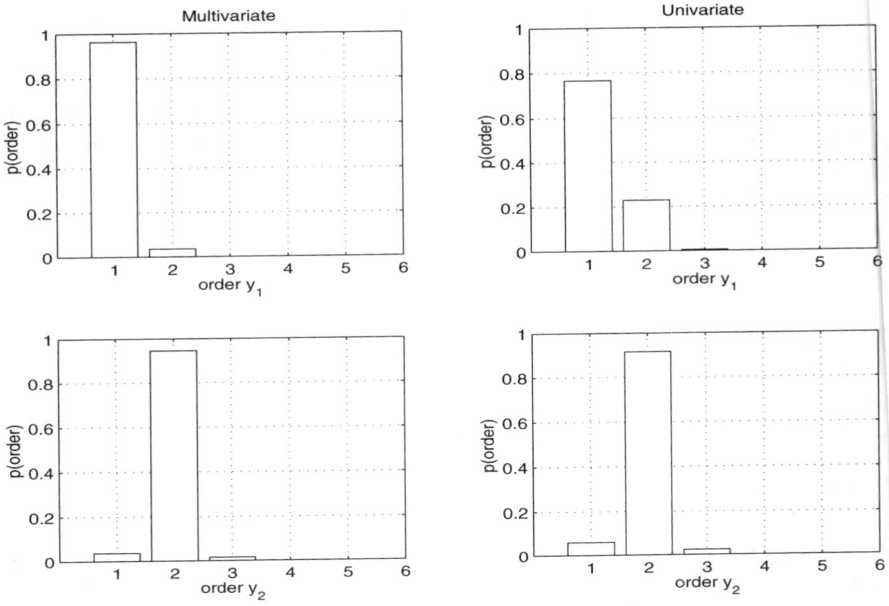

Figure 9.1 Posterior model order probabilities for the VAR example. The left-hand column of graphs shows the posterior mass $p(k_j)$ using the multivariate analysis. The right-hand column shows results where independence is assumed. The true model order is $k_1 = 1$ for the top row of graphs and $k_2 = 2$ for the bottom row.

9.6.2 Multiple curve fitting

We now consider the problem of fitting flexible curves to correlated data series. Multiple families of curves exist in many situations, see, for example, Ramsay and Silverman (1997). We use the univariate linear regression spline model of the form,

$$f(x) = \beta_0 + \sum_{j=1}^{k} \beta_j (x - t_j)_+, \tag{9.24}$$

to estimate each response and where all the knot points t_j are contained in the varying-dimensional parameter vector of model parameters $\boldsymbol{\theta}$.

As an example, we generated 10 series of 100 data points from the following two sigmoidal curves with data points $x_i \sim U[-2, 2]$,

$$\begin{pmatrix} y_{i1} \\ y_{i2} \end{pmatrix} = \begin{pmatrix} \tau_{1,1} + 10[1 + \exp\{-\tau_{1,2} z_i\}]^{-1} \\ \tau_{2,1} + 10[1 + \exp\{-\tau_{2,2} z_i\}]^{-1} \end{pmatrix} + \begin{pmatrix} \epsilon_{i1} \\ \epsilon_{i2} \end{pmatrix}, \tag{9.25}$$

where

$$\boldsymbol{\epsilon}_i = (\epsilon_{i1}, \epsilon_{i2}) \sim \mathrm{N}_2(\mathbf{0}, \boldsymbol{\Sigma}) \quad \text{and} \quad \boldsymbol{\Sigma} = \begin{pmatrix} 1.0 & 0.6 \\ 0.6 & 1.0 \end{pmatrix}$$

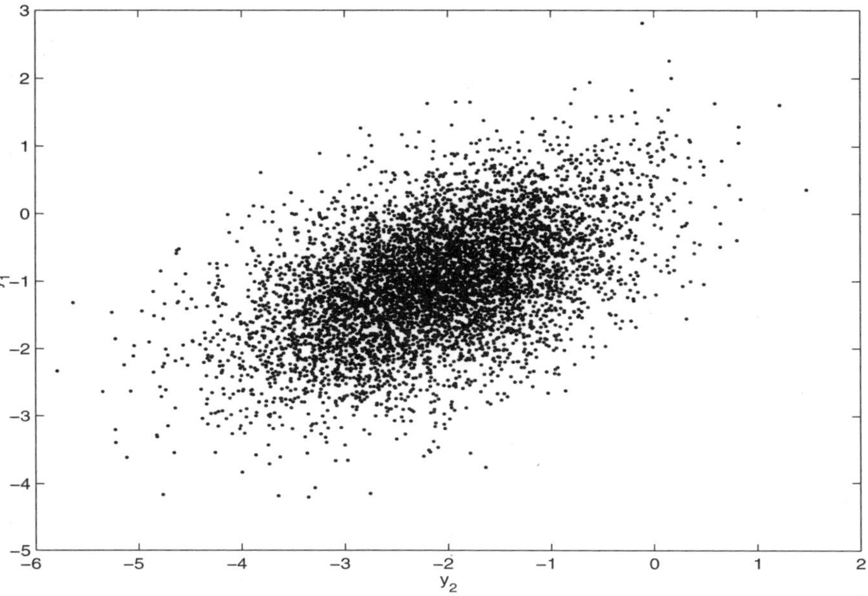

Figure 9.2 MCMC samples of the joint predictive density for $p(y_{1,20}, y_{2,20})$.

and we took $\tau_{1,1} = 10$, $\tau_{1,2} = 3$, $\tau_{2,1} = 5$ and $\tau_{2,2} = 5$ and a different draw of $\boldsymbol{\epsilon}_i$ was made for each series. One of the datasets used, together with the true curves, is shown in Figure 9.5.

We use a prior on the knot points taken to be uniform at the data points. That is, we only allow knots at the data points so that θ_j can take the value $\theta_j \in \{x_1, \ldots, x_n\}$ with $p(\theta_j = x_i) = 1/n$.

We then generated posterior mean curves for the function using both a univariate and multivariate analysis. For the univariate analysis we constrain $\boldsymbol{\Sigma}$ to be diagonal, which leads to independence between the regressions with the (j, j)th element of $\boldsymbol{\Sigma}$ then having a $\text{Ga}(\alpha, \boldsymbol{\Theta}_{jj})$ distribution, where $\boldsymbol{\Theta}_{jj}$ is the (j, j)th element of $\boldsymbol{\Theta}$.

As a performance measure we constructed an out of sample test set using 400 values of the true function at uniform grid points on $[-2, 2]$.

The sum squared error for the two methods, averaged over the 10 datasets, was 55.82 for the multivariate method and 60.03 for the univariate approach and the multivariate method had lower error rates on each of the 10 data series tested. The posterior mean curves for one of the datasets is shown in Figure 9.6 alongside the true curves.

Modelling correlation allows the fitted curves to 'borrow strength' from one another. This is illustrated by examining the posterior smoothing matrix of the model. First, we note that the expected curve $E(\boldsymbol{Y} \mid \mathcal{D}, \boldsymbol{\Sigma}, \boldsymbol{\theta})$ for any single SUR model can be

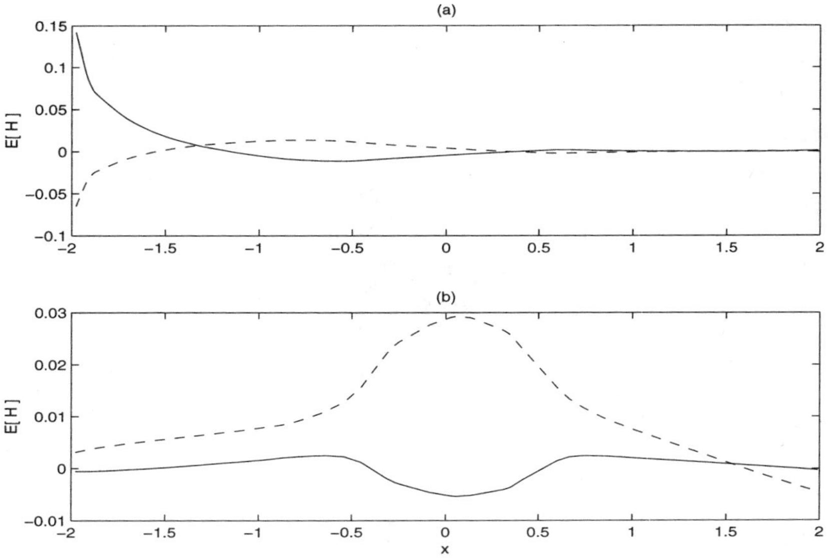

Figure 9.3 Rows of the posterior smoothing matrix for (a) $x = -1.995$ in y_1, and (b) $x = -0.005$ in y_2. The solid lines show weights given to points in y_1 and the dot dashed lines for y_2.

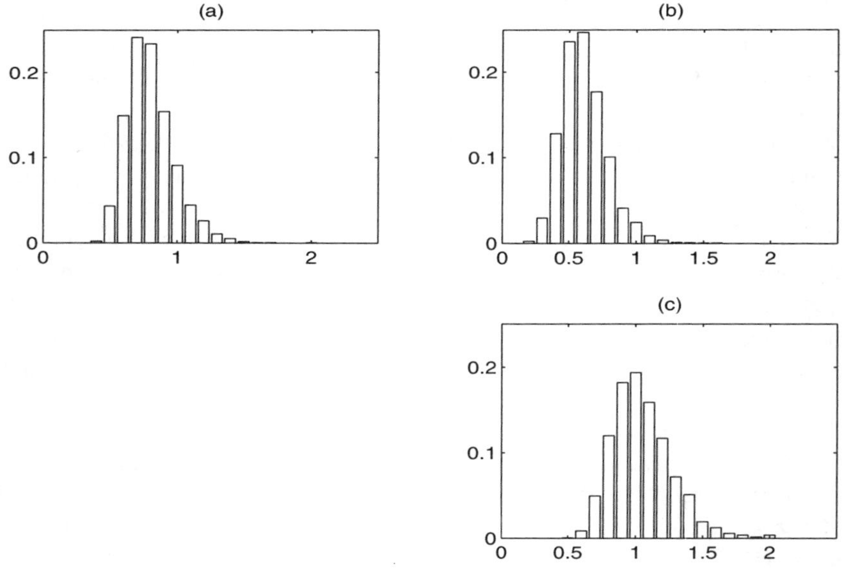

Figure 9.4 Posterior samples for the elements of $p(\boldsymbol{\Sigma} \mid \boldsymbol{Y})$. Plot (a) shows $p(\Sigma_{11} \mid \boldsymbol{Y})$; plot (b) shows $p(\Sigma_{12} \mid \boldsymbol{Y})$; plot (c) shows $p(\Sigma_{22} \mid \boldsymbol{Y})$.

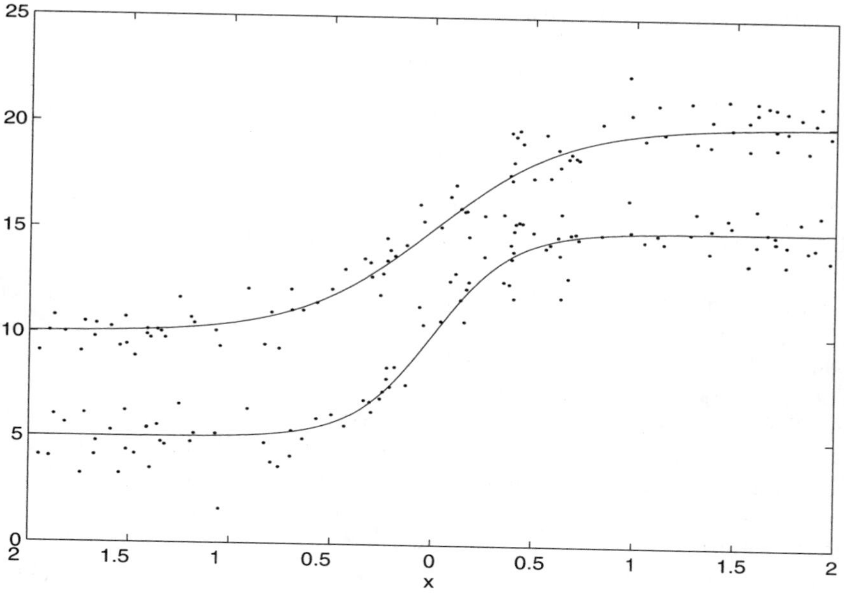

Figure 9.5 True curves plus noisy dataset used in Section 4.2.

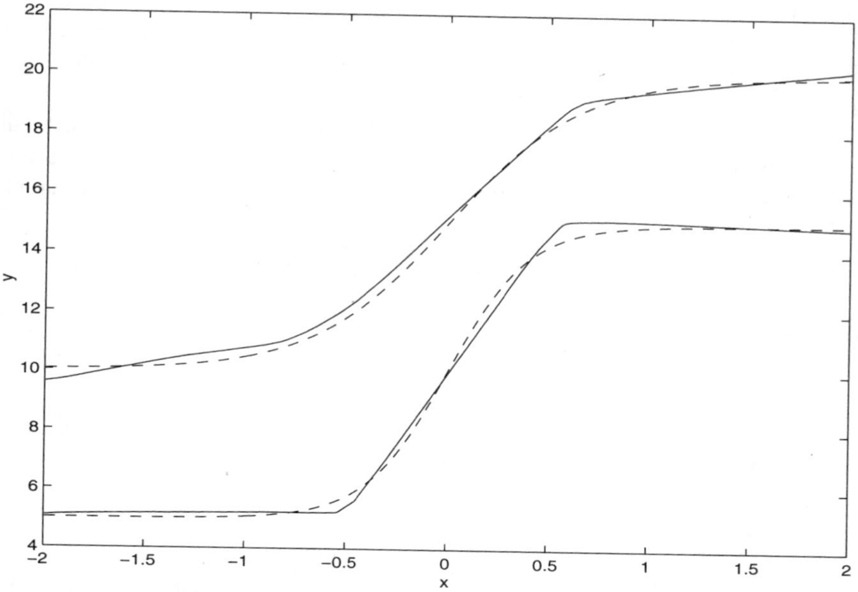

Figure 9.6 Posterior mean estimate for sigmoidal curves in Figure 9.3.

written as a smooth of the data $\boldsymbol{Y}$,

$$\begin{aligned} E(\boldsymbol{Y} \mid \mathcal{D}, \boldsymbol{\Sigma}, \boldsymbol{\theta}) &= \boldsymbol{B}\tilde{\boldsymbol{\beta}} \\ &= \boldsymbol{B}(\boldsymbol{B}'\boldsymbol{\Phi}\boldsymbol{B} + \boldsymbol{\Omega}^{-1})^{-1}\boldsymbol{B}'\boldsymbol{\Phi}\boldsymbol{Y} \\ &= \boldsymbol{S}\boldsymbol{Y}, \end{aligned} \tag{9.26}$$

where $\boldsymbol{S}$ is the $N \times N$ smoothing or hat matrix that transforms the recorded responses to their fitted values. Recall that the ith row of $\boldsymbol{S}$ gives the weight of each of the N responses when fitting the ith data point.

Each sample from the MCMC simulation will provide a basis set $\boldsymbol{B}$ from which we can calculate the smoothing matrix $\boldsymbol{S}$. The expected smooth $E(\boldsymbol{S} \mid \mathcal{D})$ is easily obtained by averaging over the individual smoothing matrices in the MCMC samples. Figure 9.3(a) shows one row of the expected smoothing matrix, $E(\boldsymbol{S} \mid \mathcal{D})$, of the SUR model for the endpoint $x = -1.995$ in the first data series Y_1. This shows how the model makes a prediction at this point, by weighting the responses in both datasets $\{Y_1, Y_2\}$ according to their proximity to x. The solid line indicates the weighting given to the first series Y_1 and the dot dashed for Y_2. Clearly, nearby points to x have the greatest effect on the fit and points in Y_1 are given more weight than those in Y_2 when forecasting on Y_1. However, it is clear that the model does borrows strength from the second series. The weights for Y_2 near x are negative, which is to be expected when a positive correlation exists between the datasets. Thus, if the data in Y_2 are unusually high at x, i.e. gives a large positive residuals, the model would lower the value in Y_1 as we expect the residuals to be correlated. In plot (b) we show a row of $E(\boldsymbol{S} \mid \mathcal{D})$ taken at $x = -0.005$ for the second series Y_2. Again local points have greater effect and there is a negative weighting for points in the other series Y_1.

The extent and uncertainty in the correlation between the two series is easily summarised by plotting the elements of $\boldsymbol{\Sigma}$ taken from the MCMC samples. These are shown in Figure 9.4 using 5000 samples. The mean covariance using these samples is

$$E(\boldsymbol{\Sigma} \mid \mathcal{D}) = \begin{pmatrix} 0.80 & 0.42 \\ 0.42 & 0.95 \end{pmatrix},$$

which compares closely to the empirical covariance

$$\hat{\boldsymbol{\Sigma}} = \begin{pmatrix} 0.81 & 0.43 \\ 0.43 & 0.95 \end{pmatrix}$$

calculated from the actual noisy data that was added to the curves.

9.7 Discussion

The analysis of seemingly unrelated regression models is, at present, not well developed when we need to have different, nonlinear basis sets to describe the relationship between each response and the covariates. In the Bayesian field there as far as we know there are only two papers to date on this topic, Smith and Kohn (2000) and Holmes *et al.* (1999a). As such, this chapter serves as a brief introduction to the field of multiple response modelling and, it is hoped, may stimulate further work in the area.

Appendix A

Probability Distributions

		Bernoulli	
X	$\sim$	$\text{Br}(\theta)$	$\theta > 0$
$p(x)$	$=$	$\theta^x(1-\theta)^{1-x}$	$x \in \{0, 1\}$
		Beta	
X	$\sim$	$\text{Be}(\alpha, \beta)$	$\alpha, \beta > 0$
$p(x)$	$=$	$\frac{\Gamma(\alpha+\beta)}{\Gamma(\alpha)\Gamma(\beta)}x^{\alpha-1}(1-x)^{\beta-1}$	$x \in (0, 1)$
		Binomial	
X	$\sim$	$\text{Bi}(n, \theta)$	$n \in \{1, 2, \ldots\},\ \theta \in (0, 1)$
$p(x)$	$=$	$\binom{n}{x}\theta^x(1-\theta)^{n-x}$	$x \in \{0, 1, \ldots, n\}$
		Dirichlet	
$\boldsymbol{X}$	$\sim$	$\text{Di}(\boldsymbol{\alpha})$	$\boldsymbol{\alpha} \in \mathbb{R}_+^k$
$p(\boldsymbol{x})$	$=$	$\frac{\Gamma(\sum_1^k \alpha_i)}{\prod_1^k \Gamma(\alpha_i)}\left(1-\sum_1^{k-1} x_i\right)^{\alpha_k-1}\prod_1^{k-1} x_i^{\alpha_i-1}$	
			$x_i \in [0, 1],\ \sum_1^{k-1} x_i < 1$
		Exponential	
X	$\sim$	$\text{Exp}(\theta)$	$\theta > 0$
$p(x)$	$=$	$\theta \exp(-\theta x)$	$x > 0$
		Gamma	
X	$\sim$	$\text{Ga}(a, b)$	$a, b > 0$
$p(x)$	$=$	$\frac{b^a}{\Gamma(a)}x^{a-1}\exp(-bx)$	$x > 0$
		Inverse-gamma	
X	$\sim$	$\text{IG}(a, b)$	$a, b > 0$
$p(x)$	$=$	$\frac{b^a}{\Gamma(a)}x^{-(a+1)}\exp(-b/x)$	$x > 0$

Multinomial

X	$\sim$	$\mathrm{Mu}(n, \boldsymbol{\phi})$	$n \in \{1, 2, \ldots\}$
			$\boldsymbol{\phi} = (\phi_1, \ldots, \phi_{k-1}),\ \sum_1^{k-1} \phi_i \leqslant 1, \phi_i \geqslant 0$
$p(x)$	$=$	$\dfrac{n!}{\prod_{i=1}^{k} x_i!} \prod_{i=1}^{k} \phi_i^{x_i}$	$x_1, \ldots, x_{k-1} \in \{0, 1, \ldots, n\}$
			$\sum_1^{k-1} x_i \leqslant n$
			$x_k = n - \sum_1^{k-1} x_i,\ \ \phi_k = 1 - \sum_1^{k-1} \phi_i$

Normal

$\boldsymbol{X}$	$\sim$	$N(\boldsymbol{\mu}, \boldsymbol{\Sigma})$	$\boldsymbol{\mu} \in \mathbb{R}^k$
			$\boldsymbol{\Sigma}$ a $k \times k$ symmetric positive-definite matrix
$p(\boldsymbol{x})$	$=$	$(2\pi)^{-k/2} \lvert\boldsymbol{\Sigma}\rvert^{-1/2} \exp\{-\frac{1}{2}(\boldsymbol{x} - \boldsymbol{\mu})' \boldsymbol{\Sigma}^{-1} (\boldsymbol{x} - \boldsymbol{\mu})\}$	$\boldsymbol{x} \in \mathbb{R}^k$

Pareto

X	$\sim$	$\mathrm{Pa}(\alpha, \beta)$	$\alpha, \beta > 0$
$p(x)$	$=$	$\alpha\beta^{\alpha} x^{\alpha-1}$	$x > \beta$

Poisson

X	$\sim$	$\mathrm{Poi}(\lambda)$	$\lambda > 0$
$p(x)$	$=$	$\dfrac{\lambda^x}{x!} \exp(-\lambda)$	$x \in \{0, 1, 2 \ldots\}$

Student

$\boldsymbol{X}$	$\sim$	$St(\boldsymbol{\mu}, \boldsymbol{\Sigma}, d)$	$\boldsymbol{\mu} \in \mathbb{R}^k, d > 0$
			$\boldsymbol{\Sigma}$ a $k \times k$ symmetric postive-definite matrix
$p(x)$	$=$	$\dfrac{\Gamma\{\frac{1}{2}(d+1)\}}{\Gamma(\frac{1}{2}d)\pi^{k/2}} \lvert\boldsymbol{\Sigma}\rvert^{-1/2} \{1 + (\boldsymbol{x} - \boldsymbol{\mu})' \boldsymbol{\Sigma}^{-1} (\boldsymbol{x} - \boldsymbol{\mu})\}^{-(d+1)/2}$	
			$x \in \mathbb{R}^k$

Uniform (discrete)

X	$\sim$	$U\{u_1, \ldots, u_m\}$	
$p(x)$	$=$	$\dfrac{1}{m}$	$x \in \{u_1, \ldots, u_m\}$

Uniform (continuous)

X	$\sim$	$U(a, b)$	$b > a$
$p(x)$	$=$	$\dfrac{1}{b-a}$	$x \in (a, b)$

Appendix B

Inferential Processes

B.1 The Linear Model

For the linear model,

$$Y = B\beta + \epsilon,$$

where $p(\epsilon) = N(\mathbf{0}, \sigma^2 I)$, $Y = (Y_1, \ldots, Y_n)'$, $\beta = (\beta_1, \ldots, \beta_p)'$ and B is an $n \times p$ fixed, conditional on the basis set $B_1, B_2, \ldots, B_p$, basis function matrix.

Let (y, x) be a general prediction point and write $b = (B_1(x), \ldots, B_p(x))'$, the output of the basis functions at predictor location x. The observed dataset is given by $\mathcal{D} = \{y_i, x_i\}_1^n$.

The prior parameters satisfy the following conditions: $m \in \mathbb{R}^p$; V is a $p \times p$ real positive-definite matrix and $a, b > 0$.

Note that the normal-inverse gamma model is a special case of this model when B is a column vector of n ones, and $\beta = \beta_1$.

Prior

$$p(\beta, \sigma^2) = \mathrm{NIG}(m, V, a, b)$$

$$= \frac{b^a (\sigma^2)^{-(a+(k/2)+1)}}{(2\pi)^{p/2} |V|^{1/2} \Gamma(a)} \exp\left\{-\frac{(\beta - m)' V^{-1} (\beta - m) + 2b}{2\sigma^2}\right\}$$

$$\beta \in \mathbb{R}^p, \; \sigma^2 > 0$$

Likelihood

$$p(\mathcal{D} \mid \beta, \sigma^2) = N(B\beta, \sigma^2 I)$$

$$= \left(\frac{1}{2\pi\sigma^2}\right)^{n/2} \exp\left\{-\frac{(Y - B\beta)'(Y - B\beta)}{2\sigma^2}\right\}$$

Posterior

$$p(\beta, \sigma^2 \mid \mathcal{D}) = \mathrm{NIG}(m^*, V^*, a^*, b^*)$$

$$m^* = (V^{-1} + B'B)^{-1}(V^{-1}m + B'Y)$$

$$V^* = (V^{-1} + B'B)^{-1}$$

$$a^* = a + n/2$$

$$b^* = b + \tfrac{1}{2}\{m'V^{-1}m + Y'Y - (m^*)'(V^*)^{-1}m^*\}$$

Marginal likelihood

$$p(\mathcal{D}) = \frac{|V^*|^{1/2} (b)^a \Gamma(a^*)}{|V|^{1/2} \pi^{n/2} \Gamma(a)} (b^*)^{-a^*}$$

Posterior predictive

$$p(y \mid x, \mathcal{D}) = \mathrm{St}(b'm^*, b^*(1 + b'V^*b), a^*)$$

$$= \frac{\Gamma\{\frac{1}{2}(a^* + 1)\}}{\Gamma(a^*/2)} \{\pi b^* (1 + b'V^*b)\}^{-1/2} \times \left\{1 + \frac{(y - b'm^*)^2}{b^*(1 + b'V^*b)}\right\}^{-(a^*+1)/2} \quad y \in \mathbb{R}$$

B.2 Multivariate Linear Model

For the multivariate linear model with q responses,

$$\boldsymbol{Y} = \boldsymbol{B}\boldsymbol{\beta} + \boldsymbol{\epsilon},$$

where $p(\boldsymbol{\epsilon}) = \mathcal{N}(I_n, \boldsymbol{\Sigma})$, $\mathcal{N}()$ denotes matrix normal, $\boldsymbol{Y} = (\boldsymbol{Y}_1, \ldots, \boldsymbol{Y}_n)'$, $\boldsymbol{Y}_i = (y_{i1}, \ldots, y_{iq})$, $\boldsymbol{\beta} = (\boldsymbol{\beta}_1, \ldots, \boldsymbol{\beta}_p)'$, $\boldsymbol{\beta}_j = (\beta_{j1}, \ldots, \beta_{jq})$ and $\boldsymbol{B}$ is an $n \times p$ fixed, conditional on the basis set $B_1, B_2, \ldots, B_p$, basis function matrix. $\mathcal{IW}$ and $\mathcal{T}$ represent the inverse-Wishart and matrix Student distributions, respectively (Brown 1993).

The prior parameters satisfy the following conditions: $\boldsymbol{m}$ is a real-valued $p \times q$ matrix; $\boldsymbol{V}$ is a $p \times p$ real positive-definite matrix and $a > 0$, $\boldsymbol{Q}$ is a $q \times q$ real positive-definite matrix.

Prior

$$p(\boldsymbol{\beta} \mid \boldsymbol{\Sigma}) = \boldsymbol{\beta} - \boldsymbol{m} \sim \mathcal{N}(\boldsymbol{V}, \boldsymbol{\Sigma})$$
$$= (2\pi)^{-pq/2}|\boldsymbol{V}|^{-q/2}|\boldsymbol{\Sigma}|^{-p/2} \times \exp\{-\tfrac{1}{2}\operatorname{tr}[\boldsymbol{V}^{-1}(\boldsymbol{\beta} - \boldsymbol{m})\boldsymbol{\Sigma}^{-1}(\boldsymbol{\beta} - \boldsymbol{m})']\}$$
$$p(\boldsymbol{\Sigma}) = \mathcal{IW}(a, \boldsymbol{Q})$$
$$= \frac{2^{-q(a+q-1)/2}}{\Gamma[(a+q-1)/2]}|\boldsymbol{Q}|^{(a+q-1)/2}|\boldsymbol{\Sigma}|^{-(a+2q)/2} \times \exp\{-\tfrac{1}{2}\operatorname{tr}(\boldsymbol{\Sigma}^{-1}\boldsymbol{Q})\}$$

Likelihood

$$p(\mathcal{D} \mid \boldsymbol{\beta}, \sigma^2) = \boldsymbol{Y} - \boldsymbol{B}\boldsymbol{\beta} \sim \mathcal{N}(I, \boldsymbol{\Sigma})$$
$$= (2\pi)^{-nq/2}|\boldsymbol{\Sigma}|^{-n/2} \times \exp\{-\tfrac{1}{2}\operatorname{tr}[(\boldsymbol{Y} - \boldsymbol{B}\boldsymbol{\beta})\boldsymbol{\Sigma}^{-1}(\boldsymbol{Y} - \boldsymbol{B}\boldsymbol{\beta})']\}$$

Posterior

$$p(\boldsymbol{\beta} - \boldsymbol{m}^* \mid \mathcal{D}, \boldsymbol{\Sigma}) = \mathcal{N}(\boldsymbol{V}^*, \boldsymbol{\Sigma})$$
$$p(\boldsymbol{\Sigma} \mid \mathcal{D}) = \mathcal{IW}(a^*, \boldsymbol{\Sigma}^*)$$
$$\boldsymbol{m}^* = (\boldsymbol{V}^{-1} + \boldsymbol{B}'\boldsymbol{B})^{-1}(\boldsymbol{V}^{-1}\boldsymbol{m} + \boldsymbol{B}'\boldsymbol{Y})$$
$$\boldsymbol{V}^* = (\boldsymbol{V}^{-1} + \boldsymbol{B}'\boldsymbol{B})^{-1}$$
$$a^* = a + n$$
$$\boldsymbol{\Sigma}^* = \boldsymbol{\Sigma} + \{\boldsymbol{m}'\boldsymbol{V}^{-1}\boldsymbol{m} + \boldsymbol{Y}'\boldsymbol{Y} - (\boldsymbol{m}^*)'(\boldsymbol{V}^*)^{-1}\boldsymbol{m}^*\}$$

Marginal likelihood

$$p(\mathcal{D}) \propto \frac{|\boldsymbol{V}|^{-q/2}}{|\boldsymbol{V}^*|^{-q/2}}|\boldsymbol{\Sigma}^*|^{-(n+a+q-1)/2}$$

Posterior predictive

$$p(\boldsymbol{y} \mid \boldsymbol{x}, \mathcal{D}) = \mathcal{T}(a^*, \tilde{\boldsymbol{V}}, \boldsymbol{\Sigma}^*),$$
$$\propto |\tilde{\boldsymbol{V}}|^{a^*/2}|\boldsymbol{\Sigma}^*|^{-1/2}\{\tilde{\boldsymbol{V}} + \boldsymbol{y}'(\boldsymbol{\Sigma}^*)^{-1}\boldsymbol{y}\}^{-(a^*+q)/2}$$
$$\tilde{\boldsymbol{V}} = \boldsymbol{I} + \boldsymbol{b}'\boldsymbol{V}^*\boldsymbol{b}$$

B.3 Exponential-Gamma Model

If $\mathcal{D} = \{y_i\}_1^n$, where each $y_i > 0$, we can use the exponential-gamma model.

The prior parameters satisfy the following conditions: $\gamma, \delta > 0$.

Prior			
$p(\phi)$	$=$	$\mathrm{Ga}(\gamma, \delta)$	
	$=$	$\dfrac{\delta^\gamma}{\Gamma(\gamma)}\phi^{\gamma-1}\exp(-\delta\phi)$	$\phi > 0$
Likelihood			
$p(\mathcal{D} \mid \phi)$	$=$	$\prod_1^n \mathrm{Exp}(y_i \mid \phi)$	
	$=$	$\phi^n \exp(-s\phi)$	
s	$=$	$\sum_1^n y_i$	
Posterior			
$p(\boldsymbol{\phi} \mid \mathcal{D})$	$=$	$\mathrm{Ga}(\gamma^*, \delta^*)$	
γ^*	$=$	$\gamma + n$	
δ^*	$=$	$\delta + s$	
Marginal likelihood			
$p(\mathcal{D})$	$=$	$\{\delta^\gamma\ \Gamma(\gamma^*)\}/\{(\delta^*)^{\gamma^*}\Gamma(\gamma)\}$	
Posterior predictive			
$p(y \mid \mathcal{D})$	$=$	$\{\gamma^*(\delta^*)^{\gamma^*}\}/\{(\delta^* + y)^{\gamma^*+1}\}$	$y > 0$

B.4 The Multinomial-Dirichlet Model

If $\mathcal{D} = \{y_i\}_1^n$ and each $y_i \in \{1, \ldots, C\}$ we can use the multinomial-Dirichlet model. Let $\boldsymbol{\phi} = (\phi_1, \ldots, \phi_{C-1})$, with $\phi_C = 1 - \sum_1^{C-1} \phi_i$ and $\phi_i \geqslant 0$ and $\sum_1^C \phi_i = 1$, then we can think of ϕ_i as representing the probability of a response being in class i. Further, let m_i be the number of observations in $\mathcal{D}$ of class i.

The prior parameters satisfy the following conditions: $\alpha_1, \ldots, \alpha_C > 0$.

Prior			
$p(\boldsymbol{\phi})$	$=$	$\mathrm{Di}(\alpha_1, \ldots, \alpha_C)$	
	$=$	$\dfrac{\Gamma(\sum_1^C \alpha_i)}{\prod_1^C \Gamma(\alpha_i)} \prod_1^C \phi_i^{\alpha_i - 1}$	$\phi_i \geqslant 0, \ \sum_1^C \phi_i = 1$
Likelihood			
$p(\mathcal{D} \mid \boldsymbol{\phi})$	$=$	$\prod_1^n \mathrm{Mu}(y_i \mid n, \boldsymbol{\phi})$	
	$=$	$\prod_1^C \phi_i^{m_i}$	
Posterior			
$p(\boldsymbol{\phi} \mid \mathcal{D})$	$=$	$\mathrm{Di}(\alpha_1^*, \ldots, \alpha_C^*)$	
α_i^*	$=$	$\alpha_i + m_i$	$i = 1, \ldots, C$
Marginal likelihood			
$p(\mathcal{D})$	$=$	$\dfrac{\Gamma(\sum_1^C \alpha_i)}{\prod_1^C \Gamma(\alpha_i)} \dfrac{\prod_1^C \Gamma(\alpha_i^*)}{\Gamma(\sum_1^C \alpha_i^*)}$	
Posterior predictive			
$p(y \mid \mathcal{D})$	$=$	$\dfrac{\alpha_y^*}{\sum_1^C \alpha_i^*} = \dfrac{m_y + \alpha_y}{n + \sum_1^C \alpha_i}$	$y = 1, \ldots, C$

B.5 Poisson-Gamma Model

If $\mathcal{D} = \{y_i\}_1^n$, where each y_i is a non-negative integer, we can use the Poisson-gamma model.

The prior parameters satisfy the following conditions: $\gamma, \delta > 0$.

Prior

$$p(\phi) = \text{Ga}(\gamma, \delta)$$

$$= \frac{\delta^\gamma}{\Gamma(\gamma)} \phi^{\gamma-1} \exp(-\delta\phi) \qquad \phi > 0$$

Likelihood

$$p(\mathcal{D} \mid \phi) = \prod_1^n \text{Poi}(y_i \mid \phi)$$

$$= \left(\prod_1^n y_i!\right)^{-1} \phi^s \exp(-n\phi)$$

$$s = \sum_1^n y_i$$

Posterior

$$p(\phi \mid \mathcal{D}) = \text{Ga}(\gamma^*, \delta^*)$$

$$\gamma^* = \gamma + s$$

$$\delta^* = \delta + n$$

Marginal likelihood

$$p(\mathcal{D}) = \{\delta\ \Gamma(\gamma^*)\}/\{(\textstyle\prod_1^n y_i!)(\delta^*)^{\gamma^*} \Gamma(\gamma)\}$$

Posterior predictive

$$p(y \mid \mathcal{D}) = \{\Gamma(\gamma^* + y)(\delta^*)^{\gamma^*}\}/\{\Gamma(\gamma^*)(\delta^* + 1)^{\gamma^*+y}\} \qquad y = 0, 1, 2 \ldots$$

B.6 Uniform-Pareto Model

If $\mathcal{D} = \{y_i\}_1^n$, where each y_i is a non-negative integer, we can use the uniform-Pareto model.

The prior parameters satisfy the following conditions: $\gamma, \delta > 0$.

Prior			
$p(\phi)$	=	$\mathrm{Pa}(\gamma, \delta)$	
	=	$\gamma \delta^{\gamma} \phi^{-(\gamma+1)}$	$\phi \geqslant \beta$
Likelihood			
$p(\mathcal{D} \mid \phi)$	=	$\prod_1^n U(y_i \mid 0, \phi)$	
	=	ϕ^{-n}	$\phi \geqslant s$
s	=	$\max\{y_1, \ldots, y_n\}$	
Posterior			
$p(\boldsymbol{\phi} \mid \mathcal{D})$	=	$\mathrm{Pa}(\gamma^*, \delta^*)$	
γ^*	=	$\gamma + n$	
δ^*	=	$\max\{\delta, s\}$	
Marginal likelihood			
$p(\mathcal{D})$	=	$\gamma/(\gamma^* \delta^n)$	$\beta^* = \beta$
$p(\mathcal{D})$	=	$(\gamma \delta^{\gamma})/(\gamma^* s^{\alpha^*})$	$\beta^* = s$
Posterior predictive			
$p(y \mid \mathcal{D})$	=	$\alpha^*/\{(\alpha^* + 1)\ \beta\}$	$0 < y \leqslant \beta^*$
$p(y \mid \mathcal{D})$	=	$\{\alpha^*(\beta^*)^{\alpha^*}\}/\{(\alpha^* + 1)\ y^{\alpha^*+1}\}$	$y > \beta^*$

References

Abramovich, F. and Steinberg, D. M. (1996) Improved inference in nonparametric regression using l_k smoothing splines. *J. Statist. Plan. Inf.* **49**, 327–341.

Abramovich, F., Bailey, T. C. and Sapatinas, T. (2000) Wavelet analysis and its statistical applications. *J. Roy. Statist. Soc.* D **49**, 1–29.

Abramovich, F., Sapatinas, T. and Silverman, B. W. (1998) Wavelet thresholding via a Bayesian approach. *J. Roy. Statist. Soc.* B **60**, 725–749.

Aha, D. W. (1997) *Lazy Learning*. Kluwer.

Ahn, H. S. (1996) Log-gamma regression modeling through regression trees. *Commun. Statist. Theor. Meth.* **25**, 295–311.

Aho, A. V., Hopcroft, J. E. and Ullman, J. D. (1983) *Data Structures and Algorithms*. Reading, MA: Addison-Wesley.

Aitkin, M. (1991) Posterior Bayes factors. *J. Roy. Statist. Soc.* B **53**, 111–142.

Akaike, H. (1974) A new look at statistical model identification. *IEEE Trans. Automatic Control* **19**, 716–727.

Albert, J. H. and Chib, S. (1993) Bayesian analysis of binary and polychotomous response data. *J. Amer. Statist. Assoc.* **88**, 669–679.

Andrews, D. F. and Mallows, C. L. (1974) Scale mixtures of normal distributions. *J. Roy. Statist. Soc.* B **36**, 99–102.

Andrieu, C. (1999) Joint Bayesian model selection and estimation of noisy sinusoids via reversible jump MCMC. *IEEE Trans. Sig. Proc.* **47**, 2667–2676.

Andrieu, C., de Freitas, N. and Doucet, A. (2001a) Robust full Bayesian learning for radial basis networks. *Neur. Comp.* **13**, 2359–2407.

Andrieu, C., Djuric, P. M. and Doucet, A. (2001b) Model selection by MCMC computation. *Sig. Proc.* **81**, 19–37.

Angers, J.-F. and Delampady, M. (1992) Hierarchical Bayesian curve fitting and smoothing. *Can. J. Statist.* **20**, 35–49.

Arjas, E. and Gasbarra, D. (1997) On prequential model assessment in life history. *Biometrika* **84**, 505–522.

Atkeson, C. G., Moore, A. W. and Schaal, S. (1997) Locally weighted learning. *AI Rev.* **11**, 11–73.

Atkinson, A. C. (1978) Posterior probabilities for choosing a regression model. *Biometrika* **65**, 39–48.

Avnimelech, R. and Intrator, N. (1999) Boosting regression estimators. *Neur. Comp.* **11**, 499–520.

Aykroyd, R. G. (1995) Partition models in the analysis of autoradiographic images. *Appl. Statist.* **44**, 441–454.

Ball, F. G. and Davies, S. S. (1995) Statistical-inference for a 2-state Markov model of a single-ion channel, incorporating time-interval omission. *J. Roy. Statist. Soc.* B **57**, 269–287.

Ball, F. G. and Rice, J. A. (1992) Stochastic models for ion channels: introduction and bibliography. *Math. Biosci.* **112**, 189–206.

Barry, D. (1983) Nonparametric Bayesian regression. PhD thesis, Yale University.

Barry, D. (1986) Nonparametric Bayesian regression. *Ann. Statist.* **14**, 934–953.

Barry, D. and Hartigan, J. A. (1992) Product partition models for change point problems. *Ann. Statist.* **20**, 260–279.

Barry, D. and Hartigan, J. A. (1993) A Bayesian analysis for changepoint problems. *J. Amer. Statist. Assoc.* **88**, 309–319.

Bartlett, M. S. (1957) A comment on D. V. Lindley's statistical paradox. *Biometrika* **44**, 533–534.

Basu, S. and Mukhopadhyay, S. (2000) Binary response regression with normal scale mixture links. In *Generalized Linear Models: A Bayesian Perspective* (ed. D. K. Dey, S. K. Ghosh and B. K. Mallick), pp. 231–241. New York: Marcel Dekker

Bauer, E. and Kohavi, R. (1999) An empirical comparison of voting classification algorithms: bagging, boosting, and variants. *Mach. Learn.* **36**, 105–139.

Bauwens, L. and Lubrano, M. (1996) Identification restrictions and posterior densities in cointegrated Gaussian VAR systems. *Adv. Econometrics* **11**, 3–28.

Bauwens, L. and Richard, J. F. (1985) A 1-1 poly-tert random variable generator with application to Monte-Carlo integration. *J. Econometrics* **29**, 19–46.

Bedrick, E. J., Christensen, R. and Johnson, W. (1996) A new perspective on priors for generalized linear models. *J. Amer. Statist. Assoc.* **91**, 1450–1460.

Bellman, R. E. (1961) *Adaptive Control Processes*. Princeton University Press.

Belsey, D. A., Kuh, E. and Welsh, R. E. (1980) *Regression diagnostics*. Wiley.

Ben-Dor, A., Bruhn, L., Friedman, N., Nachman, I., Schummer, M. and Yakhini, Z. (2000) Tissue classification with gene expression profiles. *J. Comp. Biol.* **7**, 559–583.

Berger, J. O. (1980) *Statistical Decision Theory*. Springer.

Berger, J. O. (1982) Bayesian robustness and the Stein effect. *J. Amer. Statist. Assoc.* **77**, 358–368.

Berger, J. O. and Moreno, E. (1994) Bayesian robustness in bidimensional models – prior independence. *J. Statist. Plan. Inf.* **40**, 161–171.

Bernardinelli, L., Clayton, D. G. and Montomoli, C. (1995) Bayesian estimates of disease maps – how important are priors. *Statist. Med.* **14**, 2411–2431.

Bernardo, J. M. and Smith, A. F. M. (1994) *Bayesian Theory*. Wiley.

Besag, J. E. (1974) Spatial interaction and statistical analysis of lattice systems (with discussion). *J. Roy. Statist. Soc.* B **36**, 192–236.

Besag, J. E. (1986) On the statistical analysis of dirty pictures (with discussion). *J. Roy. Statist. Soc.* B **48**, 259–302.

Besag, J. E. and Kooperburg, C. (1995) On conditional and intrinsic autoregressions. *Biometrika* **82**, 733–746.

Besag, J. E., York, J. C. and Mollié, A. (1991) Bayesian image restoration, with two applications in spatial statistics (with discussion). *Ann. Inst. Statist. Math.* **43**, 1–59.

Best, N. G., Cockings, S. and Bennett, J. (2001) Ecological regression analysis of environmental benzene exposure and childhood leukaemia. *J. Roy. Statist. Soc.* A **164**, 155–174.

Biller, C. (2000) Adaptive Bayesian regression splines in semiparametric generalized linear models. *J. Comp. Graph. Statist.* **9**, 122–140.

Bishop, C. M. (1995) *Neural Networks for Pattern Recognition*. Oxford: Clarendon Press.

Bowman, A. W. and Azzalini, A. (1997) *Applied Smoothing Techniques for Data Analysis*. Oxford: Clarendon Press.

Box, G. E. P. and Jenkins, G. M. (1970) *Time Series Analysis, Forecasting and Control*. San Francisco, CA: Holden-Day.

Box, G. E. P. and Tiao, G. C. (1968) A Bayesian approach to some outlier problems. *Biometrika* **55**, 119–129.

Breiman, L. (1993) Hinging hyperplanes for regression, classification and function approximation. *IEEE Trans. Info. Theory* **3**, 999–1013.

Breiman, L. (1996) Bagging predictors. *Mach. Learn.* **24**, 123–140.

Breiman, L., Friedman, J. H., Olshen, R. and Stone, C. J. (1984) *Classification and Regression Trees*. Belmont, CA: Wadsworth.

Breslow, N. E. and Clayton, D. G. (1993) Approximate inference in generalized linear mixed models. *J. Amer. Statist. Assoc.* **88**, 9–25.

Brockwell, P. J. and Davis, R. A. (1991) *Time Series: Theory and Methods*, 2nd edn. Springer.

Broemeling, L. D. (1974) Bayesian inferences about a changing sequence of random variables. *Commun. Statist. Theory Meth.* **3**, 243–255.

Broemeling, L. D. (1985) *Bayesian Analysis of Linear Models*. New York: Marcel Dekker.

Brooks, S. P. (1998) Quantitative convergence assessment for MCMC via CUSUMS. *Statist. Comp.* **8**, 267–274.

Brooks, S. P. and Giudici, P. (1999) Diagnosing convergence of reversible jump MCMC algorithms. In *Bayesian Statistcs 6* (ed. J. M. Bernardo, J. O. Berger, A. P. Dawid and A. F. M. Smith), pp. 733–742. Oxford: Clarendon Press

Brooks, S. P. and Roberts, G. O. (1998) Convergence assessment techniques for Markov chain Monte Carlo. *Statist. Comp.* **8**, 319–335.

Brooks, S. P. and Roberts, G. O. (1999) On quantile estimation and MCMC convergence. *Biometrika* **86**, 710–717.

Brown, P. J. (1993) *Measurement, Regression, and Calibration*. Oxford: Clarendon Press.

Brown, P. J., Vannucci, M. and Fearn, T. (1998) Multivariate Bayesian variable selection and prediction. *J. Roy. Statist. Soc.* B **60**, 627–641.

Bruce, A. and Gao, H.-Y. (1996) *Applied wavelet analysis with S-plus*. Springer.

Bruntz, S. M., Cleveland, W. S., Kleiner, B. and Warner, J. L. (1974) The dependence of ambient ozone on solar radiation, temperature, and mixing height. *Symposium on Atmospheric Diffusion and Air Pollution*, pp. 125–128. American Meteorological Society.

Bucy, R. S. and Diesposti, R. S. (1993) Decision tree design by simulated annealing. *RAIRO – Mathematical Modelling and Numerical Analysis* **27**, 515–534.

Buntine, W. L. (1992a) Learning classification trees. *Statist. Comp.* **2**, 63–73.

Buntine, W. L. (1992b) A theory of learning classification rules. PhD thesis, School of Computing Science, University of Technology, Sydney.

Buntine, W. L. and Weigend, A. S. (1991) Bayesian back-propagation. *Complex Systems* **5**, 603–643.

Byers, J. A. (1992) Dirichlet tessellation of bark beetle spatial attack points. *J. Anim. Ecol.* **61**, 759–768.

Carlin, B. P. and Chib, S. (1995) Bayesian model choice via Markov chain Monte Carlo. *J. Roy. Statist. Soc.* B **57**, 473–484.

Carlin, B. P., Gelfand, A. E. and Smith, A. F. M. (1992) Hierarchical Bayesian analysis of changepoint problems. *Appl. Statist.* **41**, 389–405.

Casdagli, M. (1989) Nonlinear prediction of chaotic time series. *Phys.* D **35**, 335–356.

Celeux, G. and Diebolt, J. (1985) The SEM algorithm: a probabilistic teacher algorithm derived from the EM algorithm for the mixture problem. *Comp. Statist. Quart.* **2**, 73–82.

Chatfield, C. (1995) Model uncertainty, data mining and statistical inference (with discussion). *J. Roy. Statist. Soc.* A **158**, 419–466.

Chatfield, C. (1996) *The Analysis of Time Series*, 5th edn. New York: Chapman & Hall.

Chen, M.-H., Shao, Q.-M. and Ibrahim, J. G. (2000) *Monte Carlo Methods in Bayesian Computation*. Springer.

Chernoff, H. and Zacks, S. (1964) Estimating the current mean of a normal distribution which is subject to changes in time. *Ann. Math. Statist.* **35**, 999–1018.

Chipman, H., George, E. I. and McCulloch, R. E. (1998a) Bayesian CART model search (with discussion). *J. Amer. Statist. Assoc.* **93**, 935–960.

Chipman, H., George, E. I. and McCulloch, R. E. (1998b) Making sense of a forest of trees. Technical Report 98-07. Department of Statistics and Actuarial Science, University of Waterloo.

Chipman, H., George, E. I. and McCulloch, R. E. (2000a) Bayesian treed models. Technical Report, Department of Statistics, University of Waterloo.

Chipman, H., George, E. I. and McCulloch, R. E. (2000b) Hierarchical priors for Bayesian CART shrinkage. *Statist. Comp.* **10**, 17–24.

Chipman, H., Kolaczyk, E. D. and McCulloch, R. E. (1997) Adaptive Bayesian wavelet shrinkage. *J. Amer. Statist. Assoc.* **92**, 1413–1421.

Clark, L. A. and Pregibon, D. (1992) Tree based models. *Statistical Models in S*. Pacific Grove: Wadsworth.

Clayton, D. G. (1996) Generalised linear mixed models. In *Markov Chain Monte Carlo in Practice* (ed. W. R. Gilks, S. Richardson and D. J. Spiegelhalter). Chapman & Hall.

Clayton, D. G. and Kaldor, J. (1987) Empirical Bayes estimates of age-standardized relative risks for use in disease mapping. *Biometrics* **43**, 671–681.

Cleveland, W. S., Devlin, S. J. and Grosse, E. H. (1988) Regression by local fitting: methods, properties and computational algorithms. *J. Econometrics* **37**, 87–114.

Clyde, M. (1999) Bayesian model averaging and model search stratergies (with discussion). In *Bayesian Statistics 6* (ed. J. M. Bernardo, J. O. Berger, A. P. Dawid and A. F. Smith). Oxford: Clarendon Press.

Clyde, M. and George, E. I. (2000) Flexible emprical Bayes estimation for wavelets. *J. Roy. Statist. Soc.* B **62**, 681–698.

Clyde, M., Parmigiani, G. and Vidakovic, B. (1998) Multiple shrinkage and subset selection in wavelets. *Biometrika* **85**, 391–402.

Cobb, G. W. (1978) The problem of the Nile: conditional solution to a changepoint problem. *Biometrika* **65**, 243–51.

Copas, J. B. (1983) Regression, prediction and shrinkage (with discussion). *J. Roy. Statist. Soc.* B **45**, 311–354.

Cowles, M. K. and Carlin, B. P. (1996) Markov chain Monte Carlo convergence diagnostics: a comparative review. *J. Amer. Statist. Assoc.* **91**, 883–904.

Cox, R. T. (1961) *The algebra of probable inference*. Baltimore: Johns Hopkins Press.

Craven, P. and Wahba, G. (1979) Smoothing noisy data with spline functions. estimating the correct degree of smoothing by the method of cross-validation. *Numer. Math.* **31**, 317–403.

Cressie, N. A. C. (1993) *Statistics for Spatial Data*, 2nd edn. London: Chapman & Hall.

Crowley, E. M. (1997) Product partition models for normal means. *J. Amer. Statist. Assoc.* **92**, 192–198.

Damien, P., Wakefield, J. C. and Walker, S. G. (1999) Gibbs sampling for Bayesian non-conjugate and hierarchical models by using auxiliary variables. *J. Roy. Statist. Soc.* B **61**, 331–344.

Dasarathy, B. V. (ed.) (1991) *Nearest Neighbor (NN) Norms: NN Pattern Classification Techniques*. IEEE Computer Society Press.

Daubechies, I. (1992) *Ten Lectures on Wavelets*. Philadelphia, PA: SIAM.

Dawid, A. P. (1981) Some matrix-variate distribution theory: notational considerations and a Bayesian application. *Biometrika* **68**, 265–274.

Dawid, A. P. (1997) Prequential analysis. In *Encylopedia of Statistical Science* (ed. S. Kotz, C. B. Read and D. L. Banks), update vol. 1. Wiley.

Dawid, A. P. and Vovk, V. G. (1999) Prequential probability: principles and properties. *Bernoulli* **5**, 125–162.

de Boor, C. (1978) *A practical guide to splines*. Springer.

de Finetti, B. (1930) Funzione caratteristica di un fenomeno aleatoria. *Men. Acad. Naz. Lincei* **4**, 86–133.

de Finetti, B. (1937/1964). La prevision: ses lois logiques, ses sources subjective. Reprinted in *Studies in Subjective Probability* (ed. H. E. Kyburg and H. E. Smokler), pp. 93–158. New York: Dover

de Finetti, B. (1963) La decision et les probabilities. *Rev. Roumaine Math. Pures Appl.* **7**, 405–413.

de Finetti, B. (1964/1972). Probabilita. Reprinted as 'Conditional probabilities and decision theory'. *Probability, Induction and Statistics*. Wiley.

DeGroot, M. H. (1970) *Optimal Statistical Decisions*. McGraw-Hill.

DeGroot, M. H. (1982) Comment on 'Lindley's paradox' by G. Shafer. *J. Amer. Statist. Assoc.* **77**, 336–339.

Dellaportas, P. and Smith, A. F. M. (1993) Bayesian inference for generalized linear and proportional hazards models via Gibbs sampling. *Appl. Statist.* **42**, 443–459.

Dellaportas, P., Forster, J. J. and Ntzoufras, I. (2002) On Bayesian model and variable selection using MCMC. *Statist. Comp.* **12**, 27–36.

Dempster, A. P., Laird, N. M. and Rubin, D. B. (1977a) Maximum likelihood from incomplete data via the EM algorithm (with discussion). *J. Roy. Statist. Soc.* B **39**, 1–38.

Dempster, A. P., Schatzoff, M. and Wermuth, N. (1977b) A simulation study of alternatives to ordinary least squares. *J. Amer. Statist. Assoc.* **72**, 77–106.

Denison, D. G. T. (1997) Simulation based Bayesian nonparamteric regression methods. PhD thesis, Department of Mathematics, Imperial College, London.

Denison, D. G. T. (2001) Bayesian motivated boosting. *Statist. Comp.* **11**, 171–178.

Denison, D. G. T., Adams, N. M., Holmes, C. C. and Hand, D. J. (2002) Bayesian partition modelling. *Comp. Statist. Data Anal.* (In the press.)

Denison, D. G. T. and George, E. I. (2001) Bayesian prediction using adaptive ridge estimators. Technical Report, Imperial College, London.

Denison, D. G. T. and Holmes, C. C. (2001) Bayesian partitioning for estimating disease risk. *Biometrics* **57**, 143–149.

Denison, D. G. T. and Mallick, B. K. (1998) A nonparametric Bayesian approach to modelling nonlinear time series. Technical Report, Imperial College, London.

Denison, D. G. T. and Mallick, B. K. (2000) Classification trees. In *Generalised Linear Models: A Bayesian Perspective* (ed. D. K. Dey, S. K. Ghosh and B. K. Mallick). New York: Marcel Dekker.

Denison, D. G. T., Mallick, B. K. and Smith, A. F. M. (1998a) Automatic Bayesian curve fitting. *J. Roy. Statist. Soc.* B **60**, 333–350.

Denison, D. G. T., Mallick, B. K. and Smith, A. F. M. (1998b) A Bayesian CART algorithm. *Biometrika* **85**, 363–377.

Denison, D. G. T., Mallick, B. K. and Smith, A. F. M. (1998c) Bayesian MARS. *Statist. Comp.* **8**, 337–346.

Devroye, L. (1986) *Non-Uniform Random Variate Generation*. Springer.

Devroye, L., Gyorfi, L. and Lugosi, G. (1996) *A Probabalistic Theory of Pattern Recognition*. Springer.

Dey, D. K., Ghosh, S. K. and Mallick, B. K. (eds) (2000) *Generalised Linear Models: A Bayesian Perspective*. New York: Marcel Dekker.

Diebolt, J. and Celeux, G. (1993) Asymptotic properties of a stochastic EM algorithm for estimating mixing proportions. *Commun. Statist. B: Stoch. Mod.* **9**, 599–613.

Diebolt, J. and Ip, E. H. S. (1996) Stochastic EM: method and application. In *Markov Chain Monte Carlo in Practice* (ed. W. R. Gilks, S. Richardson and D. J. Spiegelhalter), pp. 259–273. London: Chapman & Hall.

Dietterich, T. G. (2000) An experimental comparison of three methods for constructing ensembles of decision trees: bagging, boosting, and randomization. *Mach. Learn.* **40**, 139–157.

Diggle, P. J. (1983) *Statistical Analysis of Spatial Point Processes*. Academic.

Diggle, P. J., Tawn, J. A. and Moyeed, R. A. (1998) Model-based geostatistics (with discussion). *Appl. Statist.*

Diggle, P., Morris, S., Elliott, P. and Shaddick, G. (1997) Regression modelling of disease risk in relation to point sources. *J. Roy. Statist. Soc.* A **160**, 491–505.

Donoho, D. L. and Johnstone, I. M. (1994) Ideal spatial adaptation by wavelet shrinkage. *Biometrika* **81**, 425–55.

Donoho, D. L., Johnstone, I. M., Kerkyacharian, G. and Picard, D. (1995) Wavelet shrinkage: asymptotia? (with discussion). *J. Roy. Statist. Soc.* B **57**, 301–369.

Draper, D. (1995) Assessment and propagation of model uncertainty (with discussion). *J. Roy. Statist. Soc.* B **57**, 45–97.

Drucker, H. (1997) Improving regressors using boosting techniques. In *Proc. 14th Int. Conf. Mach. Learn.* (ed. D. H. Fisher Jr), pp. 107–115.

Duane, S., Kennedy, A. D., Pendleton, B. J. and Roweth, D. (1987) Hybrid Monte Carlo. *Phys. Lett.* B **195**, 216–222.

Durrleman, S. and Simon, R. (1989) Flexible regression models with cubic splines. *Statist. Med.* **8**, 551–561.

Edwards, W., Lindman, H. and Savage, L. J. (1963) Bayesian statistical inference for psychological research. *Psychol. Rev.* **70**, 193–242.

Efron, B. (1982) *The Jacknife, the Bootstrap and other Resampling Plans*. Philadelphia, PA: SIAM.

Efron, B. (1996) Empirical Bayes methods for combining likelihoods. *J. Amer. Statist. Assoc.* **91**, 538–550.

Efron, B. and Tibshirani, R. J. (1993) *An Introduction to the Bootstrap*. London: Chapman & Hall.

Enas, G. G. and Choi, S. C. (1986) Choice of the smoothing parameter and efficiency of k-nearest neighbor classification. *Comp. Math. Appl.* **12**, 235–244.

Eubank, R. L. (1999) *Nonparamteric Regression and Spline Smoothing*, 2nd edn, Marcel Dekker.

Everitt, B. S. (1993) *Cluster Analysis*, 3rd edn. London: Edward Arnold.

Fahrmeir, L. and Lang, S. (2001) Bayesian inference for generalized additive mixed models based on Markov random field priors. *Appl. Statist.* **50**, 201–220.

Falmer, J. D. and Sidorowich, J. J. (1987) Predicting chaotic time series. *Phys. Rev. Lett.* **59**, 845–848.

Fan, J. Q. and Gijbels, I. (1995) Data-driven bandwidth selection in local polynomial fitting – Variable bandwidth and spatial adaption. *J. Roy. Statist. Soc.* B **57**, 371–394.

Fan, J. Q. and Gijbels, I. (1996) *Local Polynomial Modelling and its Applications*. London: Chapman & Hall.

Fan, Y. and Brooks, S. P. (2000) Bayesian modelling of prehistoric corbelled domes. *J. Roy. Statist. Soc.* D **49**, 339–354.

Ferreira, J. T. A. S., Denison, D. G. T. and Holmes, C. C. (2002) Partition modelling. In *Spatial Cluster Modelling* (ed. A. B. Lawson and D. G. T. Denison). London: Chapman & Hall.

Fix, E. and Hodges, J. L. (1951) Discriminatory analysis – nonparametric discrimination: consistency properties. Technical Report, pp. 261–279. USAF School of Aviation Medicine, Randolf Field, TX.

Follman, D. A. and Lambert, D. (1989) Generalising logistic regression by nonparametric mixing. *J. Amer. Statist. Assoc.* **84**, 295–300.

Foster, D. P. and George, E. I. (1994) The risk inflated criterion for multiple regression. *Ann. Statist.* **22**, 1947–1975.

Franke, R. (1982) Scattered data interpolation: tests of some methods. *Math. Comp.* **38**, 181–200.

Freund, Y. (1995) Boosting a weak learning algorithm by majority. *Inform. Comp.* **121**, 256–285.

Freund, Y. and Schapire, R. E. (1996) Experiments with a new boosting algorithm. *Machine Learning: Proc. 13th Int. Conf.*, pp. 148–156.

Freund, Y. and Schapire, R. E. (1997) A decision-theoretic generalization of on-line learning and an application to boosting. *J. Comp. Sys. Sci.* **55**, 119–139.

Friedman, J. H. (1979) A tree-structured approach to nonparametric multiple regression. In *Smoothing Techniques for Curve Estimation* (ed. T. H. Gasser and M. Rosenblatt). Springer.

Friedman, J. H. (1991) Multivariate adaptive regression splines (with discussion). *Ann. Statist.* **19**, 1–141.

Friedman, J. H. (1993) An overview of predictive learning and function approximation. In *From Statistics to Neural Networks: Theory and Pattern Recognition Applications* (ed. V. Cherkassky, J. H. Friedman and H. Wechsler), pp. 1–61. Springer.

Friedman, J. H., Hastie, T. J. and Tibshirani, R. J. (2000) Additive logistic regression: a statistical view of boosting. *Ann. Statist.* **28**, 337–374.

Fukunaga, K. and Hostetler, L. D. (1973) Optimisation of k-nearest neighbor density estimates. *IEEE Trans. Inform. Theory* **19**, 320–326.

Furnival, G. M. and Wilson, R. W. (1974) Regression by leaps and bounds. *Technometrics* **16**, 499–511.

Gamerman, D. (1997) *Markov Chain Monte Carlo*. London: Chapman & Hall.

Geisser, S. (1975) The predictive sample reuse method with applications. *J. Amer. Statist. Assoc.* **70**, 320–328.

Geisser, S. and Eddy, W. F. (1979) A predictive approach to model selection. *J. Amer. Statist. Assoc.* **74**, 153–160.

Gelfand, A. E. and Smith, A. F. M. (1990) Sampling-based approaches to calculating marginal densities. *J. Amer. Statist. Assoc.* **85**, 398–409.

Gelman, A. and Rubin, D. B. (1992) Inference from iterative simulation using multiple sequences. *Statist. Sci.* **7**, 457–511.

Gelman, A., Carlin, J. B., Stern, H. S. and Rubin, D. B. (1995) *Bayesian Data Analysis*. Chapman & Hall.

Geman, S. and Geman, D. (1984) Stochastic relaxation, Gibbs distributions and the Bayesian restoration of images. *IEEE Trans. Pattern Anal. Mach. Intelligence* **6**, 721–740.

George, E. I. (1986a) A formal Bayes multiple shrinkage estimator. *Commun. Statist. Theory Meth.* **15**, 2099–2114.

George, E. I. (1986b) Minimax multiple shrinkage estimation. *Ann. Statist.* **14**, 188–205.

George, E. I. (1999) Discussion of 'Model averaging and model search strategies' by M. Clyde. In *Bayesian Statistics 6* (ed. J. M. Bernardo, J. O. Berger, A. P. Dawid and A. F. M. Smith), pp. 175–177. Oxford: Clarendon Press.

George, E. I. and Foster, D. P. (2000) Calibration and empirical Bayes variable selection. *Biometrika* **87**, 731–747.

George, E. I. and McCulloch, R. E. (1993) Variable selection via Gibbs sampling. *J. Amer. Statist. Assoc.* **88**, 881–889.

George, E. I. and Oman, S. D. (1996) Multiple-shrinkage principal components regression. *The Statistician* **45**, 111–124.

Gerrard, D. J. (1969) Competition quotient: a new measure of the competition affecting individual forest trees. Technical Report, Michigan State University.

Geweke, J. (1989) Bayesian-inference in econometric models using Monte-Carlo integration. *Econometrica* **57**(6): 1317–1339.

Geweke, J. (1996) Variable selection and model comparison in regression. In *Bayesian Statistics 5* (ed. J. M. Bernardo, J. O. Berger, A. P. Dawid and A. F. M. Smith), pp. 609–620. Oxford: Clarendon Press.

Geyer, C. J. and Thompson, E. A. (1995) Annealing Markov chain Monte Carlo with applications to ancestral inference. *J. Amer. Statist. Assoc.* pp. 909–920.

Ghatak, A. (1998) Vector autoregression modelling and forecasting growth of South Korea. *J. Appl. Statist.* **25**, 579–592.

Ghosh, S., Nowak, Z. and Lee, K. (1997) Tessellation-based computational methods for the characterization and analysis of heterogeneous microstructures. *Composites Sci. Tech.* **57**, 1187–1210.

Gilks, W. R. and Wild, P. (1992) Adaptive rejection sampling for Gibbs sampling. *Appl. Statist.* **41**, 337–348.

Gilks, W. R., Richardson, S. and Spiegelhalter, D. J. (1996) *Markov Chain Monte Carlo in Practice*. London: Chapman & Hall.

Giudici, P., Knorr-Held, L. and Rasser, G. (2000) Modelling categorical covariates in Bayesian disease mapping by partition structures. *Statist. Med.* **19**, 2579–2593.

Godsill, S. J. and Rayner, P. J. W. (1998) Statistical reconstruction and analysis of autoregressive signals in impulsive noise using the Gibbs sampler. *IEEE Trans. Speech Sig. Proc* **6**, 352–372.

Goldberg, D. (1989) *Genetic Algorithms in Search, Optimization, and Machine Learning*. Reading, MA: Addison-Wesley.

Goldstein, M. and Smith, A. F. M. (1974) Ridge-type estimators for regression analysis. *J. Roy. Statist. Soc.* B **36**, 284–291.

Good, I. J. (1988) The interface between statistics and philosophy of science (with discussion). *Statist. Sci.* **3**, 386–398.

Green, P. J. (1995) Reversible jump Markov chain Monte Carlo computation and Bayesian model determination. *Biometrika* **82**, 711–732.

Green, P. J. and Sibson, R. (1978) Computing Dirichlet tessellations in the plane. *Comp. J.* **21**, 168–173.

Green, P. J. and Silverman, B. W. (1994) *Nonparametric Regression and Generalised Linear Models*. London: Chapman & Hall.

Grenander, U. and Miller, M. (1994) Representations of knowledge in complex systems (with discussion). *J. Roy. Statist. Soc.* B **56**, 549–603.

Gruber, M. H. (1998) *Improving Efficiency by Shrinkage*. New York: Marcel Dekker.

Gunst, R. F. and Mason, R. L. (1977) Biased estimation in regression: an evalutation using mean squared error. *J. Amer. Statist. Assoc.* **72**, 616–628.

Gustafson, P. (2000) Bayesian regression modeling with interactions and smooth effects. *J. Amer. Statist. Assoc.* **95**, 795–806.

Gutiérrez-Peña, E. and Smith, A. F. M. (1998) Aspects of smoothing and model indequacy in generalized regression. *J. Statist. Plan. Inf.* **67**, 273–286.

Haar, A. (1910) Zur Theorie der orthogonalen Funktionen-Systeme. *Math. Ann.* **69**, 331–371.

Hajek, B. (1988) Cooling schedules for optimal annealing. *Math. Operat. Res.* **13**, 311–329.

Halpern, E. F. (1973) Bayesian spline regression when the number of knots is unknown. *J. Roy. Statist. Soc.* B **35**, 347–360.

Hamilton, J. D. (1994) *Time Series Analysis*. Princeton University Press.

Haro-Lopez, R. A. and Smith, A. F. M. (1999) On robust Bayesian analysis for location and scale parameters. *J. Multivar. Anal.* **70**, 30–56.

Harrison, D. and Rubenfeld, D. L. (1978) Hedonic housing prices and the demand for clean air. *J. Environ. Eco. and Manage.* **5**, 81–102.

Hartigan, J. A. (1975) *Clustering Algorithms*. Wiley.

Hartigan, J. A. (1990) Partition models. *Commun. Statist.* **19**, 2745–2756.

Hartigan, J. A. and Wong, M. A. (1979) A k-means clustering algorithm. *Appl. Statist.* **28**, 100–108.

Harvey, A. C. (1993) *Time Series Models*, 2nd edn. Cambridge, MA: MIT press.

Hastie, T. J. and Tibshirani, R. J. (1986) Generalized additive models (with discussion). *Statist. Sci.* **1**, 297–318.

Hastie, T. J. and Tibshirani, R. J. (1987) Generalized additive models: some applications. *J. Amer. Statist. Assoc.* **82**, 371–386.

Hastie, T. J. and Tibshirani, R. J. (1990) *Generalized Additive Models*. London: Chapman & Hall.

Hastie, T. J. and Tibshirani, R. J. (2000) Bayesian backfitting. *Statist. Sci.* **15**, 196–223.

Hastings, W. K. (1970) Monte Carlo sampling methods using Markov chains and their applications. *Biometrika* **57**, 97–109.

Heikkinen, J. and Arjas, E. (1998) Non-parametric Bayesian estimation of a spatial Poisson intensity. *Scan. J. Statist.* **25**, 435–450.

Heikkinen, J. and Arjas, E. (1999) Modeling a Poisson forest in variable elevation: A nonparametric Bayesian approach. *Biometrics* **55**, 738–745.

Hodges, J. S. (1987) Uncertainty, policy analysis, and statistics. *Statist. Sci.* **2**, 259–291.

Hodgson, M. E. A. (1999) A Bayesian restoration of an ion channel signal. *J. Roy. Statist. Soc.* B **61**, 95–114.

Hodgson, M. E. A. and Green, P. J. (1999) Bayesian choice among Markov models of ion channels using Markov chain Monte Carlo. *Proc. Roy. Soc. London* A **455**, 3425–3448.

Hoerl, A. E. and Kennard, R. W. (1970a) Ridge regression: applications to nonorthogonal problems. *Technometrics* **12**, 69–82.

Hoerl, A. E. and Kennard, R. W. (1970b) Ridge regression: biased estimation for nonorthogonal problems. *Technometrics* **12**, 55–67.

Hoeting, J. A., Madigan, D., Raftery, A. E. and Volinsky, C. T. (1999) Bayesian model averaging: a tutorial (with discussion). *Statist. Sci.* **14**, 382–417.

Hoggart, C. J. and Griffin, J. E. (2001) A Bayesian partition model for survival data with a survival fraction. In *ISBA Proceedings* (ed. E. I. George).

Holland, J. H. (1975) *Adaption in Natural and Artificial Systems*. Ann Arbor, MI: University of Michigan Press.

Holmes, C. C. and Adams, N. M. (2002) A probabilistic nearest-neighbor method for statistical pattern recognition. *J. Roy. Statist. Soc.* B **64**, 1–12.

Holmes, C. C. and Denison, D. G. T. (1999) Bayesian wavelet analysis with a model complexity prior. In *Bayesian Statistcs 6* (ed. J. M. Bernardo, J. O. Berger, A. P. Dawid and A. F. M. Smith), pp. 769–776. Oxford: Clarendon Press.

Holmes, C. C. and Denison, D. G. T. (2002) A Bayesian MARS classifier. *Mach. Learn.* (In the press.)

Holmes, C. C. and Mallick, B. K. (1998) Bayesian radial basis functions of variable dimension. *Neural Comp.* **10**, 1217–1233.

Holmes, C. C. and Mallick, B. K. (2000a) Bayesian wavelet networks for nonparametric regression. *IEEE. Trans. Neural Networks* **11**, 27–35.

Holmes, C. C. and Mallick, B. K. (2000b) Generalised nonlinear modelling with multivariate regression splines. Technical Report, Imperial College, London.

Holmes, C. C. and Mallick, B. K. (2001) Bayesian regression with multivariate linear splines. *J. Roy. Statist. Soc.* B **63**, 3–17.

Holmes, C. C., Denison, D. G. T. and Mallick, B. K. (1999a) Accounting for model uncertainty in seemingly unrelated regressions. Technical Report, Imperial College, London.

Holmes, C. C., Denison, D. G. T. and Mallick, B. K. (1999b) Bayesian partitioning for classification and regression. Technical Report, Imperial College, London.

Husmeier, D., Penny, W. D. and Roberts, S. J. (1999) An empirical evaluation of Bayesian sampling with hybrid Monte Carlo for training neural network classifiers. *Neural Networks* **12**, 677–705.

Ibrahim, J. G. and Laud, P. W. (1991) On Bayesian analysis of generalized linear models using Jeffrey's prior. *J. Amer. Statist. Assoc.* **86**, 981–986.

Ishwaran, H. (1999) Applications of hybrid Monte Carlo to Bayesian generalized linear models: Quasicomplete separation and neural networks. *J. Comp. Graph. Statist.* **8**, 779–799.

Isik, C. and Ammar, S. (1992) Fuzzy optimal search methods. *Fuzzy Sets and Systems* **46**, 331–337.

Izenman, A. J. (1983) J. R. Wolf and J. A. Wolfer: an historical note on the Zurich sunspot relative numbers. *J. Roy. Statist. Soc.* A **146**, 311–318.

Jain, A. K., Zhong, Y. and Dubuisson-Jolly, M. P. (1998) Deformable template models: a review. *Sig. Proc.* **71**, 109–129.

James, W. and Stein, C. M. (1961) Estimation with quadratic loss. *Proc. 4th Berkeley Symposium 1* pp. 361–379.

Jandhyala, V. K. and Fotopoulos, S. B. (1999) Capturing the distributional behaviour of the maximum likelihood estimator of a changepoint. *Biometrika* **86**, 129–140.

Jeffreys, W. H. and Berger, J. O. (1992) Ockham's razor and Bayesian analysis. *American Scientist* **80**, 64–72.

Kass, R. E. and Raftery, A. E. (1995) Bayes factors. *J. Amer. Statist. Assoc.* **90**, 773–795.

Kass, R. E. and Wasserman, L. (1996) The selection of prior distributions by formal rules. *J. Amer. Statist. Assoc.* **91**, 1343–1370.

Kato, Z., Pong, T. C. and Lee, J. C. M. (2001) Color image segmentation and parameter estimation in a markovian framework. *Pattern Recognition Letters* **22**, 309–321.

Kennedy, M. C. and O'Hagan, A. (2001) Bayesian calibration of computer models. *J. Roy. Statist. Soc.* B **63**, 425–450.

Key, J. T., Pericchi, L. R. and Smith, A. F. M. (1999) Bayesian model choice: what and why? In *Bayesian Statistics 6* (ed. J. M. Bernardo, J. O. Berger, A. P. Dawid and A. F. M. Smith), pp. 343–370. Oxford: Clarendon Press.

Kirkpatrick, S., Gelatt, C. D. and Vecchi, M. P. (1983) Optimization by simulated annealing. *Science* **220**, 671–680.

Kitagawa, G. (1987) Non-Gaussian state-space modelling of nonstationary time series. *J. Amer. Statist. Assoc.* **82**, 1032–1063.

Knorr-Held, L. and Besag, J. E. (1998) Modelling risk from a disease in time and space. *Stat. Med.* **17**, 2045–2060.

Knorr-Held, L. and Rasser, G. (2000) Bayesian detection of clusters and discontinuities in disease maps. *Biometrics* **56**, 13–21.

Knott, G. D. (1999) *Interpolating Cubic Splines*. Boston: Birkhauser.

Knuiman, M. W. and Speed, T. P. (1988) Incorporating prior information into the analysis of contingency tables. *Biometrics* **44**, 1061–1071.

Kohn, R., Smith, M. and Chan, D. (2001) Nonparametric regression using linear combinations of basis functions. *Statist. Comp.* pp. 313–322.

Kwok, S. W. and Carter, C. (1990) Multiple decision trees. In *Uncertainty in Artificial Intelligence 4* (ed. R. D. Schachter, T. S. Levitt, L. N. Kanal and J. F. Lemmer), pp. 327–335. Elsevier.

Laird, N. M. (1978) Nonparametric maximum likelihood estimation of a mixing distribution. *J. Amer. Statist. Assoc.* **73**, 805–811.

Lampinen, J. and Vehtari, A. (2001) Bayesian approach for neural networks – review and case studies. *Neural Networks* **14**, 257–274.

Laud, P. W. and Ibrahim, J. G. (1995) Predictive model selection. *J. Roy. Statist. Soc.* B **57**, 247–262.

Lauritzen, S. L. (1996) *Graphical Models*. Oxford University Press.

Lavine, M. (1992) Some aspects of Polya tree distributions for statistical modelling. *Ann. Statist.* **20**, 1222–1235.

Lavine, M. (1994) More aspects of Polya tree distributions for statistical modelling. *Ann. Statist.* **22**, 1161–1176.

Lavine, M. and Schervish, M. J. (1999) Bayes factors: what they are and what they are not. *Amer. Statist.* **53**, 119–122.

Lavine, M., Wasserman, L. and Wolpert, R. L. (1991) Bayesian inference with specified prior marginals. *J. Amer. Statist. Assoc.* **86**, 964–971.

Lawless, J. F. (1981) Mean squared error properties of generalized ridge estimators. *J. Amer. Statist. Assoc.* **76**, 462–466.

Lawson, A., Biggeri, A. and Bohning, D. (eds) (1999) *Disease Mapping and Risk Assessment for Public Health*. Wiley.

Leamer, E. E. (1978) *Specification Searches*. Wiley.

Leblanc, M. and Crowley, J. (1993) Survival trees by goodness of split. *J. Amer. Statist. Assoc.* **88**, 457–467.

Lee, C. B. (1997) Estimating the number of change points in exponential families distributions. *Scan. J. Statist.* **24**, 201–210.

Lee, C. B. (1998) Bayesian analysis of a change-point in exponential families with applications. *Comp. Statist. Data Anal.* **27**, 195–208.

Lewis, P. A. W. and Ray, B. K. (1997) Modeling long-range dependence, nonlinearity and periodic phenomena in sea surface temperatures using TSMARS. *J. Amer. Statist. Assoc.* **92**, 881–893.

Lewis, P. A. W., Ray, B. K. and Stevens, J. G. (1991) Nonlinear modeling of time series using multivariate adaptive regression splines. *J. Amer. Statist. Assoc.* **86**, 864–877.

Lim, T. S., Lo, W. Y. and Shih, Y. S. (2000) A comparison of prediction accuracy, complexity, and training time of thirty-three old and new classification algorithms. *Mach. Learn.* **40**, 203–228.

Lindley, D. V. (1957) A statistical paradox. *Biometrika* **45**, 533–534.

Lindley, D. V. (1980) Approximate Bayesian methods. In *Bayesian Statistics* (ed. J. M. Bernardo, M. H. DeGroot, D. V. Lindley and A. F. M. Smith), pp. 223–245. Valencia: University Press.

Lindley, D. V. (1995) Discussion of 'Assessment and propagation of uncertainty' by D. Draper. *J. Roy. Statist. Soc.* B **57**, 75.

Lindley, D. V. and Smith, A. F. M. (1972) Bayes estimates for the linear model (with discussion). *J. Roy. Statist. Soc.* B **34**, 1–41.

Liu, C. L. and Nakagawa, M. (2001) Evaluation of prototype learning algorithms for nearest-neighbor classifier in application. *Pattern Recognition* **34**, 601–615.

Lovasz, L., Naor, M., Newman, I. and Wigderson, A. (1995) Search problems in the decision tree model. *SIAM J. Disc. Math.* **8**, 119–132.

McCullagh, P. and Nelder, J. A. (1989) *Generalized Linear Models*, 2nd edn. London: Chapman & Hall.

McCulloch, W. S. and Pitts, W. (1943) A logical calculus of ideas eminent in neural activity. *Bull. Math. Biophys.*, pp. 115–133.

MacKay, D. J. C. (1992a) Bayesian interpolation. *Neural Computation* **4**, 415–447.

MacKay, D. J. C. (1992b) A practical Bayesian framework for backpropagation networks. *Neural Computation* **4**, 448–472.

MacKay, D. J. C. (1995) Probable networks and plausible predictions – a review of practical Bayesian methods for supervised neural networks. *Network: Computation in Neural Systems* **6**, 1053–1062.

Mackey, M. C. and Glass, L. (1977) Oscillations and chaos in physiological control systems. *Science* **287**, 1087–1091.

McLachlan, G. J. (1992) *Discriminant Analysis and Statistical Pattern Recognition*. Wiley.

Madigan, D. and Raftery, A. E. (1994) Model selection and accounting for model uncertainty in graphical models using Occam's window. *J. Amer. Statist. Assoc.* **89**, 1535–1546.

Madigan, D. and York, J. C. (1995) Bayesian graphical models for discrete data. *Int. Statist. Rev.* **63**, 215–232.

Mallat, S. (1989) A theory for multiresolution signal decomposition: the wavelet representation. *IEEE Trans. Pattern Anal. Machine Intell.* **11**, 674–693.

Mallick, B. K. (1998) Bayesian curve estimation by polynomial of random order. *J. Statist. Plan. Inf.* **70**, 91–109.

Mallick, B. K. and Gelfand, A. E. (1994) Generalised linear models with unknown link functions. *Biometrika* **81**, 237–245.

Mallick, B. K., Denison, D. G. T. and Smith, A. F. M. (1999) Bayesian survival analysis using a MARS model. *Biometrics* **55**, 1071–1077.

Mallows, C. L. (1973) Some comments on C_p. *Technometrics* **15**, 661–675.

Mertens, B. J. A. and Hand, D. J. (1999) Adjusted estimation for the combination of classifiers. *Adv. Intell. Data Anal.* **1642**, 317–330.

Metropolis, N., Rosenbluth, A. W., Rosenbluth, M. N., Teller, A. H. and Teller, E. (1953) Equations of state calculations by fast computing machines. *J. Chem. Phys.* **21**, 1087–91.

Michie, D., Spiegelhalter, D. J. and Taylor, C. C. (eds) (1994) *Machine Learning, Neural and Statistical Classification*. Ellis Horwood.

Mitchell, M. (1996) *An Introduction to Genetic Algorithms*. Cambridge, MA: MIT Press.

Mitchell, T. J. and Beauchamp, J. J. (1988) Bayesian variable selection in linear regression. *J. Amer. Statist. Assoc.* **83**, 1023–1036.

Modha, D. S. and Masry, E. (1998) Prequential and cross-validated regression estimation. *Mach. Learn.* **33**, 5–39.

Møller, J. and Skare, Ø. (2001) Bayesian image analysis with coloured Voronoi tessellations and a view to applications in reservoir modelling. Technical Report, Department of Mathematical Sciences, Aalborg University.

Morgan, J. N. and Sonquist, J. A. (1963) Problems in the analysis of survey data and a proposal. *J. Amer. Statist. Assoc.* **58**, 415–434.

Moulin, P. and Liu, J. (1999) Analysis of multiresolution image denoising schemes using generalized Gaussian and complexity priors. *IEEE Trans. Inf. Theor.* **45**, 909–919.

Mukhopadhyay, S. and Gelfand, A. E. (1997) Dirichlet process mixed generalized linear models. *J. Amer. Statist. Assoc.* **92**, 633–639.

Müller, P. and Rios Insua, D. (1998a) Feedforward neural networks for nonparametric regression. In *Practical nonparametric and semiparametric Bayesian statistics* (ed. D. Dey, P. Müuller and D. Sinha). Springer.

Müller, P. and Rios Insua, D. (1998b) Issues in Bayesian analysis of neural network models. *Neural Comp.* **10**, 571–592.

Müller, P. and Vidakovic, B. (1999) *Bayesian inference in wavelet-based models*. Springer.

Murthy, S. K., Kasif, S. and Salzberg, S. (1994) A system for induction of oblique decision trees. *J. Artificial. Intell.* **2**, 1–32.

Nason, G. P. and von Sachs, R. (1999) Wavelets in time-series analysis. *Phil. Trans. Roy. Soc.* A **357**, 2511–2526.

Neal, R. M. (1996) *Bayesian learning for Neural Networks*. Springer.

Neal, R. M. (1999) Regression and classification using Gaussian process priors. In *Bayesian Statistics 6* (ed. J. M. Bernardo, J. O. Berger, A. P. Dawid and A. F. M. Smith), pp. 475–502. Oxford: Clarendon Press.

Nelder, J. A. and Wedderburn, R. W. M. (1972) Generalised linear models. *J. Roy. Statist. Soc.* A **135**, 370–384.

Nelson, L. M., Bloch, D. A., Longstreth, W. T. and Shi, H. (1998) Recursive partitioning for the identification of disease risk subgroups: a case-control study of subarachnoid hemorrhage. *J. Clin. Epidem.* **51**, 199–209.

Newton, M. A., Czado, C. and Chappell, R. (1996) Bayesian inference for semiparametric binary regression. *J. Amer. Statist. Assoc.* **91**, 142–153.

Nicholls, G. K. (1997) Coloured continuum triangulation models in the Bayesian analysis of two dimensional change point problems. In *The Art and Science of Bayesian Image Analysis: Proceeedings of the Leeds Annual Statistics Research Workshop 1997* (ed. K. V. Mardia and R. G. Aykroyd). Leeds University Press.

Nicholls, G. K. (1998) Bayesian image analysis with Markov chain Monte Carlo and coloured continuum triangulation models. *J. Roy. Statist. Soc.* B **60**, 643–659.

Nikolaev, N. I. and Slavov, V. (1997) Inductive genetic programming with decision trees. *Mach. Learn.: ECML-97* **1224**, 183–190.

Nowak, R. D. and Baraniuk, R. G. (1999) Wavelet-based transformations for nonlinear signal processing. *IEEE Trans. Sig. Proc.* **47**, 1852–1865.

Ntzoufras, I., Dellaportas, P. and Forster, J. J. (2001) Bayesian variable and link determination for generalised linear models. Technical Report, Athens University of Economics and Business.

Oh, M. S. (1997) A Gibbs sampling approach to Bayesian analysis of generalized linear models for binary data. *Commun. Statist.* **12**, 431–445.

O'Hagan, A. (1978) On curve fitting and optimal design for regression (with discussion). *J. Roy. Statist. Soc.* B **40**, 1–42.

O'Hagan, A. (1979) On outlier rejection in Bayes inference. *J. Roy. Statist. Soc.* B **41**, 358–367.

O'Hagan, A. (1988) Modelling with heavy tails. In *Bayeisan Statistics 3* (ed. J. M. Bernardo, M. H. DeGroot, D. V. Lindley and A. F. M. Smith), pp. 345–359. Oxford: Clarendon Press.

O'Hagan, A. (1994) *Kendall's Advanced theory of Statistics: Bayesian Inference*, vol. 2b. Cambridge: Arnold.

Okabe, A., Boots, B., Sugihara, K. and Chiu, S.-N. (2000) *Spatial Tessellations: Concepts and Applications of Voronoi Diagrams*, 2nd edn. Wiley.

Oliver, J. J. and Hand, D. J. (1994) Averaging over decision stumps. In *Machine Learning: ECML-94* (ed. F. Bergadano and L. de Raedt). Springer.

Oliver, J. J. and Hand, D. J. (1995) On pruning and averaging decision trees. *Proc. Int. Machine Learning Conf.*, pp. 430–437.

Oliver, J. J. and Hand, D. J. (1996) Averaging over decision trees. *J. Classification* **13**, 281–297.

Otten, R. H. J. M. and van Ginneken, L. P. P. P. (1989) *The Annealing Algorithm*. Boston, MA: Kluwer.

Pascutto, C., Wakefield, J. C., Best, N. G., Richardson, S., Bernardinelli, L., Staines, A. and Elliott, P. (2000) Statistical issues in the analysis of disease mapping data. *Statist. Med.* **19**, 2493–2519.

Peng, F., Jacobs, R. A. and Tanner, M. A. (1996) Bayesian inference and hierarchical mixtures-of-experts models with an application to speech recognition. *J. Amer. Statist. Assoc.* **91**, 953–960.

Penny, W. D. and Roberts, S. J. (1999) Bayesian neural networks for classification: how useful is the evidence framework? *Neural Networks* **12**(6): 877–892.

Percival, D. B. (1995) On estimation of the wavelet variance. *Biometrika* **82**, 619–631.

Percival, D. B. and Walden, A. T. (2000) *Wavelet Methods for Time Series Analysis*. Cambridge University Press.

Percy, D. F. (1992) Prediction for seemingly unrelated regressions. *J. Roy. Statist. Soc.* B **54**, 243–252.

Percy, D. F. (1996) Zellner's influence on multivariate linear models. In *Bayesian Analysis in Statistics and Econometrics* (ed. D. A. Berry, K. M. Chaloner and J. K. Geweke), pp. 203–213. Wiley.

Pericchi, L. R. (1984) An alternative to the standard Bayesian procedure for discrimination between normal linear models. *Biometrika* **71**, 575–586.

Pericchi, L. R. and Smith, A. F. M. (1992) Exact and approximate posterior moments for a normal location parameter. *J. Roy. Statist. Soc.* B **54**, 793–804.

Phillips, D. B. and Smith, A. F. M. (1994) Bayesian faces via hierarchical template modeling. *J. Amer. Statist. Assoc.* **89**, 1151–1163.

Phillips, D. B. and Smith, A. F. M. (1996) Bayesian model comparison via jump diffusions. In *Markov Chain Monte Carlo in Practice* (ed. W. R. Gilks, S. Richardson and D. J. Spiegelhalter), pp. 215–239. London: Chapman & Hall.

Poiner, I., Blaber, S., Brewer, D., Burridge, C., Caeser, D., Connell, M., Dennis, D., Dew, G., Ellis, A., Farmer, M., Fry, G., Glaister, J., Gribble, N., Hill, B., Long, B., Milton, D., Pitcher, C., Proh, D., Salini, J., Thomas, M., Toscas, P., Veronise, S., Wang, Y. and Wassenberg, T. (1997) The effects of prawn trawling in the far northern section of the Great Barrier Reef. Technical Report, CSIRO Vision of Marine Research, CSIRO, Queensland, Australia.

Press, S. J. (1972) *Applied Multivariate Analysis*. New York: Holt, Rinehart and Winston.

Press, W. H. (1992) *Numerical Recipes in C: the Art of Scientific Computing*, 2nd edn. Cambridge University Press.

Priestley, M. B. (1988) *Nonlinear and Nonstationary Time Series Analysis*. Academic.

Punska, O., Andrieu, C., Doucet, A. and Fitzgerald, W. J. (1999) Bayesian segmentation of piecewise constant autoregressive processes using MCMC. Technical Report TR 344, Engineering Department, University of Cambridge.

Quinlan, J. R. (1986) Induction of decision trees. *Machine Learning* **1**, 81–106.

Quinlan, J. R. (1993) Combining instance-based and model-based learning. *Machine Learning: Proc. 10th Int. Conf., Amherst, MA*. Morgan Kaufmann.

Quinlan, J. R. and Rivest, R. L. (1989) Inferring decision trees using the minimum description length principle. *Inf. Comp.* **80**, 227–248.

Quintana, F. A. and Iglesias, P. L. (2001) Bayesian clustering and product partition models. Technical Report, Departamento de Estadística, Pontificia Universidad Católica de Chile.

Raftery, A. E. and Akman, V. E. (1986) Bayesian analysis of a Poisson process with a change-point. *Biometrika* **73**, 85–89.

Raftery, A. E., Madigan, D. and Hoeting, J. A. (1997) Model selection and accounting for model uncertainty in linear regression models. *J. Amer. Statist. Assoc.* **92**, 179–191.

Raftery, A. E., Madigan, D. and Volinsky, C. T. (1996) Accounting for model uncertainty in survival analysis improves predictive performance (with discussion). In *Bayesian Statistics 5* (ed. J. M. Bernardo, J. O. Berger, A. P. Dawid and A. F. M. Smith), pp. 323–349. Oxford: Clarendon Press.

Ramsay, J. O. and Silverman, B. W. (1997) *Functional Data Analysis*. Springer.

Rasmussen, C. E. (1996) Evaluation of Gaussian processes and other methods for non-linear regression. PhD thesis, University of Toronto.

Rekaya, R., Carabano, M. J. and Toro, M. A. (2000) Assessment of heterogeneity of residual variances using changepoint techniques. *Genetics Selection Evolution* **32**, 383–394.

Richardson, S. and Green, P. J. (1997) On Bayesian analysis of mixtures with an unknown number of components (with discussion). *J. Roy. Statist. Soc.* B **59**, 731–792.

Ripley, B. D. (1994) Neural networks and related methods for classification (with discussion). *J. Roy. Statist. Soc.* B **56**, 409–456.

Ripley, B. D. (1996) *Pattern Classification and Neural Networks*. Cambridge University Press.

Robert, C. P. (1995a) Convergence control techniques for MCMC algorithm. *Statist. Sci.* **10**, 231–253.

Robert, C. P. (1995b) A note on Jeffrey's–Lindley's paradox. *Statist. Sin.* **3**, 601–608.

Robert, C. P. (1995c) Simulation of truncated normal variables. *Statist. Comp.* **5**, 121–125.

Robert, C. P. and Casella, G. (1999) *Monte Carlo Statistical Methods*. Springer.

Roberts, G. O. and Tweedie, R. L. (1996) Geometric convergence and central limit theorems for multidimensional Hastings and Metropolis algorithms. *Biometrika* **83**, 95–110.

Roberts, H. V. (1965) Probabilistic prediction. *J. Amer. Statist. Assoc.* **60**, 50–62.

Rosales, R., Stark, J. A., Fitzgerald, W. J. and Hladsky, S. B. (1998) Bayesian estimation of ion channel hidden markov models. *Biophys. J.* **74**, A321–A321.

Rosenblatt, M. (1956) Remarks on some nonparametric estimates of a density function. *Ann. Math. Statist.* **27**, 832–836.

Rosenblatt, M. (1958) The perceptron: a probabalistic model for information storage and organization in the brain. *Psychol. Rev.* **65**, 386–408.

Rue, H. and Hurn, M. (1999) Bayesian object identification. *Biometrika* **86**, 649–660.

Ruggeri, F. and Vidakovic, B. (1999) A Bayesian decision theoretic approach to the choice of thresholding parameter. *Statist. Sin.* **9**, 183–197.

Rumelhart, D. E., Hinton, G. E. and Williams, R. J. (eds) (1986) *Parallel Distributed Processing: Explorations in the Microstructure of Cognition*, vol. 1. *Foundations*. Cambridge, MA: MIT Press.

Safavian, S. R. and Landgrebe, D. (1991) A survey of decision tree classifier methodology. *IEEE Trans. Sys. Man. Cybernet.* **21**, 660–674.

Samaniego, F. J. and Neath, A. A. (1996) How to be a better Bayesian. *J. Amer. Statist. Assoc.* **91**, 733–742.

Sambridge, M. (1999a) Geophysical inversion with a neighbourhood algorithm. I. Searching a parameter space. *Geophys. J. Int.* **138**, 479–494.

Sambridge, M. (1999b) Geophysical inversion with a neighbourhood algorithm. II. Appraising the ensemble. *Geophys. J. Int.* **138**, 727–746.

Savage, L. J. (1971) Elicitation of personal probabilities and expectations. *J. Amer. Statist. Assoc.* **66**, 781–801.

Schapire, R. E. (1990) The strength of weak learnability. *Machine Learning* **5**, 197–227.

Schapire, R. E. and Singer, Y. (1999) Improved boosting algorithms using confidence-rated predictions. *Mach. Learn.* **37**, 297–336.

Schapire, R. E., Freund, Y., Bartlett, P. and Lee, W. S. (1998) Boosting the margin: a new explanation for the effectiveness of voting methods. *Ann. Statist.* **26**, 1651–1686.

Schiff, S. J., Aldroubi, A., Unser, M. and Sato, S. (1994) Fast wavelet transformation of EEG. *Electroen. Clin. Neuro.* **91**, 442–455.

Schlattmann, P. and Bohning, D. (1993) Mixture-models and disease mapping. *Statist. Med.* **12**, 1942–1950.

Schwarz, G. (1978) Estimating the dimension of a model. *Ann. Statist.* **6**, 461–464.

Serroukh, A., Walden, A. T. and Percival, D. B. (2000) Statistical properties and uses of the wavelet variance estimator for the scale analysis of time series. *J. Amer. Statist. Assoc.* **95**, 184–196.

Shafer, G. (1982) Lindley's paradox (with discussion). *J. Amer. Statist. Assoc.* **77**, 325–351.

Shannon, W. D. and Banks, D. (1999) Combining classification trees using MLE. *Statist. Med.* **18**, 727–740.

Shively, T. S., Kohn, R. and Wood, S. (1999) Variable selection and function estimation in additive nonparamteric regression. *J. Amer. Statist. Assoc.* **94**, 777–794.

Silverman, J. F. and Cooper, D. B. (1988) Bayesian clustering for unsupervised estimation of surface and texture models. *IEEE Trans. Pattern Anal. Mach. Intell.* **10**, 482–495.

Skouras, K. and Dawid, A. P. (1998) On efficient point prediction systems. *J. Roy. Statist. Soc.* B **60**, 765–780.

Smith, A. F. M. (1975) A Bayesian approach to inference about a changepoint in a sequence of random variables. *Biometrika* **62**, 407–416.

Smith, A. F. M. (1980) Change-point problems: approaches and applications. In *Bayesian Statistics* (ed. J. M. Bernardo, M. H. DeGroot, D. V. Lindley and A. F. M. Smith), pp. 83–89. Valencia University Press.

Smith, A. F. M. and Roberts, G. O. (1993) Bayesian computation via the Gibbs sampler and related Markov chain Monte Carlo methods. *J. Roy. Statist. Soc.* B **55**, 3–23.

Smith, A. F. M. and Spiegelhalter, D. J. (1980) Bayes factors and choice criteria for linear models. *J. Roy. Statist. Soc.* B **42**, 213–220.

Smith, M. and Kohn, R. (1996) Nonparametric regression using Bayesian variable selection. *J. Econometrics* **75**, 317–344.

Smith, M. and Kohn, R. (1997) A Bayesian approach to nonparametric bivariate regression. *J. Amer. Statist. Assoc.* **92**, 1522–1535.

Smith, M. and Kohn, R. (2000) Nonparametric seemingly unrelated regression. *J. Econometrics* **98**, 257–281.

Smith, M., Wong, C. M. and Kohn, R. (1998) Additive nonparametric regression with autocorrelated errors. *J. Roy. Statist. Soc.* B **60**, 311–331.

Smith, P. L. (1982) Curve fitting and modeling with splines using statistical variable selection techniques. Technical Report, NASA.

Snapp, R. R. and Venkatesh, S. S. (1998) Asymptotic expansions of the *k* nearest neighbor risk. *Ann. Statist.* **26**, 850–878.

Spiegelhalter, D. J. and Smith, A. F. M. (1982) Bayes factors for linear and log-linear models with vague prior information. *J. Roy. Statist. Soc.* B **44**, 377–387.

Spyers-Ashby, J. M. (1996) The recording and analysis of tremor in neurological disorders. PhD thesis, Imperial College, London University.

Stark, J. A., Rosales, R., Fitzgerald, W. J. and Hladsky, S. B. (1999) Representing idealizations of ion channel records. *Biophys. J.* **76**, A199–A199.

Stephens, D. A. (1994) Bayesian retrospective multiple-changepoint identification. *Appl. Statist.* **43**, 159–178.

Stone, C. J. (1985a) Additive regression and other nonparametric models. *Ann. Statist.* **13**, 689–705.

Stone, C. J. (1985b) The dimensionality reduction principle for generalized additive models. *Ann. Statist.* **14**, 590–606.

Sun, D., Speckman, P, L. and Tsutakawa, R. K. (2000) Random effects in generalized linear mixed models. In *Generalised Linear Models: A Bayesian Perspective* (ed. D. K. Dey, S. K. Ghosh and B. K. Mallick). Marcel Dekker.

Thisted, R. A. (1988) *Elements of Statistical Computing*. New York: Chapman & Hall.

Thodberg, H. H. (1995) A review of Bayesian neural networks with an application to near infrared spectroscopy. *IEEE Trans. Neural Networks* **7**, 56–72.

Tierney, L. (1994) Markov chains for exploring posterior distributions (with discussion). *Ann. Statist.* **22**, 1701–1762.

Titterington, D. M., Smith, A. F. M. and Makov, U. E. (1985) *Statistical Analysis of Finite Mixture Distributions*. Wiley.

Tjelmeland, H. and Besag, J. E. (1998) Markov random fields with higher-order interactions. *Scan. J. Statist.* **25**, 415–433.

Tong, H. (1983) *Threshold Models in Nonlinear Time Series Analysis*. Springer.

Upsdell, M. P. (1996) Choosing an appropriate covariance function in Bayesian smoothing. In *Bayesian Statistics 5* (ed. J. M. Bernardo, J. O. Berger, A. P. Dawid and A. F. M. Smith), pp. 747–756. Oxford University Press.

Vehtari, A. and Lampinen, J. (2000) Bayesian MLP neural networks for image analysis. *Pattern Recognition Letters* **21**, 1183–1191.

Venables, W. N. and Ripley, B. D. (1997) *Modern Applied Statistics with S-PLUS*, 2nd edn. Springer.

Verdinelli, I. and Wasserman, L. (1991) Bayesian analysis of outlier problems using the Gibbs sampler. *Statist. Comp.* **1**, 105–117.

Vidakovic, B. (1998) Nonlinear wavelet shrinkage with Bayes rules and Bayes factors. *J. Amer. Statist. Assoc.* **93**, 173–179.

Vidakovic, B. (1999) *Statistical Modelling by Wavelets*. Wiley.

Vivarelli, F. and Williams, C. K. I. (2001) Comparing Bayesian neural network algorithms for classifying segmented outdoor images. *Neural Networks* **14**, 427–437.

Volinsky, C. T. (1997) Bayesian model averaging for censored survival data. PhD thesis, University of Washington, Seattle.

Volinsky, C. T., Madigan, D., Raftery, A. E. and Kronmal, R. A. (1997) Bayesian model averaging in proportional hazards models. Assessing the risk of a stroke. *Appl. Statist.* **46**, 433–448.

Voronoi, M. G. (1908) Nouvelles applications des paramètres continus à la théorie des formes quadratiques. *J. Reine Angew. Math.* **134**, 198–287.

Wahba, G. (1978) Improper priors, spline smoothing, and the problem of guarding against model errors in regression. *J. Roy. Statist. Soc.* B **40**, 364–372.

Wahba, G. (1983) Bayesian 'confidence intervals' for the cross-validated smoothing spline. *J. Roy. Statist. Soc.* B **45**, 133–150.

Walker, S. G. and Mallick, B. K. (1997) Hierarchical generalized linear models and frailty models with Bayesian nonparametric mixing. *J. Roy. Statist. Soc.* B **59**, 845–860.

Walker, S. G., Damien, P., Laud, P. W. and Smith, A. F. M. (1999) Bayesian nonparametric inference for random distributions and related functions. *J. Roy. Statist. Soc.* B **61**, 485–527.

Wallace, C. S. and Freeman, P. R. (1987) Estimation and inference by compact coding (with discussion). *J. Roy. Statist. Soc.* B **49**, 240–251.

Wallace, C. S. and Patrick, J. D. (1993) Coding decision trees. *Mach. Learn.* **11**, 7–22.

Waller, L. A., Turnbull, B. W., Clark, A. L. and Nasca, P. (1994) Spatial pattern analyses to detect rare disease clusters. In *Case Studies in Biometry* (ed. N. Lange, L. Ryan, L. Billard, D. Brillinger, L. Conquest and J. Greenhouse), pp. 1–23. Wiley.

Watson, I. and Marir, F. (1994) Case-based reasoning – a review. *Knowledge Engng Rev.* **9**, 327–354.

Wedderburn, R. W. M. (1976) On the existence and uniqueness of the maximum likelihood estimates for certain generalized linear models. *Biometrika* **63**, 27–32.

Weigend, A. S., Huberman, B. A. and Rumelhart, D. E. (1992) Predicting sunspots and exchange rates with connectionist networks. In *Nonlinear Modeling and Forecasting* (ed. C. Casdagli and S. Eubank). Addison-Wesley.

West, M. (1987) On scale mixtures of normal distributions. *Biometrika* **74**, 646–648.

Whittaker, E. (1923) On a new method of graduation. *Proc. Edinburgh Math. Soc.* **41**, 63–75.

Widrow, B. and Hoff, M. E. (1960) Adaptive switching circuits. *IRE WESCON Convention Record* **4**, 94–104.

Williams, C. K. I. (1998) Prediction with Gaussian processes: from linear regression to linear prediction and beyond. In *Learning in Graphical Models* (ed. M. I. Jordan). Amsterdam: Kluwer.

Williams, C. K. I. and Rasmussen, C. E. (1996) Gaussian processes for regression. In *Advances in Neural Information Processing Systems 8* (ed. D. S. Touretzky, M. C. Mozer and M. E. Hasselmo). Boston, MA: MIT Press.

Wong, C. M. and Kohn, R. (1996) A Bayesian approach to additive semiparamteric regression. *J. Econometrics* **74**, 209–235.

Wood, S. and Kohn, R. (1998) A Bayesian approach to robust binary nonparametric regression. *J. Amer. Statist. Assoc.* **93**, 203–213.

Wu, M. and Fitzgerald, W. J. (1995) Analytical approach to changepoint detection in Laplacian noise. *IEE Proc. Vis. Image Sig. Proc.* **142**, 174–180.

Yao, Y. C. (1984) Estimation of a noisy discrete-time step function: Bayes and empirical Bayes approaches. *Ann. Statist.* **12**, 1434–1447.

Yau, P., Kohn, R. and Wood, S. (2002) Bayesian variable selection and model averaging in high dimensional multinomial nonparametric regression. *J. Comp. Graph. Statist.* (In the press.)

Yule, G. U. (1927) On a method of investigating periodicities in disturbed series with special reference to Wolfer's sunspot numbers. *Phil. Trans. Roy. Soc. Lond.* A **226**, 267–298.

Zacks, S. (1982) Classical and Bayesian approaches to the changepoint problem. *Statistique et Analysé des Donnees* **1**, 48–81.

Zeger, S. L. and Karim, L. (1991) Generalized linear models with random effects: a Gibbs sampling approach. *J. Amer. Statist. Assoc.* **86**, 79–86.

Zellner, A. (1962) An efficient method of estimating seemingly unrelated regressions and tests of aggregation bias. *J. Amer. Statist. Assoc.* **57**, 500–509.

Zellner, A. (1986) On assessing prior distributions and Bayesian regression analysis with g-prior distributions. In *Bayesian Inference and Decision Techniques: Essays in Honor of Bruno de Finetti* (ed. P. K. Goel and A. Zellner). Amsterdam: North-Holland.

Zellner, A. and Vandaele, W. (1974) Bayes Stein estimators for K means, regression and simultaneous equations models. In *Studies in Bayesian Econometrics and Statistics* (ed. S. E. Feinberg and A. Zellner), pp. 628–653. Amsterdam: North-Holland.

Zhang, Q. (1997) Using wavelet networks in nonparametric estimation. *IEEE Trans. Neur. Net.* **8**, 227–236.

Zheng, Z. J. and Webb, G. I. (2000) Lazy learning of Bayesian rules. *Mach. Learn.* **41**, 53–84.

Index

Author Index